Lecture Notes in Mechanical Engineering

Lecture Notes in Mechanical Engineering (LNME) publishes the latest developments in Mechanical Engineering—quickly, informally and with high quality. Original research reported in proceedings and post-proceedings represents the core of LNME. Volumes published in LNME embrace all aspects, subfields and new challenges of mechanical engineering.

To submit a proposal or request further information, please contact the Springer Editor of your location:

Europe, USA, Africa: Leontina Di Cecco at Leontina.dicecco@springer.com
China: Ella Zhang at ella.zhang@springer.com
India: Priya Vyas at priya.vyas@springer.com
Rest of Asia, Australia, New Zealand: Swati Meherishi at swati.meherishi@springer.com

Topics in the series include:

- Engineering Design
- Machinery and Machine Elements
- Mechanical Structures and Stress Analysis
- Automotive Engineering
- Engine Technology
- Aerospace Technology and Astronautics
- Nanotechnology and Microengineering
- Control, Robotics, Mechatronics
- MEMS
- Theoretical and Applied Mechanics
- Dynamical Systems, Control
- Fluid Mechanics
- Engineering Thermodynamics, Heat and Mass Transfer
- Manufacturing
- Precision Engineering, Instrumentation, Measurement
- Materials Engineering
- Tribology and Surface Technology

Indexed by SCOPUS, EI Compendex, and INSPEC.

All books published in the series are evaluated by Web of Science for the Conference Proceedings Citation Index (CPCI).

To submit a proposal for a monograph, please check our Springer Tracts in Mechanical Engineering at https://link.springer.com/bookseries/11693.

Said Aniss · Miloud Rahmoune ·
Somia Mordane · Mohamed Elamine Ait Ali ·
Rabha Khatyr · Fakher Chaari · Mohamed Haddar
Editors

Advances in Mechanics

Selected Contributions to the 15th International Moroccan Congress of Mechanics, May 24–27, 2022, CMM 2022, Casablanca, Morocco

Springer

Editors
Said Aniss
Faculty of Sciences Ain Chock
University of Hassan II Casablanca
Casablanca, Morocco

Somia Mordane
Faculty of Sciences Ben M'Sik
University of Hassan II Casablanca
Casablanca, Morocco

Rabha Khatyr
Faculty of Science Ain Chock
University of Hassan II Casablanca
Casablanca, Morocco

Mohamed Haddar
National School of Engineers of Sfax
University of Sfax
Sfax, Tunisia

Miloud Rahmoune
Superior School of Technology
Moulay Ismail University of Meknes
Meknes, Morocco

Mohamed Elamine Ait Ali
Mechanical Engineering Department
Mohammed V University in Rabat -
Mohammadia School of Engineering
Rabat, Morocco

Fakher Chaari
National School of Engineers of Sfax
University of Sfax
Sfax, Tunisia

ISSN 2195-4356 ISSN 2195-4364 (electronic)
Lecture Notes in Mechanical Engineering
ISBN 978-3-031-46972-5 ISBN 978-3-031-46973-2 (eBook)
https://doi.org/10.1007/978-3-031-46973-2

This Springer imprint is published by the registered company Springer Nature Switzerland AG
The registered company address is: Gewerbestrasse 11, 6330 Cham, Switzerland

Paper in this product is recyclable.

Preface

The Moroccan Congress of Mechanics is a scientific event held every two years during the April–May period, initiated by the Moroccan Society of Mechanical Sciences (SMSM). SMSM is an academic society that brings together Moroccan researchers and their foreign colleagues working in the field of mechanical sciences. The 15th edition of this congress—CMM 2022—was held from May 24–27, 2022, at Hassan II University in Casablanca, Morocco, and was organized by the SMSM and the Faculty of Sciences Ain Chok of Hassan II University in collaboration with the France-Maghreb Association of Mechanics and Materials (AF3M).

The main objective of the congress is to provide an overview of research and advancements in the field and to highlight forthcoming challenges in scientific research within the mechanical sciences. This helps strengthen research capabilities, foster the growth of existing inter-university connections, and encourage collaboration among various laboratories involved in mechanical sciences research. The congress brought together researchers, academicians, and industry professionals from around the world to exchange knowledge, ideas, and advancements in various areas of mechanics.

We are delighted to present this collection of 23 research articles, carefully selected from a pool of over 170 oral communications delivered at the 15th Congress of Mechanics CMM 2022. This collection represents a diverse range of topics and showcases the innovative and cutting-edge research conducted in the field of mechanics. The rigorous two-phase selection process ensured that only the most significant and impactful contributions were included in this compilation. In the first stage of this process, the identification of the most significant oral communications from the congress occurred, overseen by session chairs and members of the scientific committee. Following this, authors of 34 selected papers were invited to submit extended versions of their articles. These manuscripts then advanced to the next phase, where two reviewers from an international group of senior scientists assessed each of them. We extend our gratitude to these reviewers for their dedication and availability in evaluating the proposals. The favorable evaluations from both reviewers were requested for each paper's acceptance after necessary revisions, ultimately resulting in the selection of 23 articles. This procedure guarantees that the compilation at hand comprises solely the most influential and pioneering contributions, solidifying their place within this esteemed collection.

Covering various mechanical engineering disciplines, these papers explore numerous captivating subjects.

Readers will find here two papers within the domain of dynamics, vibrations, and acoustics. The first is about calculating dispersion curves in graphite-epoxy composite plates using a hybrid analytic approach, while the second addresses the area of active vibration control in non-uniform section beams.

In the field of solid and structural mechanics modeling, there are here a few studies offering valuable insights into: numerical modeling of stress concentrations in welded tubular Y-joints under deflected bending load; the coupling of the method of fundamental solutions with domain decomposition techniques to solve the Laplace equation; and dynamic modeling of multilayered piezoelectric beams employing the pseudo-Stroh formalism.

The integrated design and manufacturing processes section of this book features a comprehensive study investigating the effects of endurance and take-off distance on pre-sizing unmanned aerial vehicles, which may translate in some practical implications for the aerospace industry.

In the porous media and granular media domains, a compelling study probes the infiltration phenomenon in unsaturated porous media, addressing phenomenology and modeling techniques.

The realm of fluid mechanics, rheology, hydrodynamics instabilities, and transition showcases diverse research subfields, encompassing topics like Faraday instability in Hele-Shaw cells, bifurcation analysis of incompressible fluid flow, the effect of load on minimum film thickness in non-Newtonian MEHD parabolic slider bearings, and linear stability analysis of viscoelastic fluids in plane channel flow.

Regarding heat transfer, mass transfer, renewable energy, and the environment, several investigations shed light on key aspects such as: the numerical study of latent heat discharge in phase change material shell-and-tube thermal energy storage systems, the numerical evaluation of petroleum coke and biomass co-combustion in rotary cement kilns, and the impact of improved thermoelectric module performance on efficiency. This part also reports on new analytical and numerical solutions of steady convection–diffusion around a circular cylinder heated with uniform heat flux.

The studies described in this book, which aptly demonstrate the ongoing innovation and advancement in diverse subfields of mechanics, are expecting to contribute significantly to the progress of engineering and applied sciences.

We extend our gratitude to all the researchers who submitted their work for consideration. The quality and quantity of submissions posed a challenge during the selection process, and we commend their dedication and scholarly contributions. Our sincere appreciation also goes to the members of the scientific committee, who diligently reviewed each submission to ensure the highest standards of quality and relevance. Furthermore, we extend our thanks to the organizing committee of the 15th Congress of Mechanics for providing a platform that facilitated this intellectual exchange. Their efforts in organizing the conference and creating an environment conducive to collaboration and learning have been instrumental in the success of this endeavor.

We hope that this collection of research articles will serve as a valuable resource for researchers, scholars, and practitioners in the field of mechanics. It is our firm belief that these contributions will inspire further exploration, eventually resulting in advancements addressing the challenges of our ever-changing world.

Contents

Porous Media and Granular Media

Fluids Mechanics, Rheology, Hydrodynamics Instabilities and Transition

Heat Transfer, Mass Transfer, Renewable Energy and Environment

Dynamics, Vibrations and Acoustics

Calculation of Dispersion Curves in the Composite Graphite-Epoxy Plate by a Hybrid Analytic Method

I. Zitouni(✉), H. Rhimini, and A. Chouaf

Laboratory of Mechanics, Engineering and Innovation, National School of Electricity and Mechanics, Hassan II University, Casablanca, Morocco
{Ismaine.zitouni.doc20,h.rhimini,a.chouaf}@ensem.ac.ma

Abstract. In recent years, graphite-epoxy materials have been widely used in the space industry due to their ability to satisfy design demands (minimum weight and high structural rigidity). However, radiation resulting from the space environment can create imperfections in these structures. The ultrasonic non-destructive testing by guided wave presents an alternative for the detection of these defects. The knowledge of the dispersion curves is an essential step in order to know the adequate frequencies susceptible to propagate in our material for a successful control. In this manuscript, we present a hybrid analytical method to plot the dispersion curves of a graphite-epoxy composite material. The advantage of this method over the classical analytical methods has been studied. The accuracy of the obtained results is controlled using the DISPERSE software. Next, the profiles of the displacements and stresses normalized by the acoustic power flow of the fundamental modes are plotted at a given frequency. A discussion of the vibratory states of the plate is established.

Keywords: Ultrasonic guided waves · dispersion curve · graphite-epoxy composite · hybrid analytical method · displacements and stresses

1 Introduction

Aeronautical structures are always exposed to solicitations caused by the radiation from the space environment and atomic oxygen [1]. The need for a choice of material is fundamental to have good functionality and high durability of aeronautical pieces and avoid space disasters. Graphite-epoxy composites were able to meet all these requirements. They have been used to reduce thermal distortion and control the effects of solar radiation. In addition, this type of material is characterized by a high structural rigidity while having a low weight [1, 2].

Despite all these advantages, the non-adhesion between the fibers and the matrix may lead to the appearance of imperfections that are not detectable in some cases.

The non-destructive testing by Ultrasonic Guided Waves (UGW) presents an advantageous tool for the detection of these defects. In fact, UGW has the ability to generate a vibration of the whole piece while having a large propagation distance. The knowledge

S. Aniss et al. (Eds.): CMM 2022, LNME, pp. 3–12, 2024.
https://doi.org/10.1007/978-3-031-46973-2_1

of the dispersion curves is a major phase in the ultrasonic control. They allow us to know the modes susceptible to propagate in the piece and can detect the defects. The study of dispersion curves has been widely developed by several methods. Analytical roots finding algorithms methods applied to characteristic equations were the first used to obtain these curves [3, 4]. However, the computation time and a large number of iterations are still important limitations for these methods. Several other methods have been introduced subsequently such as the semi-analytical method [5, 6], the spectral method [7, 8], the method of the Legendre polynomial [9, 10] and also the matrix methods (TMM: Transfer Matrix Method, GMM: Global Matrix Method and SSTM: Stable Transfer Matrix Method [11, 12].

In this paper, we introduce a hybrid analytical method for plotting the dispersion curves of a graphite-epoxy composite material. We compare our curves with those provided by the Disperse software [11], considered as the leader in the field of NDT by UGW. A perfect concordance of the results of the two techniques was observed. Our hybrid method allowed us to have dispersion curves at the neighborhood of the cut-off frequencies of the higher order modes. Then, with the help of these dispersion curves, we plot the displacement and stress profiles of the fundamental S_0 and A_0 modes of the composite for couples of wave numbers frequency.

2 Guided Waves Theory

We consider the propagation of guided waves in an orthotropic homogeneous plate of thickness e = 2h along the x_3 direction and assumed infinite in the x_1 and x_2 (Fig. 1).

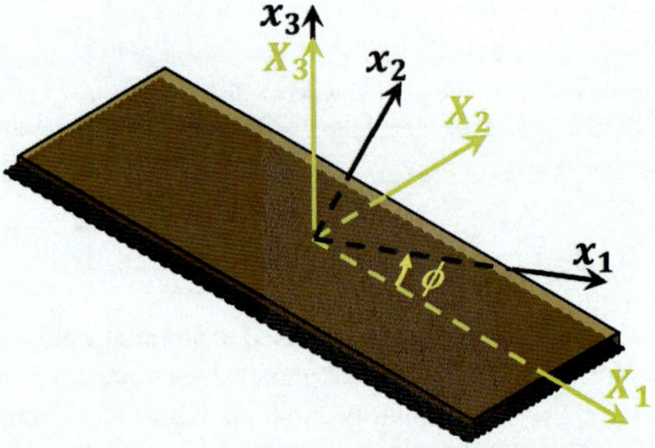

Fig. 1. Geometry of the composite plate plate, (x_1, x_2, x_3): Global Cartesian coordinate system. (X_1, X_2, X_3): Cartesian reference system. φ: angle who describe the orientation of the fiber.

The composite plate is governed by the following three equations:

$$\frac{\partial \sigma_{ij}}{\partial x_i} = \rho \frac{\partial^2 u_i}{\partial t^2} \quad (1)$$

$$\sigma_{ij} = C_{ijop} \varepsilon_{op} \quad (2)$$

$$\varepsilon_{op} = \frac{1}{2}(\frac{\partial u_o}{\partial x_p} + \frac{\partial u_p}{\partial x_o}) \tag{3}$$

The equations in (1) represent the equations of motion of an elastic wave in an orthotropic plate in the global Cartesian coordinate system (x_1, x_2, x_3). With σ_{ij} representing the stress tensor, ρ the density and u_i the displacement field. The second equation represents Hooke's law of behavior. With C_{ijop} representing the elastic stiffness tensor (defined by 9 distinct constants) and ε_{op} representing the deformation tensor. Finally, the third equation describe the deformation-displacement relationship.

Replacing (2) and (3) in (1) we obtain the Christofell system. Based on these equations and considering the propagation of a harmonic elastic wave, we can approximate the displacement field as follows:

$$(u_1, u_2, u_3) \sum_{m=1}^{6} (1, V_m, W_m) U_{1m} e^{ik(x_1 + \alpha_q x_3 - ct)} \tag{4}$$

where U_{1m}, V_m and W_m are the displacement amplitudes in the x_1, x_2 and x_3 directions. c is the phase velocity, k is the wave number and m is a summation index.

The stress field expression is obtained by replacing (4) in (3) and (2).

The plate is considered free in stresses, the stresses σ_{33}, σ_{13} and σ_{23} are null at the upper and lower surfaces. The annulation of these stresses leads to a system of 6 equations on the amplitudes. This system is as follows [13]:

$$[D]_{(6,6)} \left\{ U_{ij} \right\}_{(6,1)} = \{0\}_{(6,1)} \tag{5}$$

The non-trivial solutions are derived from the cancellation of the determinant of this system. The components of the matrix D are written as a function of the engineering constants (E_i: Young's modulus, G_{ij}: shear modulus, ν_{ij}: Poisson's ratio (Table 1)), of the frequency f, of the wave number and of the density.

At this stage, we need to find solutions that allow the determinant of the matrix *[D]* to be cancelled. Analytical methods for finding the zeros of a function will be presented and applied in the following section.

3 Hybrid Method

Roots finding algorithms of a function have the advantage of being robust and are often convergent. One of these analytical methods is the bisection method which has been successfully used to plot the dispersion curves of several isotropic and anisotropic materials [3]. But this method finds problems for multilayer structures, moreover, the calculation time of this method is very high. The Newton-Raphson method [4] also allows the obtaining of these curves, it presents a great saving in time compared to the dichotomy method but finds problems for the obtaining of all the roots because of the weak approximation of the initial values. The hybrid method that we present in this paper comes to overcome the limitations of these two methods. For our case, we seek to find the couple frequency-wave number (f,k) that allow the cancellation of the UGW dispersion equations of the composite plate. To do this, we first fix the wave number and we carry out our search on a range of frequency, each time the characteristic equations change sign we apply the bisection method to reduce the interval where the root is assumed, then we apply the Newton-Raphson method to take advantage of its convergence speed until the solution does not differ from the previous one ($|f_i - f_{i-1}| \leq \varepsilon$). The study being developed for a fixed value of k, we repeat the same procedure for the whole range of the wave number (Fig. 2).

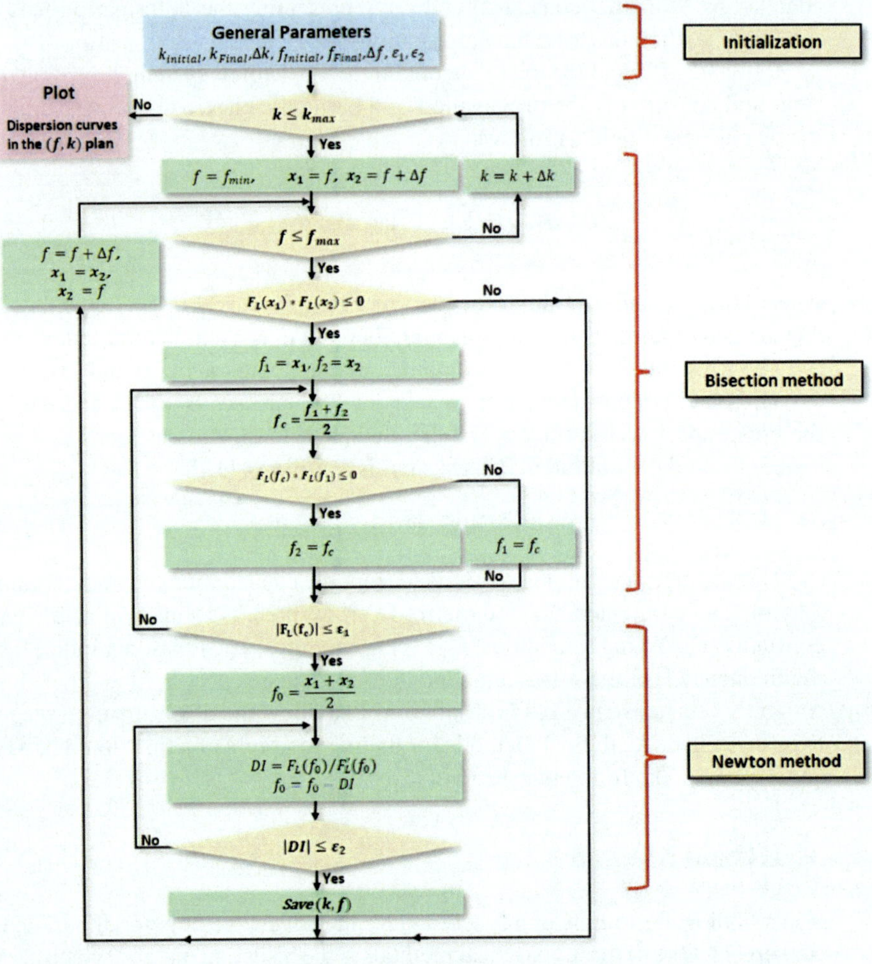

Fig. 2. Hybrid method algorithm.

4 Dispersion Curves of Graphite-Epoxy Composite

For our application, we considered a composite plate in graphite-epoxy of $e = 4$ mm thickness and density $\rho = 1600\ kg/m^3$, the mechanical characteristics are given in Table 1 with the Young's modulus and shear modulus are in GPa. For more details, see references [13, 14].

Table 1. Mechanical properties of a graphite-epoxy composite plate [13].

Young's modulus			Shear modulus			Poisson ratio		
E_1	E_2	E_3	G_{12}	G_{13}	G_{23}	υ_{12}	υ_{13}	υ_{23}
154.25	14.72	14.72	7.46	5.82	32	0.18	0.18	0.26

We take the wave number and frequency ranges as follows:
$k = 10^{-5}$:100:12000 (m^{-1})
f = 10:50:6 10^6 (Hz)

Figure 3 shows the dispersion curves of the first 12 modes of the graphite-epoxy composite in the (fe,k) plan for values of 0°, 30°, 45° and 90° of the φ angle that relates the global coordinate system to the reference coordinate system. The curves produced by the hybrid method are compared with those obtained by the DISPERSE software.

Table 2. Root square error (RSE = $(\sqrt{}(fe_{Hybrid} - fe_{Disperse})^2)$

Methods	Frequency of the Hybrid method (MHz.mm)	Frequency of the Disperse Software (MHz.mm)	Error
Frequency of the S_1 mode at k = 1000 m^{-1}	4.7532	4.7542	9.5245 10^{-4}
Frequency of the A_0 mode at k = 1000 m^{-1}	1.2984	1.2987	2.9157 10^{-4}

Table 2 shows the calculation of the error between the solutions obtained by the hybrid method and those derived from Disperse software for a graphite-epoxy composite plate with fiber direction φ = 0° for the two fundamental modes (S_1 and A_0) at k = 1000 m^{-1}.

The precision evaluated between the results obtained by the hybrid method and the DISPERSE software is around 10^{-4}, which shows the accuracy of the solutions obtained by the hybrid method. We observe that the its succeeded in defining the first 12 modes in the whole range of the wave number, which is not the case of the DISPERSE software that finds problems to generate the S_0 mode (at 0° and 90°) as well as to define completely the modes for very small values of the wave number.

Table 3. Comparison of the computation time between the analytical methods for obtaining the first 12 modes.

Methods	Bisection	Newton-Raphson	Hybrid
Calculation time (s)	1481	1068	927

In Table 3 we have reported the computation time of the Bisection, Newton and hybrid methods for the case of a graphite-epoxy composite plate at $\varphi = 0^{\circ}$ (the frequency interval is 10:50:6.10^6 (Hz), the wave number interval is 10^{-5}:100:12000 (m^{-1}). From Table 3, we notice that the computation time of our method is much shorter than the other classical methods. Moreover, with this method, we manage to detect all the modes existing in the defined search of zeros space and more particularly in the proximity of the cut-off frequencies of these modes. This proves the speed and accuracy of our method in determining the dispersion curves of a composite material with different degrees of anisotropy.

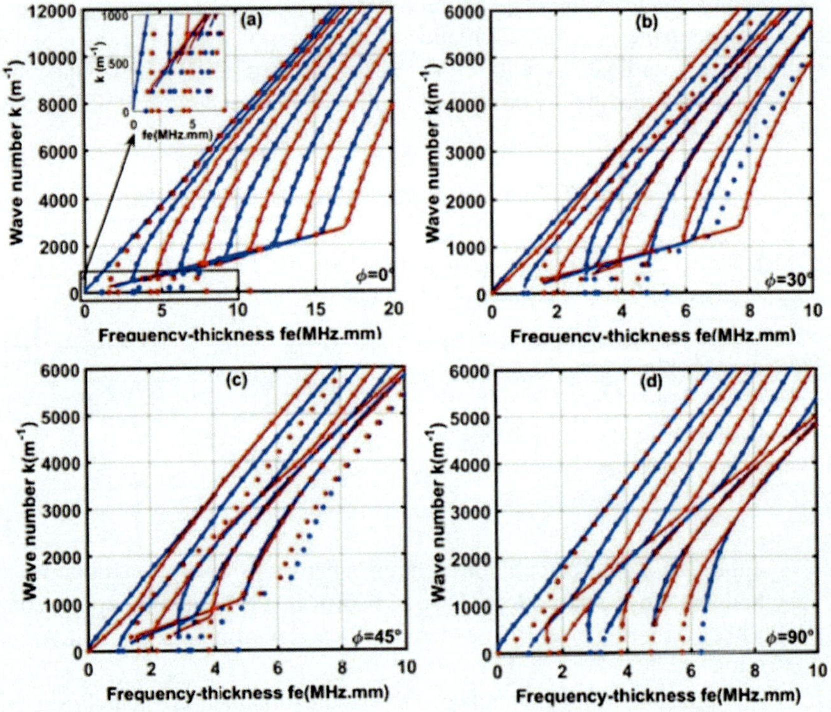

Fig. 3. Dispersion curves in the (fe,k) plane plotted by the hybrid method (solid circles) and Disperse software (continuous lines) for angles 0° (a), 30° (b), 45° (c) and 90° (d).

To carry out a proper non-destructive inspection using ultrasonic guided waves, two steps are essential. The first is the knowledge of the frequencies susceptible to generate guided waves (dispersion curve explained above) in the waveguide. These curves depend essentially on the waveguide geometry and the type of material.

The second step is to determine the shape of the mode generated. Indeed, the shape of the mode provides information on the type of defect being controlled [15, 16]. In the following section, we will carry out a study of the analytical displacements and stresses of the graphite-epoxy composite plate.

5 Displacement and Stress Profiles of the Graphite-Epoxy Composite Plate at 0°

The dispersion curves show the pairs of frequencies-wave numbers that are susceptible to propagate in the piece and generate vibrations to control the health of the plate. These vibratory states depend on the chosen generation mode. In fact, for a graphite-epoxy composite plate at $\varphi = 0°$(Fig. 4), Lamb waves and transverse waves polarized along the x_2 direction can propagate in this type of waveguide. Moreover, these modes can be symmetric or antisymmetric. In this section, we present the displacement and stress profiles of the graphite-epoxy composite plate at $\varphi = 0°$. For this purpose, we will choose to follow the two fundamental modes S_0 and A_0 (the most used modes [17] in the industry for UGW inspection) for a frequency of 500 kHz.

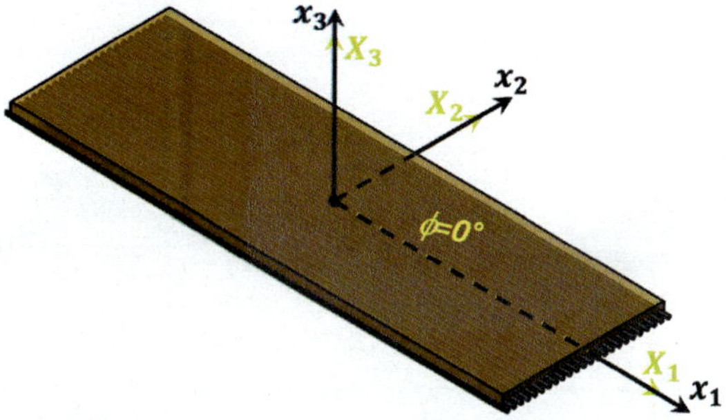

Fig. 4. Graphite-epoxy composite plate at $\varphi = 0°$.

The expressions for the analytical displacement fields are known in function of a constant [18] this makes no physical sense. To remedy this, several authors proposed the normalization by the acoustical power [19]. It allows having exact profiles of the displacements for each value of the frequency-wave number couples. Figures 5 and 6 shows the profiles of displacements and stresses normalized by the acoustic power. The frequency and wave number pairs taken are: (f = 500 kHz, k = 1046 m^{-1}) for the S_0 mode and (f = 500 kHz, k = 1516 m^{-1}) for the A_0 mode.

We note that the fundamental mode S_0 propagates symmetrically through the thickness of the composite plate. On the other hand, the A_0 mode propagates in an antisymmetric way. At a frequency of 500 kHz, the S_0 mode allows a greater vibration of the waveguide than the A_0 mode. In fact, the displacement u_1 vibrates between values of ±2 nm and u_3 between ±5 nm, while for the A_0 mode the longitudinal displacement is between ±1 nm and the transverse displacement between ±4 nm. Similarly, this will imply a high stress concentration for the S_0 mode. This information will help us in the characterization of the defects present in the graphite-epoxy composite plate under inspection.

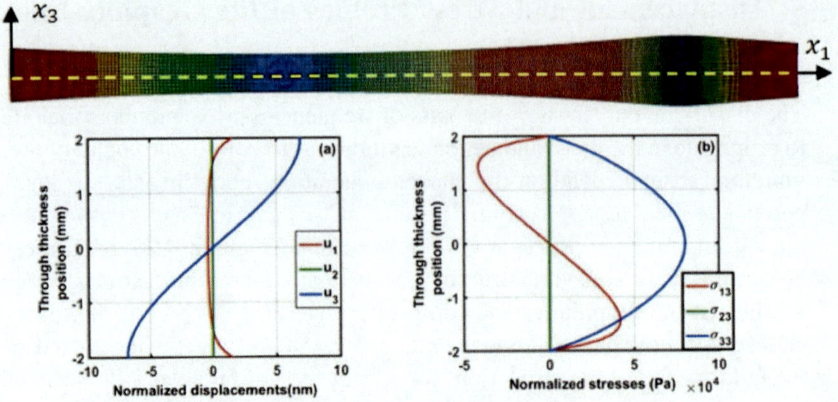

Fig. 5. Vibratory state of the composite plate for the S_0 mode. (a–b) Analytical displacements and stresses normalized to the power of the S_0 mode.

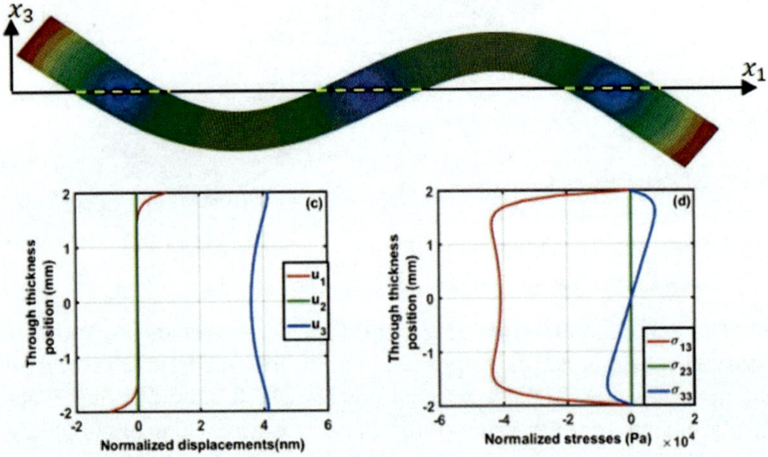

Fig. 6. Vibratory state of the composite plate for the A_0 mode. (c–d) Analytical displacements and stresses normalized by the power of the A_0 mode.

6 Conclusion

The hybrid method offers a very competitive way to plot the dispersion curves of a graphite-epoxy composite material for different degrees of anisotropy (transverse isotropic behavior for angles of 0° and 90° as well as monoclinic behavior for angles of 30° and 45°). This method overcomes the limitations of the bisection method (speed) and of the Newton-Raphson method (root jump). Adding to this a considerable gain in computation time. Comparing it with the DISPERSE software, the hybrid method was able to detect even modes close to the cut-off frequencies, which is not the case for the DISPERSE software. Following this, a discussion of the displacement profiles and the normalized acoustic power stresses of the fundamental modes was set up.

During experimental non-destructive testing in industry, the controllers use the DISPERSE software as a way of generating the dispersion curves, but as we have shown in the paper, the disperse software may not give accurate results for certain types of material (notably composites with arbitrary fiber directions). Our method did not have such problems (Sect. 4), and the hybrid method also showed a gain in computation time. So, with a view to the future, we're aiming to set up a digital interface based on the hybrid method, which will make it easier for industry to choose the control frequency.

References

1. Lukez, R.: The use of graphite/epoxy composite structures in space applications (1987)
2. Mayer, N.J.: Carbon composites in space vehicle structures (1974)
3. Dayal, V., Kinra, V.K.: Leaky lamb waves in an anisotropic plate. I: an exact solution and experiments. J. Acoust. Soc. Am. **85**(6), 2268–2276 (1989)
4. Neau, G.: Ondes de Lamb dans des plaques anisotropes viscoélastiques: étude des fronts d'onde et atténuation (Doctoral dissertation, Bordeaux 1) (2003)
5. Seyfaddini, F., Nguyen-Xuan, H., Nguyen, V.H.: A semi-analytical isogeometric analysis for wave dispersion in functionally graded plates immersed in fluids. Acta Mech. **232**(1), 15–32 (2021)
6. Barazanchy, D., Giurgiutiu, V.: A comparative convergence and accuracy study of composite guided-ultrasonic wave solution methods: comparing the unified analytic method, safe method and disperse. Proc. Inst. Mech. Eng. Part C J. Mech. Eng. Sci. **231**(16), 2961–2973 (2017)
7. Adamou, A.T.I., Craster, R.V.: Spectral methods for modelling guided waves in elastic media. J. Acoust. Soc. Am. **116**(3), 1524–1535 (2004)
8. Quintanilla, F.H., Lowe, M.J.S., Craster, R.V.: Modeling guided elastic waves in generally anisotropic media using a spectral collocation method. J. Acoust. Soc. Am. **137**(3), 1180–1194 (2015)
9. Zheng, M., He, C., Lu, Y., Wu, B.: State-vector formalism and the Legendre polynomial solution for modelling guided waves in anisotropic plates. J. Sound Vib. **412**, 372–388 (2018)
10. Zheng, M., He, C., Lyu, Y., Wu, B.: Guided waves propagation in anisotropic hollow cylinders by Legendre polynomial solution based on state-vector formalism. Compos. Struct. **207**, 645–657 (2019)
11. Pavlakovic, B., Lowe, M., Alleyne, D., Cawley, P.: Disperse: a general purpose program for creating dispersion curves. In: Thompson, D.O., Chimenti, D.E. (eds.) Review of Progress in Quantitative Nondestructive Evaluation, pp. 185–192. Springer, Boston (1997). https://doi.org/10.1007/978-1-4615-5947-4_24
12. Kamal, A.M., Gresil, M., Giurgiutiu, V.: Comparative study of several methods for the calculation of ultrasonic guided waves in composites. In: 54th AIAA/ASME/ASCE/AHS/ASC Structures, Structural Dynamics, and Materials Conference, pp. 1901 (2013)
13. Nayfeh, A.H.: Wave propagation in layered anisotropic media: with application to composites. Elsevier (1995)
14. Nayfeh, A.H., Chimenti, D.E.: Free wave propagation in plates of general anisotropic media (1989)
15. Ismaine, Z., Hassan, R., Abdelkerim, C.: New spectral displacement post-processing approach for modeling ultrasonic guided waves. In: 2023 3rd International Conference on Innovative Research in Applied Science, Engineering and Technology (IRASET), pp. 1–8. IEEE, May 2023
16. Zitouni, I., Rhimini, H., Chouaf, A.: Modeling the propagation of ultrasonic guided waves in a composite plate by a spectral approximation method. Eng. Trans. **71**(2), 213–227 (2023)

17. Mori, N., Biwa, S.: Transmission characteristics of the S0 and A0 Lamb waves at contacting edges of plates. Ultrasonics **81**, 93–99 (2017)
18. Zitouni, I., Rhimini, H., Chouaf, A.: Comparative study of the spectral method, DISPERSE and other classical methods for plotting the dispersion curves in anisotropic plates. J. Appl. Comput. Mech. **9**(4), 955–973 (2023)
19. Royer, D., Dieulesaint, E.: Elastic Waves in Solids I: Free and Guided Propagation. Springer, Heidelberg (1999)

Active Vibration Control of Non-uniform Section Beam

K. El Harti[1(✉)], M. Sanbi[2], I. Kadiri[1], and M. Rahmoune[1]

[1] Laboratory of Advanced Materials Studies and Applications, FS-EST, Moulay Ismail University of Meknes, Meknes, Morocco
elhartikhalid@gmail.com

[2] National School of Applied Sciences, Abdelmalek Essaadi University, Tétouan, Morocco

Abstract. This work presents the dynamics and AVC of a non-uniform FGM beam. The beam is equipped with four layers of piezoelectric sensors/actuators, on different finite elements to see the influence of its location on the dynamics and AVC. In this study, the classical EB theory combined with FEM is applied to a beam dividing it into a finite number of elements. Principle of Hamilton is applied to generate the equation of motion. The structure is modeled analytically then numerically and the simulation results are presented at the end.

Keywords: Vibration Control · Piezoelectricity · Functionally Graded Materials · Non-Uniform Beam · Euler-Bernoulli Beam Theory

1 Introduction

The field of smart materials is a relatively virgin and unexplored area. It is therefore undeniable that the characteristics and applications of smart materials have yet to be explored. In order to facilitate product design innovation and open up new opportunities, designers and manufacturers must be able to understand the essential characteristics of smart materials. Since the beginning, materials have been classified into two main categories: structural materials that are mainly used for their mechanical properties (building construction, armor…) and functional materials whose ability to conduct electrical current, for example, or transparency find many applications. Smart materials also celebrate the growing role of biological models in the design of new products.

Copying living systems, molecular or cellular micromachines, active or selective membranes, allows the exploration of many new applications in the medical or computing fields. Today, smart materials are used in a large number of sectors, from construction to sports equipment, biomedicine, robotics and the military, because a smart material is sensitive, adaptable and evolutive. For example, it is possible to detect structural weaknesses in the skin of an airplane, cracks in a building or a concrete dam, reduce vibrations in helicopter blades, or insert filters into arteries that will dilate to reduce the risk of blood clots.

S. Aniss et al. (Eds.): CMM 2022, LNME, pp. 13–22, 2024.
https://doi.org/10.1007/978-3-031-46973-2_2

In the mid-1980s, a new class of composites, called functional gradient materials (FGM), was born by a group of Japanese scientists [1]. That year, Japanese researchers were faced with a problem that necessitated a specific type of composite material that could withstand a very large temperature difference in a space project. These materials exhibit a progressive change in composition with volume [2]. These composites are largely used in various engineering applications due to their resistance to high temperatures.

Mechanical structures are often subjected to vibrations from a variety of sources. These vibrations are generally sources of problems affecting the proper functioning of many systems and processes in different industrial sectors and many engineering fields such as automotive, aeronautics, naval.... These systems are increasingly integrating composite materials in the manufacture of structures. The amplitude of vibration can lead to a significant number of problems such as structural damage or fatigue. One of the solutions to such a problem is active vibration control (AVC) [3].

Born in the thirties, active control only really developed with the advent of digital signal processors in the eighties. In addition, mechanical systems are increasingly incorporating composite materials into the fabrication of structures. In the case of thin structures, the use of piezoelectric transducers bonded to the structure is often used.

Piezoelectric materials, for example, have the ability to produce an electrical voltage when compressed. And conversely, when subjected to an electric current, they can also be mechanically deformed. The frequency of the electrical signal and its amplitude vary directly with the mechanical deformation they undergo. These materials are generally made of ceramic and more recently of polymers. The best known are the quartz crystals used in quartz watches to maintain the basic vibrations used to measure time. Piezoelectric materials are also used to dampen vibrations and reduce noise. For example, piezoelectric materials can be used to surround a rotating shaft to significantly reduce vibration. Piezoelectric polymers are also used for medical applications or to capture ultrasound. An interesting application of piezoelectric materials is monitoring the health of certain materials used in the construction of aircraft hulls or cement buildings. A piezoelectric sensor can detect localized defects, such as cracks, holes or impacts. A fracture will generate noise or vibrations that can be analyzed by the sensor. When carbon fibers break, they will change the resistance of the electrical circuit they constitute. Here is an application: the "intelligent cement".

This cement has a kind of "nervous system" that allows it to detect internal changes and transmit information to the outside. With this type of cement, we can build bridges or dams capable of warning engineers of areas of fragility at the very places where cracks or fractures may appear. Subjected to various stresses, weight, vibrations, frost, earthquakes, cement constructions can fail suddenly without any sign being detected during preventive visits. The classical method consists in defining and identifying beforehand the supposedly fragile points to apply monitoring probes. But these points must be selected at the time of construction and their usefulness may diminish over time. This is why researchers at the University of New York at Buffalo had the idea of creating a real nervous system in cement based on carbon fibers. These fibers, 10 microns in diameter and a few centimeters long, are mixed into the cement during its preparation. Even if they represent only 0.05% of its volume, they increase its electrical conductivity by 10%. These fibers protrude on the outside, which ensures good electrical contact. It is therefore possible to place electrodes anywhere on the surface of a "smart cement" construction and detect a change in stress. The electrical resistance of the cement can be measured. An alarm can now be triggered long before a wall cracks or a bridge shaken by an earthquake threatens to collapse.

It is important to note that thin walled composite beams have a wide range of applications in aero-space engineering and other sectors. Therefore, the reduction of noise and vibration of these structures is a priority research topic in many industrial sectors.. Also, the additional weight and size of the sensors and actuators required for active vibration control have slowed down the development of many applications. It is only recently, in the last decades, that the use of transducers based on piezoelectric materials has allowed advances in applications that were previously forbidden. These materials, attached to the structure to be damped, are compressed/stretched when the structure vibrates and converting a part of the mechanical energy of the structure into electrical energy. The aim is to transfer or degrade this energy, the mechanical energy of the structure is reduced and the vibrations are therefore damped.

Examples of systems where AVC can be applied are numerous, such as fans, vehicle interiors, precision equipment, engines, electric and hydraulic drives, noise barriers and enclosures… [4–6]. There have been extensive theoretical and experimental studies on FGM structures that have been published, especially beams and are still of interest to researchers because of their applications. El Harti et al. studied the AVC of an FGM beam with a piezo-materials with a new geometry [5]. They also addressed the active control of the porous Euler-Bernoulli FGM beam in a thermal environment with symmetrically bonded piezo materials [7], as well as the dynamic analysis and AVC of the distributed piezo-thermo-elastic porous FGM beam modeled by the FEM [8]. New numerical methods as robust as the FEM can be proposed for the structural dynamics analysis and vibration control of an FGM beam, such as the "Differential Quadrature" method and the "Bezier" method.

Non-uniform variable cross-section structures are used commonly in different applications of engineering, including wind turbine, helicopter blade, ship propeller, space and marine structures, because they respond to specific architectural requirements due to their capacity, also they present an optimized weight and strength distribution., etc. Sahabni and Cunedioglu studied the free vibration analysis of cracked FG non-uniform beams [9]. Ebrahimi and Hashemi investigated the vibration analysis of non-uniform FG porous beams under thermal loading [10]. Kumar et al. analyzed the geometrically nonlinear free vibration of axially FG taper beams [11]. Chen studied the vibration analysis of axially FG Timoshenko beams with non-uniform section-cross [12]. Huang et al. examined the free vibration of axially functionally graded Timoshenko beam with non-uniform cross section [13]. Ozdemir et al. presented a study of the vibration analysis of a rotating tapered beam using DTM [14].

This paper is an extension of our previous works [5, 7, 8]. The novelty of this work is manifested in its study of the active vibration control of functionally graded beams with non-uniform section using Euler-Bernoulli's beam theory combined with FEM.

2 Mathematical Model

A non-uniform FG beam with dimension L with constant height h, and width $b(x)$ is considered variable in this study.

The general equations for the cross-sectional area, width and moment of inertia of the beam are given by [10]:

$$
\begin{aligned}
b(x) &= b_0\left(1 - C_b\frac{x}{L}\right)^m \\
A(x) &= A_0\left(1 - C_b\frac{x}{L}\right)^m \\
I_y(x) &= I_{y0}\left(1 - C_b\frac{x}{L}\right)^m
\end{aligned}
\tag{1}
$$

The width taper ratio (C_b) must satisfy $0 \leq C_b < 1$ to ensure that $A(x)$ and $I_y(x)$ are positive. Knowing the width quantities at the embedding and at the free end of the beam, the width taper ratio can be calculated as follows:

$$
C_b = 1 - \frac{b}{b_0}
\tag{2}
$$

In this study, the beam shrinks linearly ($m = 1$). The quantities of cross sectional area, moment of inertia at the beam embedment can be written as follows:

$$
\begin{aligned}
A_0 &= b_0 h \\
I_{y0} &= \frac{b_0 h^3}{12}
\end{aligned}
\tag{3}
$$

To describe the variation of material properties from pure metal on the bottom side to pure ceramic on the top side, we consider the power law presented as follows [7]:

$$
V_c = \left(\frac{z}{h}\right)^k = 1 - V_m
\tag{4}
$$

Considering the Euler-Bernoulli beam theory, the axial and transverse displacements, u and w, of any point on the beam are given by [7]:

$$
\begin{aligned}
u(x, y, z, t) &= z\frac{\partial w(x, t)}{\partial x} \\
w(x, y, z, t) &= w(x, t) \\
v(x, y, z, t) &= 0
\end{aligned}
\tag{5}
$$

Under EB' assumptions and we use the boundary conditions for 2d.d.l per node, the final form of becomes [5]:

$$[w(x,t)] = \left[f_1(x)\, f_2(x)\;\; f_3(x)\, f_4(x) \right] \times \left[w_1\; \theta_1\;\; w_2\; \theta_2 \right] = \left[n^T \right][q] \tag{6}$$

With

$$[w\prime(x,t)] = \left[n_2^T \right][q] \tag{7}$$

$$\left[w''(x,t) \right] = \left[n_1^T \right][q] \tag{8}$$

$$[\dot{w}(x,t)] = \left[n_1^T \right][\dot{q}] \tag{9}$$

The motion' equation of the 2-node element is obtained via application of Lagrange principle [7]:

$$M\ddot{q} + Kq = f(t) \tag{10}$$

3 Dynamic Equation and State Space Model

To focus on the control of the first vibration modes, we used the transformation $q = Tg$. . The state space (SS) model in Multi-In Multi-Out (MIMO) mode of the beam, for the first vibrational modes is given by [7]:

$$\dot{x} = Ax(t) + Bu(t) + Er(t) \tag{11}$$

$$y(t) = C^T x(t) + Du(t) \tag{12}$$

4 Results and Discussion

In the aim to validate the control procedure, we consider an embedded-free FGM beam with a non-uniform section, partially covered by four layers of piezoelectric materials. A pulse of amplitude (1N) is applied as external force. The figures below show the response of the beam to the pulse. In all results, we compare responses with and without control, varying the location of piezoelectric actuators as well as the taper ratio, and the power index.

Figure 1 shows the variation of the actuators location. Figure 2 and Fig. 3 show the variation of the power index, for width ($b = 2$; $b = 3$) successively, keeping the actuators on FE2. The Fig. 4 show Variation of vibration frequencies (Hz) as a function of power index k, location of piezoelectric actuators and taper ratio.

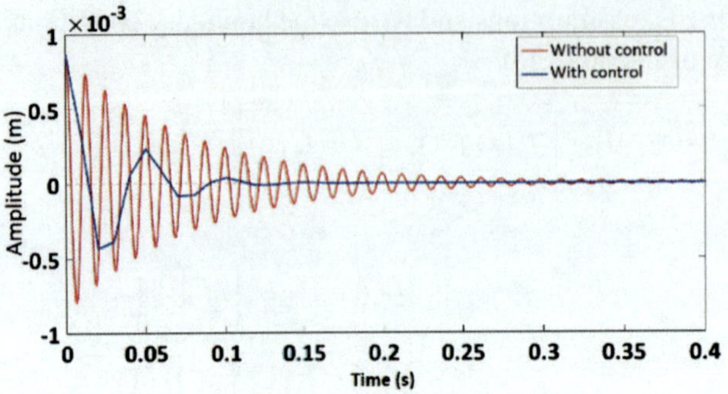

(a) Actuators on FE2

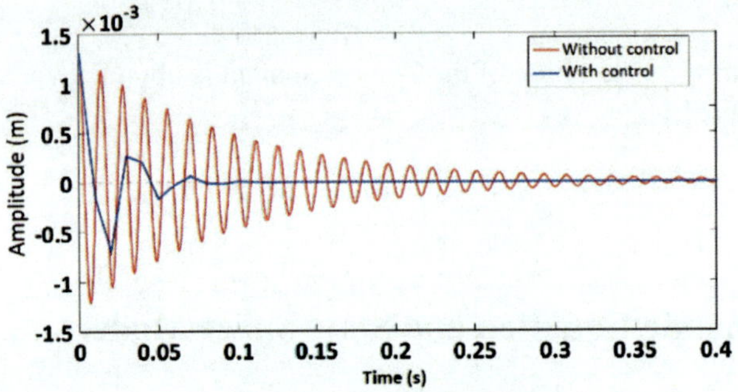

(a) Actuators on FE3

Fig. 1. Impulse response, (k = 1 and b = 3)

The analysis of results shows that the increase in the power index, implies increases in vibration amplitudes (see Fig. 2 and Fig. 3), also decreases in vibration frequencies (see Fig. 4). Increasing the taper ratio C_b, , implies increases in vibration amplitudes (see Fig. 2 and Fig. 3), also decreases in vibration frequencies (see Fig. 4). Moving the actuators towards the free end involves increases in vibration amplitudes (see Fig. 1), as well as decreases in vibration frequencies (see Fig. 4).

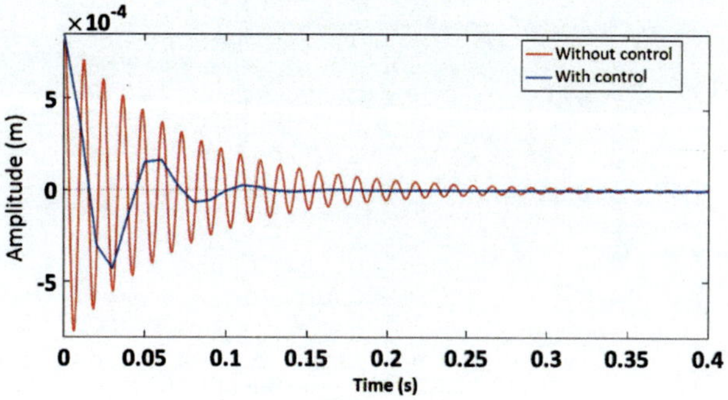

(a) k=0.5

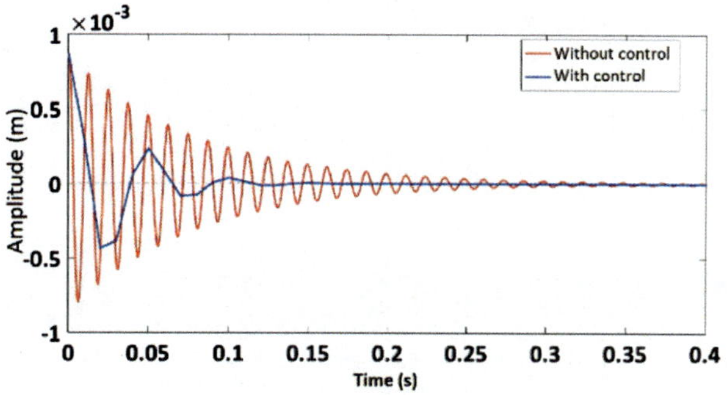

(b) k=1

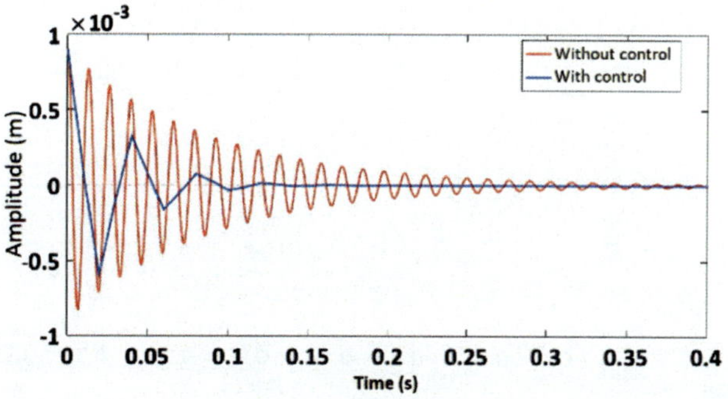

(c) k=2

Fig. 2. Impulse response, with power index variation (Actuators on FE2, and b = 2)

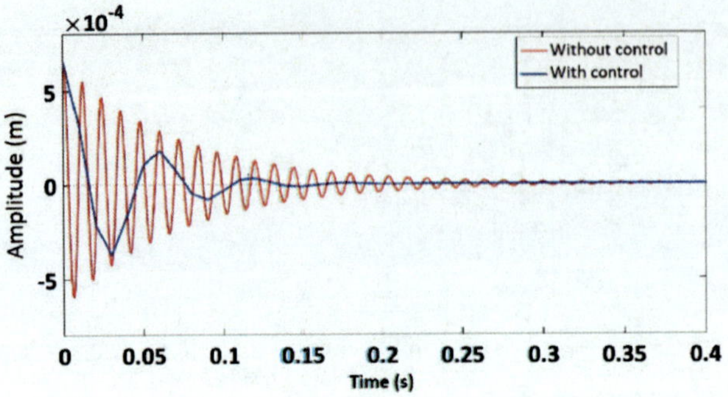

(a) k=0.5

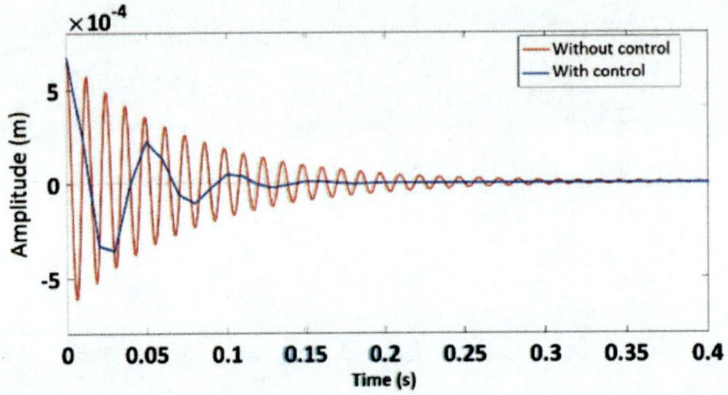

(b) k=1

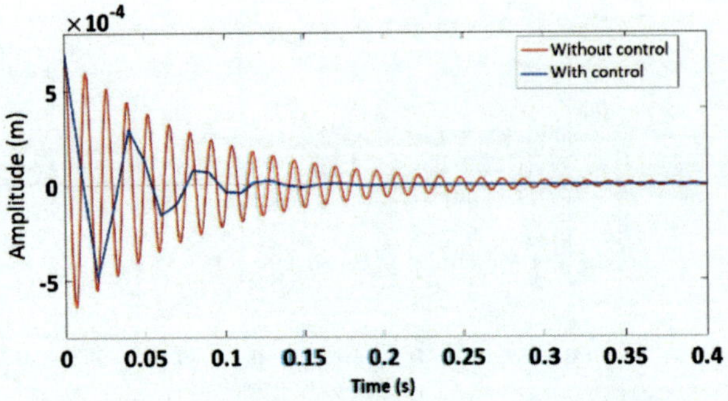

(c) k=2

Fig. 3. Impulse response, with power index variation (Actuators on FE2 and b = 3)

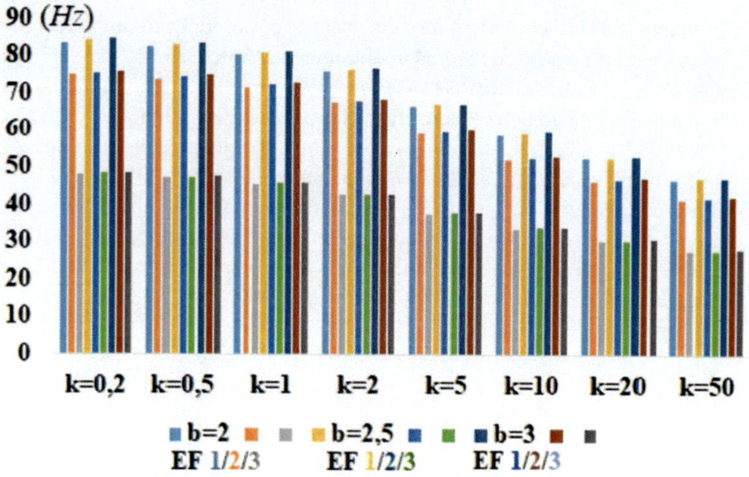

Fig. 4. Variation of vibration frequencies (Hz) as a function of power index k, location of piezoelectric actuators and taper ratio.

5 Conclusion

In this paper, we have studied the theoretical analysis of the AVC of a smart beam with a non-uniform cross-section, based on the Euler-Bernoulli beam theory combined with the finite element method. During the study, we examined the influence of the power index, the taper ratio and the location of the actuators on the vibrations and the active control. We also showed the reliability of our LQG optimal control procedure with Kalman filter.

References

1. Koizumi, M.: FGM activities in Japan. Compos. Part B Eng. **28**, 1–4 (1997)
2. Bharti, I., et al.: Novel applications of functionally graded nano, optoelectronic and thermoelectric materials. Int. J. Mater. Mech. Manuf. **1**(3), 221–224 (2013)
3. Beards, C.E.: Structural Vibration: Analysis and Damping, 1st edn. London (1996)
4. Bendine, K., et al.: Active vibration control of functionally graded beams with piezoelectric layers based on higher order shear deformation theory. Earthq. Eng. Eng. Vib. Eng. Eng. Vib. **15**(4), 611–620 (2016)
5. El Harti, K., et al.: Active vibration control of sandwich FGM beam with piezoelectric sensor/actuator. Int. J. Appl. Eng. Res. **12**(20), 9338–9345 (2017)
6. Panda, R.K., et al.: Active vibration control of smart functionally graded beams. Procedia Eng. **144**, 551–559 (2016)
7. El Harti, K., et al.: Dynamic control of Euler-Bernoulli FG porous beam under thermal loading with bonded piezoelectric materials. Ferroelectrics **558**(1), 104–116 (2020)
8. El Harti, K. et al.: Dynamic analysis and active control of distributed piezothemoelastic FGM composite beam with porosities modeled by the finite element method. Compos. Mech. Computations Appl. Int. J. **12**(1), 57–74 (2021)
9. Shabani, S., Cunedioglu, Y.: Free vibration analysis of cracked functionally graded non-uniform beams. Mater. Res. Express **7**(1), 015707 (2020)

10. Ebrahimi, F., Hashemi, M.: Vibration analysis of non-uniform imperfect functionally graded beams with porosities in thermal environment. J. Mech. **33**(6), 739–757 (2017)
11. Kumar, S., Mitra, A., Roy, H.: Geometrically nonlinear free vibration analysis of axially functionally graded taper beams. Eng. Sci. Technol. Int. J. **18**(4), 579–593 (2015)
12. Chen, W. R.: Vibration analysis of axially functionally graded Timoshenko beams with non-uniform cross-section. Lat. Am. J. Solids Struct. **18**, e397 (2021)
13. Huang, Y., Yang, L.E., Luo, Q.Z.: Free vibration of axially functionally graded Timo-shenko beams with non-uniform cross-section. Compos. Part B Eng. **45**(1), 1493–1498 (2013)
14. Ozdemir Ozgumus, O., Kaya, M.O.: Vibration analysis of a rotating tapered Timoshenko beam using DTM. Meccanica **45**(1), 33–42 (2010)

Solid and Structural Mechanics Modeling

Numerical Study of the Effect of the Brace Inclination on the Stress Concentration Factor for Three Different Welded Tubular Y-Joints Subjected to Deflected Bending Load

E. Chouha(✉), SE. Jalal, Z. El Maskaoui, and A. Chouaf

Mechanics, Engineering and Innovation Laboratory, Higher School of Electricity and Mechanics ENSEM, Hassan II University, Casablanca, Morocco
{h.chouha,se.jalal,z.elmaskaoui,a.chouaf}@ensem.ac.ma

Abstract. The tubular metal constructions find their applications in the following industrial sectors: metal tanks, industrial hangars, metal bridges, commercial buildings, pipelines, support for oil platforms etc. The phenomenon of stress concentration, born in the welded zones, contributes to the damage of these structure.

In the literature, the semi-parametric formulas give information about the stress concentration factor value, but gives no idea about its distribution.

In this study, three different Y-joints have been studied. A finite element modeling was used to show the inclination brace effect ($\theta = 45°$, $\theta = 60°$ and $\theta = 90°$) on the distribution of stress concentration factor SCF and on the tubular Y-joint strength under deflected bending loads.

Keywords: Tubular Joint · Stress Concentration Factor · Hot Spot · Linear extrapolation method · Y-joint · Finite Elements Methods

1 Introduction

Metal construction is a broader field, it is used in a wide of industrial fields, including: metal bridges, offshore platform supports, pipelines, pressure vessels, handling cranes, electric power transmission pylons, lighting masts, metal frames, etc.

These structures are designed based on elements of circular sections assembled by welding. The preparation operations for the assembly and the welding process give rise to places of geometric discontinuity at intersections of the welded elements. In service, these structures undergo various loading; out-of-plane bending, tension, in-plane bending, deflected bending and combined loads etc. Which generates two damage mechanisms depending on the type of applied loading. When this loading is increasing and monotonous, it leads to damage to the assembly by plastic deformation due to the stress concentration in the vicinity of the welded zones. But, when it is a cyclic loading, we speak then about the damage by fatigue.

S. Aniss et al. (Eds.): CMM 2022, LNME, pp. 25–36, 2024.
https://doi.org/10.1007/978-3-031-46973-2_3

In the literature, some others Chen et al.[1], Shao et al.[2] have found parametric formulas allowing the prediction of the maximum value of the stress concentration factor on simple uniplanar joints form Y, X, K, T and KT-joints. In order to find the stress concentration factor in overlapped joints, several studies [3, 4] have been carried out.

The structural stress approach, has used [5] to study the distribution of the stress concentration factor along the welded zone in uniplanar joints with various shapes.

Some authors have preferred the study of multi-planar joints with different shapes such as XX, KK etc. Chiew et al. [6] and Karamanos et al. [7] have studied muti-planar XX-joints. The authors Woghiren et al. [8] have studied the stress concentration at the weld toe for the reinforced multi-planar KK -joint under out-of-plane bending and tensile load.

Unfortunately, the semi-parametric formulas that we find in the literature, can only inform us about the value of the factor of stress concentration, but gives no idea about its distribution around the weld toe.

For this reason, in the present study, a FEM modeling was carried out for three tubular Y-joints but with a different brace angle ($\theta = 45°$, $\theta = 60°$ and $\theta = 90°$), this in the objective to find the effect of this angle θ on the stress concentration factor distribution and on the joint mechanical resistance under a series of deflected bending loads.

2 Dimensions and Mechanicals Properties

In this study, we worked on three Y-joints of the same dimensions, the only difference being the angle inclination brace; the brace of the first Y-joint is perpendicular ($\theta = 90°$) to the chord. The brace of the second Y-joints is inclined by $\theta = 60°$ to the chord, while the last joint has a brace inclined at $\theta = 45°$ to the chord.

The chord of all shapes studied has the same length L = 720 mm and the same diameter D = 88,7 mm with the same thickness T = 3 mm. All the braces of these joints have the same diameter d = 48 mm, the same length l = 350 mm and the same thickness t = 2 mm.

The three Y-joints were made of mild steel with a yield strength $\sigma_y = 240$ MPa, a Young's modulus E = 210 GPa and a Poisson's ratio $\nu = 0{,}3$. Some details are given in Fig. 1 below:

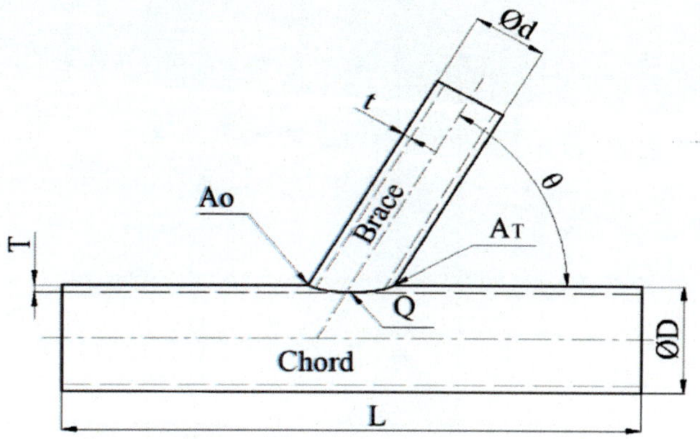

Fig. 1. Details of joints studied

3 Finite Element Analysis

In this study, the three models studied were created by the CATIA V5 design software, and subsequently the numerical simulation of these models was carried out by the Patran-Nastran software.

3.1 Numerical Model

For numerical simulations, we used the finite element method to evaluate the mechanical behavior of structures. However, the simulation accuracy results are affected by several factors. The type of finite elements used, the quality of mesh refinement in stress concentration zones and the integration scheme considered.

Two types of finite elements "thin shell" and "solid element" can be used. The authors Chang et al. [9] and Hamed et al. [10] have used a shell element with 8 nodes. But Chiew et al. [6] recommend the use of solid elements such hexahedral with 8 or 20 nodes.

According to the latter authors, the use of shell elements gives an inaccurate result and the use of the linear extrapolation (Fig. 2) recommended by IIW [11] is not possible to find the local stress in the weld toe vicinity.

In this study, the mesh used is carried out using solid hexahedral elements with eight nodes according to the ARSEM recommendations [12]. This choice was made so that the weld bead could be taken into consideration. In order to apply the extrapolation method, the mesh enhancement around the weld toe is performed (Fig. 3).

3.2 Mesh Convergence Studies

In order to have a good accuracy of the FE results with a reasonable calculation time, a better size of finite elements must be considered. For this reason, a mesh convergence studies were carried out. Indeed, we considered a T-joint ($\theta = 90°$) in simple tension loading, and four sizes of elements adjacent to the intersection line: 0, 5; 1; 2 et 5 mm, the number of elements adjacent to this line is: 420; 264; 128 and 56 elements respectively. The results obtained are reported in the Table 1 and in Fig. 4.

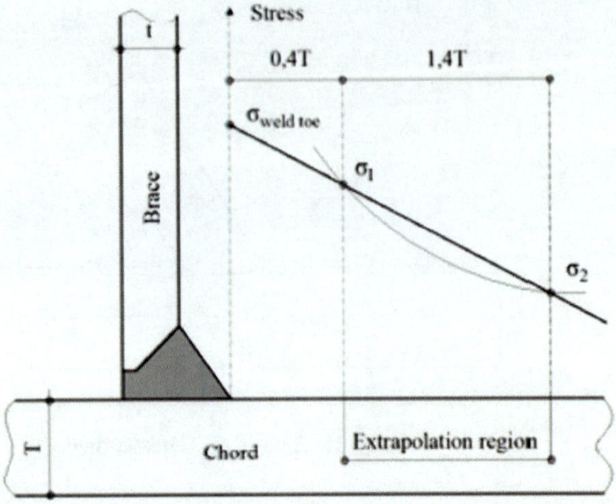

Fig. 2. Linear extrapolation method

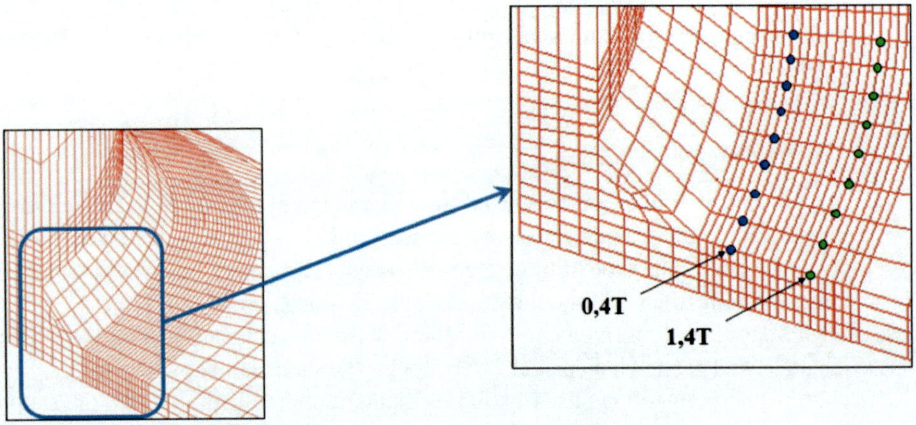

Fig. 3. Mesh refinement and lincar extrapolation point

Table 1. Mesh convergence studies result

Element size (mm)	Number of element adjacent	Stress (MPa)
0,5	420	235
1	264	231
2	128	185
5	56	160

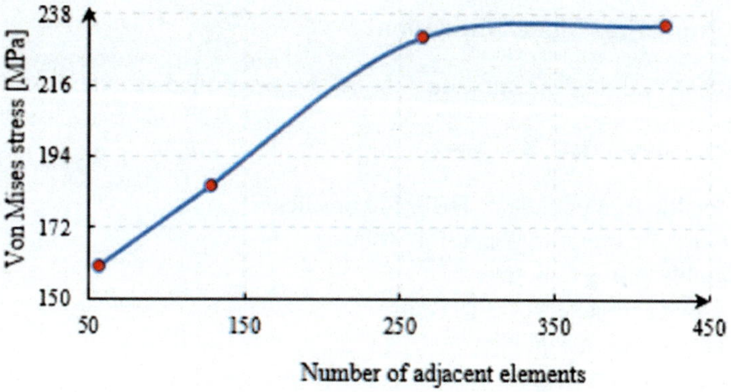

Fig. 4. Mesh convergence study

Figure 4 show that using a mesh size of 1 mm for refinement at the intersection area between brace and chord is so sufficient to have a good accuracy.

3.3 Loading and Boundary Conditions

After fixed the both ends of the chord (Fig. 5), a force F = 1647 N has been applied in (Oxy) plane to the free brace end by using a rigid body. In nineteen angular orientations ψ, this force was applied with an increment of 10° counted from Crown point (Fig. 5).

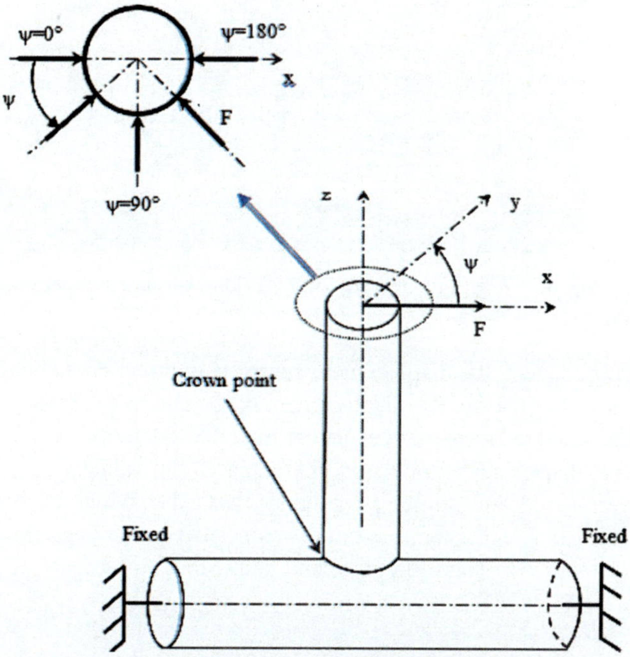

Fig. 5. Boundary and load conditions

3.4 Numerical Model Validation

In order to validate the numerical model, we performed an experiment on the Y90° joint under tensile load.

The test rig (Fig. 6 and 7) is composed of the following elements:

- Metallic frame made of HEB 240 profiles
- Hydraulic pressure generator 600 bar;
- Double acting cylinder;
- Displacement sensor ± 50 mm;
- Force sensor (200 kN);
- Data acquisition card;

The chord ends of the specimen were rigidly fixed on the rigid frame by eight bolts and the force is applied at the free end brace. The force sensor is fixed between the cylinder and the brace (Fig. 7b) using a pivot connection.

Fig. 6. View of the test rig for Y90°-joint specimen under tension loading.

During the test, the displacement sensor is in direct contact with the flat part of the force sensor to measure the displacement of the end of the brace.

Force and displacement sensors are linked to a laptop via a data acquisition card to record and display real-time force and displacement values.

Experimental and numerical results (elastic behavior) are reported on the graph, Fig. 8. We can see that the maximum error between the experiment and the numerical model is about 7%. Therefore, we can assume that our numerical model is valid.

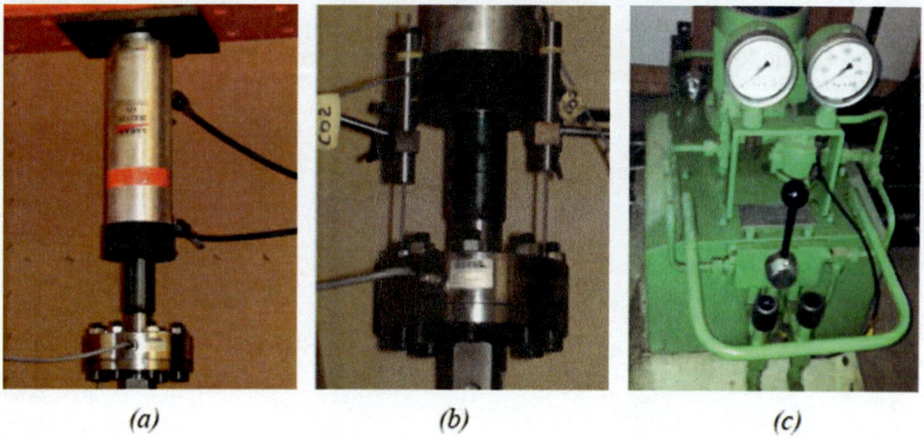

Fig. 7. Main elements of the test rig. (a) Double acting cylinder, (b) Force sensor with two displacement sensors, (c) Hydraulic pressure generator 600 bar.

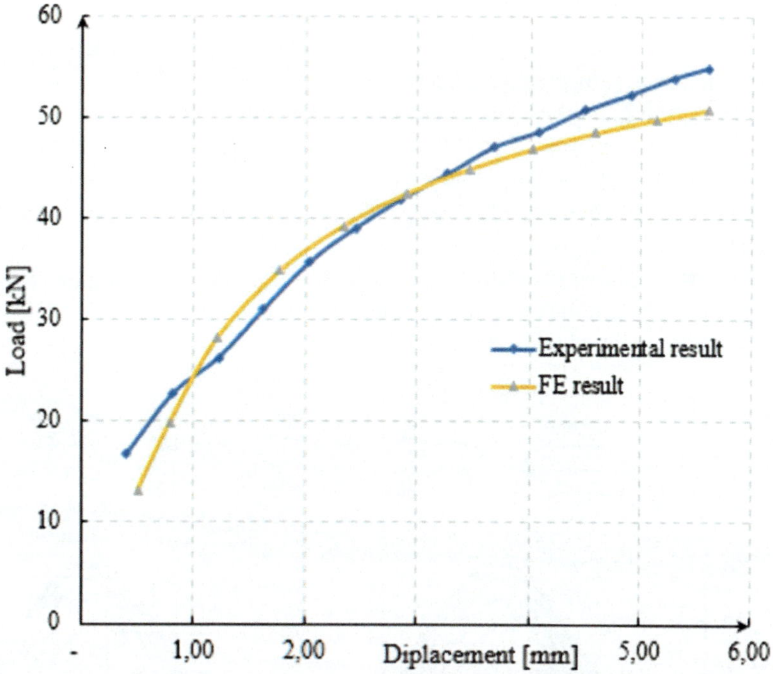

Fig. 8. Experimental and finite element results for axial tension loading.

4 Results and Discussions

To circumvent the issue of geometric singularities, for each load orientation and for each joint, the local stress evaluation around the welded zone is made by linear extrapolation method (Fig. 2). The goal of this method is to evaluate the stress on the contours spaced at 0,4T and 1,4T from the welded zone. This method is applied to all the nodes forming the weld toe. Therefore, local stress is calculated as follows:

$$\sigma_{Local} = 1,4\sigma_1 - 0,4\sigma_2 \tag{1}$$

Under the moment (M_i), The nominal stress can be calculated from the following formula:

$$\sigma_{nominal} = \frac{32dM_i}{\pi\left[d^4 - (d-2t)^4\right]} \tag{2}$$

Once the nominal stress is found by Eq. (2), the SCF of these joints can be determined as follow:

$$SCF = \frac{\sigma_{Local}}{\sigma_{nominal}} \tag{3}$$

where the σ_{Local} is the hot spot stress around the weld toe.

Figures 9, 10 and 11 show the SCF at the weld zone for Y90°, Y60° and Y45° joints respectively. When $\psi = 0°$, bending occurs in plane defined by joint axis. On the other hand, if $\psi = 90°$, the bending is out-of-plane bending. When $\psi \in]0°,90°[\ U]90°,180°[$, the junction is subjected to deflected bending.

From the plane bending to out-of-plane bending, the hot spot moves towards the Saddle point. For ($\psi = 0°$ or 180°) the extrapolated local stress is minor, and it is highest in the case of out-of-plane bending ($\psi = 90°$ or 270°).

In the service state, these joints support combined loading: in and out of plane bending.

Figure 12 shows the ultimate safe limit for combined loadings discussed in this study.

In the Fig. 13, the maximum local stress of the three Y-joints, has been represented depending on the loading direction.

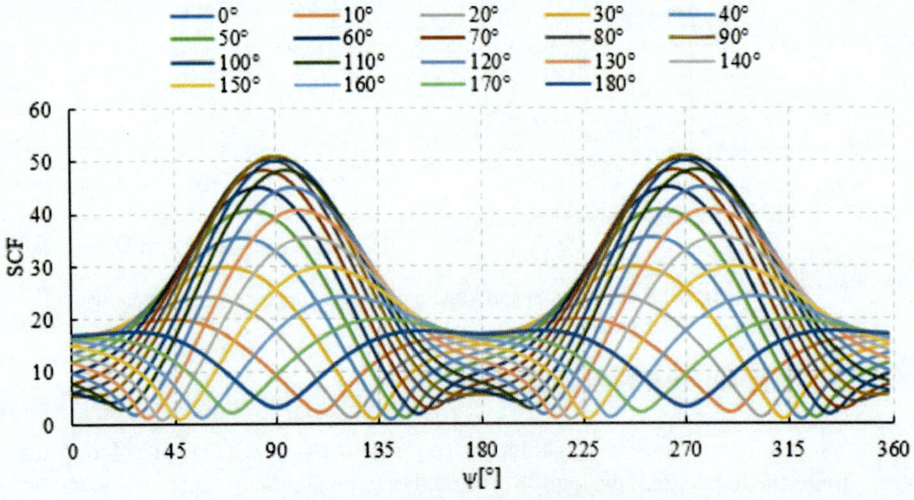

Fig. 9. Distribution of SCF at the welded zone for Y90°.

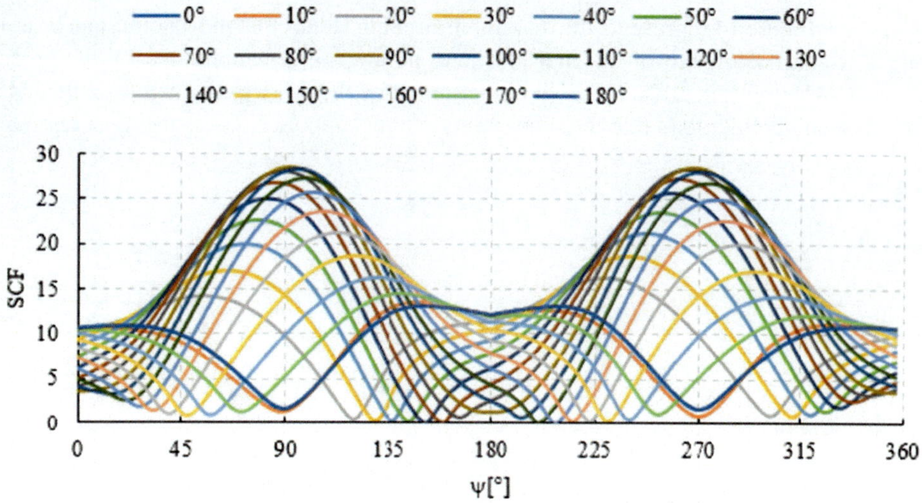

Fig. 10. Distribution of SCF at the weld toe for Y60°.

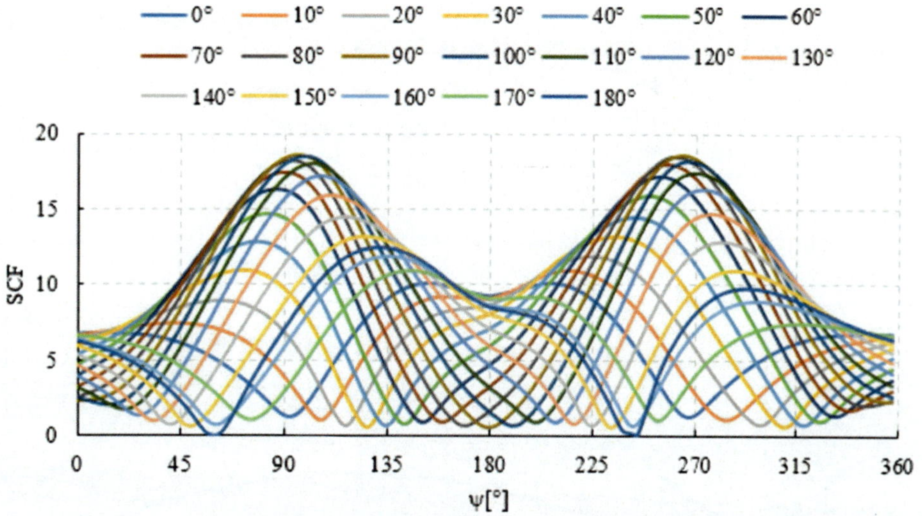

Fig. 11. Distribution of SCF at the weld toe for Y45°.

The maximum stress is lower in the case of the small brace inclination angle than those of the others.

In the deflected loading case, the maximum stress is equal to 327 MPa (Y90°). For this configuration, the maximum stress exceeds far the yielding stress, which locally generates a plastic deformation. While the Y45° joint has a maximum stress of 166 MPa that's a ratio of 1,97. The Saddle points ($\psi = 90°$ and $\psi = 270°$) correspond to stress concentration zone.

Figure 14 represents the maximum strength values that the Y-joints can bear, leading to the occurrence of the first plasticity point. The area delimited by the green curve (Y45°-joint) is greater than those delimited by the other two. Therefore, the Y45° joint supports high load that the other joints cannot support. So, it is the most resistant.

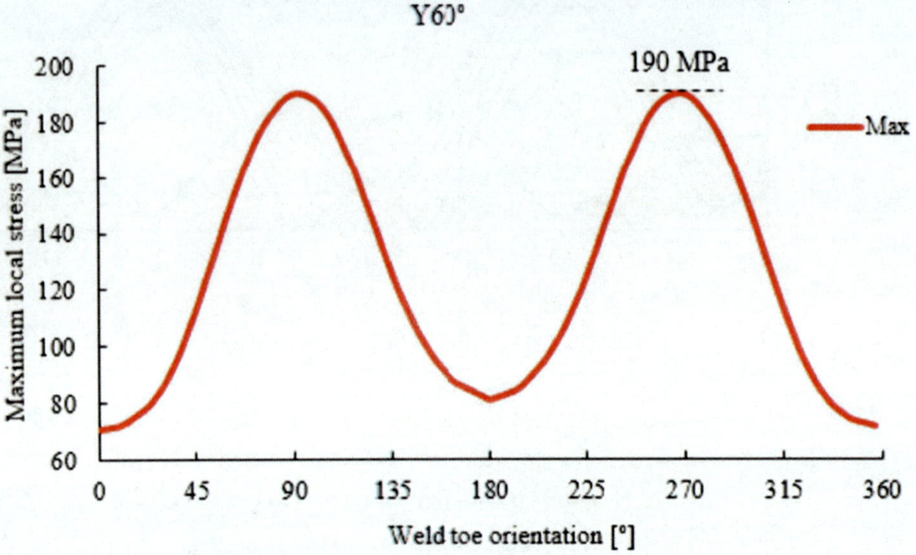

Fig. 12. Safety limit load.

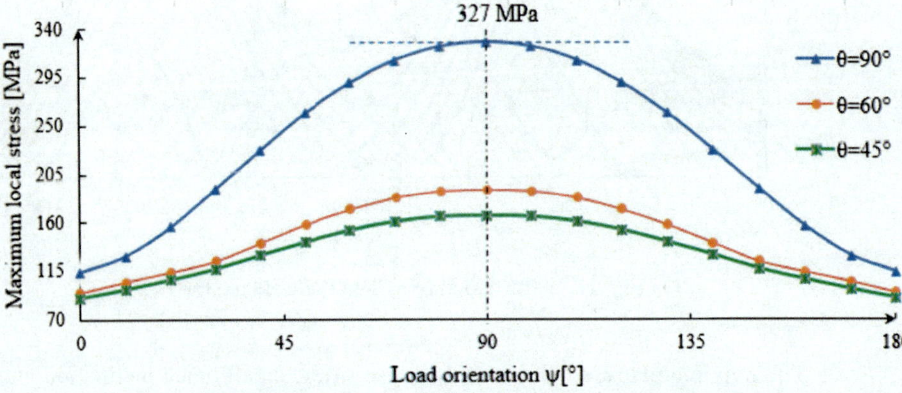

Fig. 13. Maximum local stress distribution vs load orientation.

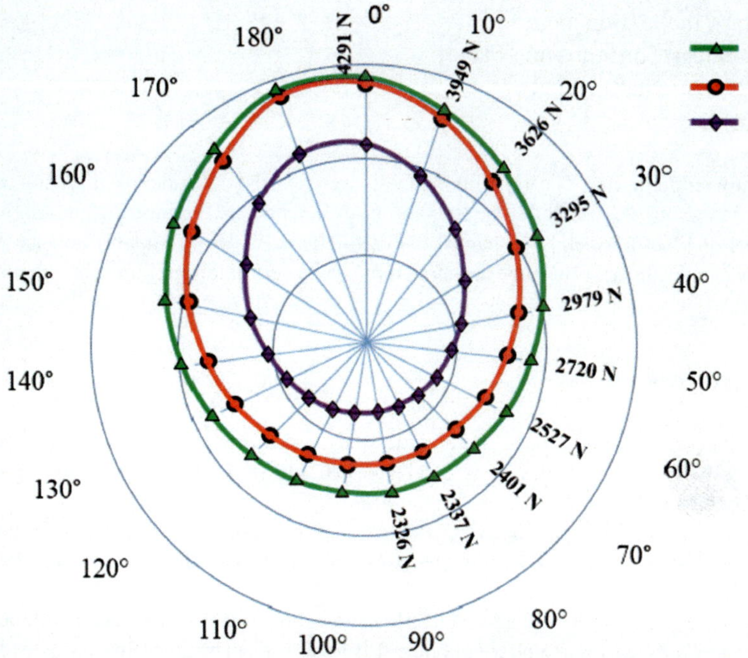

Fig. 14. Limit load before yielding for each angular direction.

5 Conclusion

In this study, the response of the Y45°, Y60° and Y90°-joint to deflected bending loads was treated. The obtained results allowed us to conclude that:

- Compared to the Y90°-joint, the Y45°-joint resists 1,97 times more;
- Compared to the Y60°-joint, the Y45°-joint resists 1,14 times more;
- The Y45° joint is the strongest of all the configurations studied;
- For the three joints, the Saddle point ($\psi = 90°$) is always the position of the highest local stress, on the other hand, the Toe and Heel points are always safe;
- The out-plane bending load is the most harmful of all the situations that were evaluated.
- Each joint's limit load surface has been drawn, serving as a tool for designers to make decisions;
- Finally, the decrease in θ leads to increase Y-joint strength.

Notations

L: Chord length
A_T: Crown Toe ($\psi = 0°$)
A_O: Crown Heel ($\psi = 180°$)
t : Brace wall thickness
T: Chord wall thickness
D: Brace diameter

θ: Brace inclination angle
SCF: Stress Concentration Factor
ψ: Load orientation angle
Q: Saddle

Acknowledgements. The experimental tests were carried out under the direction of Pr S.E. Jalal, Z. El Maskaoui, A. Chouaf at the Laboratory of Mechanics, Engineering and Innovation, Higher School of Electricity and Mechanics ENSEM, Hassan II University, Casablanca Morocco. The authors gratefully acknowledge the anonymous reviewers of this preliminary version of this article.

References

1. Chen, Y., Yang, J., Hu, K.: Parametric study and formulae of SCFs for positive large eccentricity CHS N-joints. J. Constr. Steel Res. **120**, 117–131 (2016). https://doi.org/10.1016/j.jcsr.2016.01.005
2. Shao, Y.-B., Du, Z.-F., Lie, S.-T.: Prediction of hot spot stress distribution for tubular K-joints under basic loadings. J. Constr. Steel Res. **65**(10), 2011–2026 (2009). https://doi.org/10.1016/j.jcsr.2009.05.004
3. Gao, F., Shao, Y.B., Gho, W.M.: Stress and strain concentration factors of completely overlapped tubular joints under lap brace IPB load. J. Constr. Steel Res. **63**(3), 305–316 (2007). https://doi.org/10.1016/j.jcsr.2006.05.007
4. Yang, J., Chen, Y., Hu, K.: Stress concentration factors of negative large eccentricity tubular N-joints under axial compressive loading in vertical brace. Thin-Walled Struct. **96**, 359–371 (2015). https://doi.org/10.1016/j.tws.2015.08.027
5. Liu, G., Zhao, X., Huang, Y.: Prediction of stress distribution along the intersection of tubular T-joints by a novel structural stress approach. Int. J. Fatigue **80**, 216–230 (2015). https://doi.org/10.1016/j.ijfatigue.2015.05.021
6. Chiew, S.P., Soh, C.K., Wu, N.W.: General SCF design equations for steel multiplanar tubular XX-joints. Int. J. Fatigue **22**(4), 283–293 (2000). https://doi.org/10.1016/S0142-1123(99)00130-9
7. Karamanos, S.A., Romeijn, A., Wardenier, J.: Stress concentrations in multi-planar welded CHS XX-connections. J. Constr. Steel Res. **50**(3), 259–282 (1999). https://doi.org/10.1016/S0143-974X(98)00244-2
8. Woghiren, C.O., Brennan, F.P.: Weld toe stress concentrations in multi-planar stiffened tubular KK joints. Int. J. Fatigue **31**(1), 164–172 (2009). https://doi.org/10.1016/j.ijfatigue.2008.03.039
9. Chang, E., Dover, W.D.: Parametric equations to predict stress distributions along the intersection of tubular X and DT-joints. Int. J. Fatigue **21**(6), 619–635 (1999). https://doi.org/10.1016/S0142-1123(99)00018-3
10. Hamed, A.F., Khalid, Y.A., Sahari, B.B., Hamdan, M.M.: Finite element and experimental analysis for the effect of elliptical chord shape on tubular T-joint strength. Proc. Inst. Mech. Eng. Part E J. Process Mech. Eng. **215**(2), 123–131 (2001). https://doi.org/10.1243/0954408011530370
11. Hobbacher, A., I. I. of Welding. Joint Working Group XIII-XV. Fatigue Design of Welded Joints and Components: Recommendations of IIW Joint Working Group XIII-XV, XIII-1539–96/XV-845–96. Abington (1996)
12. Welded tubular joints (1987 edition) | Open Library. https://openlibrary.org/works/OL18922254W/Welded_tubular_joints

Coupling of the Method of Fundamental Solutions and the Domain Decomposition Technique for Solving Laplace Equation

Y. Kharbaoui[1(✉)], O. Askour[2], B. Braikat[1], A. Tri[3,4], H. Zahrouni[5,6], and M. Potier-Ferry[5,6]

[1] Laboratoire d'Ingénierie et Matériaux, Faculté des Sciences Ben M'Sik, Hassan II University of Casablanca, Casablanca, Morocco
yassinekharbaoui@gmail.com

[2] Mathematics, Computer Science and Communication Systems Laboratory, National School of Applied Sciences of Safi, Cadi Ayyad University, BP 63, Safi, 46000 Sidi Bouzid, Morocco

[3] Laboratory of Mechanics, Faculty of Sciences Ain Chock, Hassan II University Casablanca, Casablanca, Morocco

[4] Higher Institute of Maritimes Studies (ISEM), Km 7 Route d'El Jadida, Casablanca, Morocco

[5] Université de Lorraine, CNRS, Arts et Métiers ParisTech, LEM3, 57000 Metz, France

[6] DAMAS, Laboratory of Excellence on Design of Alloy Metals for low-mAss Structures, Université de Lorraine, Metz, France

Abstract. In this work, we are interested in combining the method of fundamental solutions (MFS) with the domain decomposition technique for solving Laplace equations. This coupling allows a good control of the resulting system generated by the MFS. Our main objective is to stabilize numerically the MFS by controlling the ill-conditioning of the system matrices. Several examples are considered to analyse the robustness and efficiency of this coupling.

Keywords: Meshless Method · Domain Decomposition Technique · Method of Fundamental Solutions

1 Introduction

In the last few decades, meshless methods have shown great promises due to their simplicity and potential to solve many physic problems. These methods consist to replace the study domain into randomly distributed points, unlike methods with mesh in which each node has a connectivity to its neighbouring nodes. These nodes, in meshless methods, do not have connectivity and therefore can handle complex geometries without any difficulty. There are many meshless methods, such as the method of fundamental solutions (MFS). The general concept of this technique consists in approaching the solution by a linear combination of

S. Aniss et al. (Eds.): CMM 2022, LNME, pp. 37–45, 2024.
https://doi.org/10.1007/978-3-031-46973-2_4

fundamental solutions taken into account on source points located on a fictitious boundary [1–3].

The MFS was originally introduced by Kupradze and Aleksidze in 1964 [10] and has been widely used effectively for the numerical resolution of a wide variety of partial differential equations and has become a robust tool for solving a large number of physical problems. In 2009, Li and Zhu [4] combined the Method of Fundamental Solutions (MFS) with two other methods, the Radial Basis Functions (RBF) and the Ana-log Equation Method (AEM), to solve nonlinear elliptic problems. After two years, Tri et al. [5] have associated the Method of Fundamental Solutions with the Asymptotic Numerical Method (ANM) to solve the nonlinear Poisson problem. In 2018, Askour et al. [6–8] have coupled the Method of Fundamental Solutions (MFS) with the ANM to solve nonlinear elastic problem. They also used regularization methods, like the SVD and TSVD methods associated with the Tikonov and L-curve methods to stabilize the global system.

In this work, we will combine the MFS with the domain decomposition technique which is based on a very simple principle, which is to divide the domain into several sub-domains [9]. The reason for using this method is to minimize the numerical instabilities of the full matrices of the system to be solved. To test its robustness and efficiency, we have applied it to Laplace's equation for different boundary conditions.

2 Coupling of the Method of Fundamental Solutions with the Domain Decomposition Technique for the Laplace Equation

In general, Laplace equation is written in the following form:

$$\begin{cases} \Delta u = 0 & \text{in} \quad \Omega \\ u \quad = g & \text{on} \quad \partial\Omega \end{cases} \tag{1}$$

where Δ is the Laplacian operator, u is the unknown and g is a given function. By using the MFS, the approximation of the solution u of the problem (1) in $2D$ is sought in the following form:

$$u(x,y) = \sum_{j=1}^{N_s} \alpha_j G_j(r) \tag{2}$$

where α_j are unknown coefficients, N_s is the number of source points and $G_j(r)$ are the fundamental solutions of the Laplace operator which are illustrated as:

$$G_j(r) = \frac{1}{2\pi} log(r) \tag{3}$$

with $r = \sqrt{(x_j - x_{sj})^2 + (y_j - y_{sj})^2}$, (x_j, y_j) are the coordinates of the collocation points and (x_{sj}, y_{sj}) are the coordinates of the source points. By applying the domain decomposition technique, the problem is written in the case of two subdomains without covering, in the following way:

$$\begin{cases} \Delta u_1 = 0 & \text{in} \quad \Omega_1 \\ u_1 \quad = g & \text{on} \quad \partial\Omega_1 \\ \Delta u_2 = 0 & \text{in} \quad \Omega_2 \\ u_2 \quad = g & \text{on} \quad \partial\Omega_2 \\ u_1 \quad = u_2 & \text{on} \quad \partial\Omega_1 \cap \partial\Omega_2 \\ \frac{\partial u_1}{\partial n} = -\frac{\partial u_2}{\partial n} & \text{on} \quad \partial\Omega_1 \cap \partial\Omega_2 \end{cases} \tag{4}$$

where u_1 and u_2 represent the solutions in the domains Ω_1 and Ω_2 respectively and n is the outward unit normal vector to $\partial\Omega_1$. In the following, the indices 1 and 2 are introduced to differentiate the parameters between the two domains.

The approximation by MFS allows us to obtain the following system of equations:

$$\begin{cases} \sum_{j=1}^{N_{s1}} \alpha_{1j} \Delta G_j(r) = 0 & \text{in} \quad \Omega_1 \\ \sum_{j=1}^{N_{s1}} \alpha_{1j} G_j(r) \quad = 0 & \text{on} \quad \partial\Omega_1 \\ \sum_{j=1}^{N_{s2}} \alpha_{2j} \Delta G_j(r) = 0 & \text{in} \quad \Omega_2 \\ \sum_{j=1}^{N_{s2}} \alpha_{2j} G_j(r) \quad = 0 & \text{on} \quad \partial\Omega_2 \\ \sum_{j=1}^{N_{s1}} \alpha_{1j} G_j(r) \quad = \sum_{j=1}^{N_{s2}} \alpha_{2j} G_j(r) & \text{on} \quad \partial\Omega_1 \cap \partial\Omega_2 \\ \sum_{j=1}^{N_{s1}} \alpha_{1j} \frac{\partial G_j(r)}{\partial n_1} \quad = -\sum_{j=1}^{N_{s2}} \alpha_{2j} \frac{\partial G_j(r)}{\partial n_2} & \text{on} \quad \partial\Omega_1 \cap \partial\Omega_2 \end{cases} \tag{5}$$

where N_{s1} and N_{s2} are respectively the number of source points on the fictitious boundary of the first and second sub-domains, α_{1j} and α_{2j} are the unknowns of the problems in first and second subdomains. In matrix form, the system (5) becomes:

$$\begin{bmatrix} [G_1]_{N_{b1},N_{s1}} & [0]_{N_{b1},N_{s2}} \\ [0]_{N_{b2},N_{s1}} & [G_2]_{N_{b2},N_{s2}} \\ [G_1]_{N_r,N_{s1}} & -[G_2]_{N_r,N_{s2}} \\ [\frac{\partial G_j(r)}{\partial n_1}]_{N_r,N_{s1}} & [\frac{\partial G_j(r)}{\partial n_2}]_{N_r,N_{s2}} \end{bmatrix} \begin{Bmatrix} \alpha_1 \\ \alpha_2 \end{Bmatrix} = \begin{Bmatrix} \{g\}_{N_{b1},1} \\ \{g\}_{N_{b2},1} \\ \{0\}_{N_r,1} \\ \{0\}_{N_r,1} \end{Bmatrix} \tag{6}$$

where N_r is the number of connection points between the two sub-domains, N_{b1} and N_{b2} represent respectively the number of points in each sub-domain.

3 Numerical Applications and Discussions

in this section, we consider two geometrical domains; the first is square and the second is circular.

3.1 Square Domain

The Laplace equation is defined in a square domain of side $L = 11m$. This domain is replaced by a distribution of 21 nodes along the x-axis and of 11 nodes along the y-axis. In this example, we study three cases of decomposition; two, three and four sub-domains. The fictitious boundary is taken in circular form of radius $R_{FS} = 20m$ for each sub-domain and $N_s = 60$.

In Fig. 1, we depict the solution U as a function of the spatial variable x for two, three, and four sub-domains. This figure shows the comparison between the numerical and analytical results for the three cases of sub-domain. From this figure, we remark that the numerical solutions are in good agreement with analytical one.

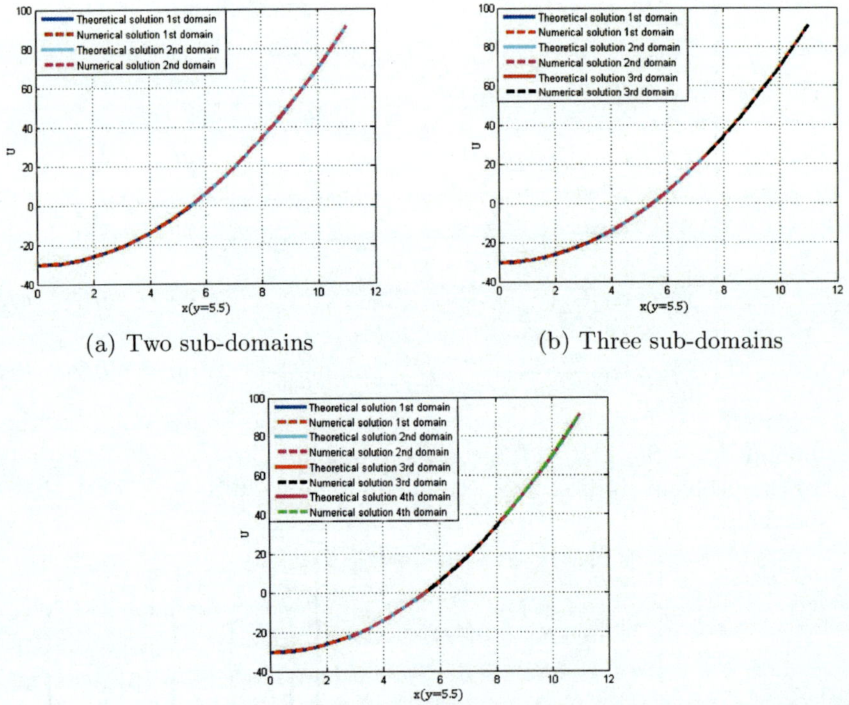

(a) Two sub-domains

(b) Three sub-domains

(c) Four sub-domains

Fig. 1. Comparison between the numerical solutions for the three cases of decomposition with that analytical for $(0 \leq x \leq L, y = 5.5m)$

Table 1. Mean squared errors and conditioning for two subdomains

	RMSE and COND	Two sub-domains	One sub-domain
$g(x, y) = \frac{(x-100)-(y-100)}{1000}$	RMSE1	2.1473×10^{-7}	2.6016×10^{-9}
	RMSE2	2.1476×10^{-7}	
	COND	7.4727×10^{16}	3.4277×10^{18}
$g(x, y) = (x^2 - 1) - (y^2 - 1)$	RMSE1	1.5397×10^{-5}	2.7203×10^{-7}
	RMSE2	1.5397×10^{-5}	
	COND	7.4727×10^{16}	3.4277×10^{18}

Table 2. Mean squared errors and conditioning for three sub-domains

	RMSE and COND	Three sub-domains	One sub-domain
$g(x, y) = \frac{(x-100)(y-100)}{1000}$	RMSE1	9.8529×10^{-9}	2.6016×10^{-9}
	RMSE2	3.3275×10^{-8}	
	RMSE3	2.7881×10^{-8}	
	COND	5.2615×10^{16}	3.4277×10^{18}
$g(x, y) = (x^2 - 1) - (y^2 - 1)$	RMSE1	1.0132×10^{-6}	2.7203×10^{-7}
	RMSE2	3.4292×10^{-6}	
	RMSE3	2.8718×10^{-6}	
	COND	5.2615×10^{16}	3.4277×10^{18}

In Tables 1-3, we report the Mean Square Errors $RMSE1 = \sqrt{\frac{1}{N_{b1}} \sum_{i=1}^{N_{b1}} (\tilde{u}_{1,i} - u_{1,i})^2}$ and $RMSE2 = \sqrt{\frac{1}{N_{b2}} \sum_{i=1}^{N_{b2}} (\tilde{u}_{2,i} - u_{2,i})^2}$; $\tilde{u}_{1,i}$ and $\tilde{u}_{2,i}$ represent the numerical solutions of the first and the second domain while $u_{1,i}$ and $u_{2,i}$ are their analytical solutions. Note that the conditioning (COND) is calculated directly from the software library. Three cases of the domain decomposition are investigated in this examples.

In Table 1, we notice that the error obtained by doing the domain decomposition technique is larger than the one obtained by doing a single subdomain. But, the conditioning obtained by the domain decomposition technique is smaller than that obtained by a single domain.

In Table 2, we observe that the error remains constant. However, the conditioning obtained by the domain decomposition technique looks similar that of Table 1.

the results of Table 3 confirm the results obtained previously.

Table 3. Mean squared errors and conditioning for four sub-domains

	RMSE and COND	Four sub-domains	One sub-domain
$g(x,y) = \frac{(x-100)(y-100)}{1000}$	RMSE1	8.3678×10^{-10}	2.6016×10^{-9}
	RMSE2	2.6585×10^{-9}	
	RMSE3	5.2142×10^{-9}	
	RMSE4	4.0364×10^{-9}	
	COND	9.7067×10^{16}	3.4277×10^{18}
$g(x,y) = (x^2-1)-(y^2-1)$	RMSE1	6.4021×10^{-7}	2.7203×10^{-7}
	RMSE2	1.9593×10^{-6}	
	RMSE3	5.3296×10^{-6}	
	RMSE4	4.5514×10^{-6}	
	COND	9.7067×10^{16}	3.4277×10^{18}

In Fig. 2, we represent the variations of conditioning and of error with respect to number of sub-domains. From this figure, the numerical solutions are obtained with very small mean square errors varying between 10^{-6} and 10^{-10}. Nevertheless, a slight decreases in conditioning is observed going from 10^{18} for a single domain to 10^{16} in the case of two sub-domains and decreases further in the case of a decomposition of three and four sub-domains.

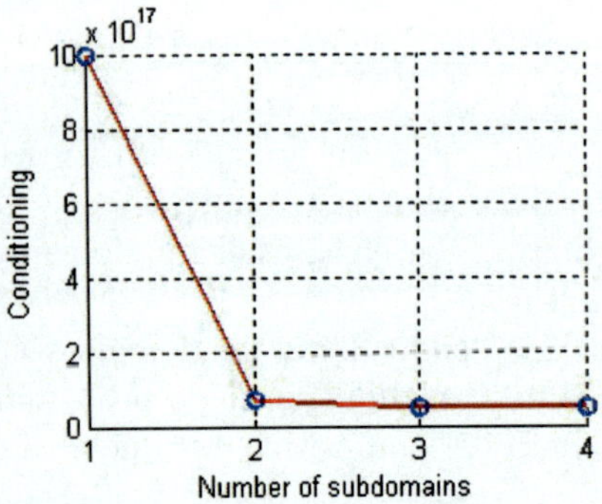

(a) Conditioning variation

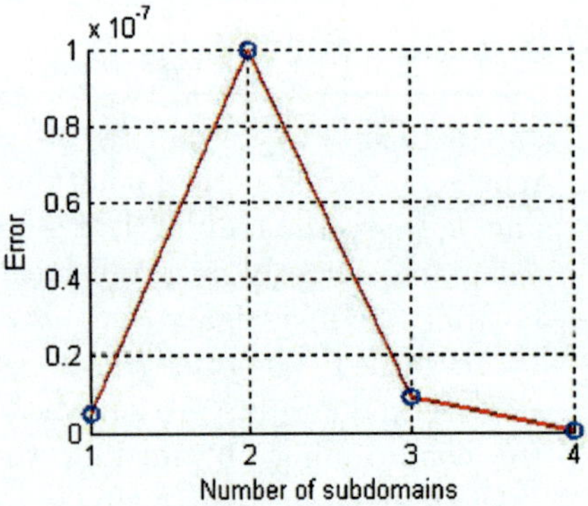

(b) Error variation

Fig. 2. Variations of the conditioning and of the error with respect to number of sub-domains

3.2 Circular Domain

For the second example, we consider a circular domain of radius $R = 1m$ which we have subdivided down into sub-domains. For the four sub-domains, the fictitious boundary is taken in circular form of radius $R_{FS} = 5m$. In Figs. 3, we represent the solution obtained by the domain decomposition technique with respect to x and y for two, three and four sub-domains and the analytical solution for functions $g(x, y) = e^x sin(y)$ and $g(x, y) = (x^2 - 1) - (y^2 - 1)$. From this figure, we remark that the solution obtained by the domain decomposition technique is in good agreement with that analytical.

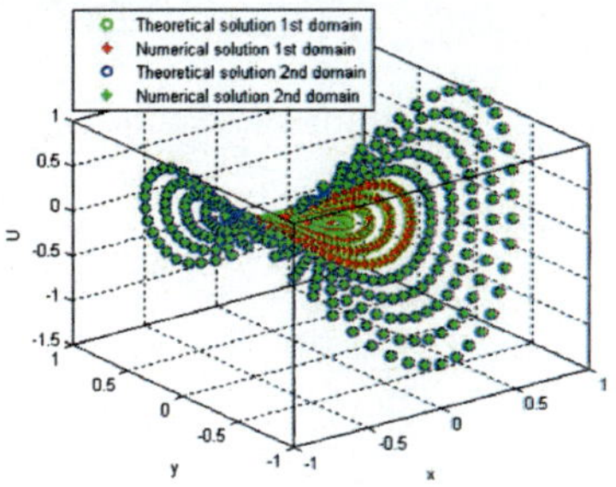

(a) Two sub-domains

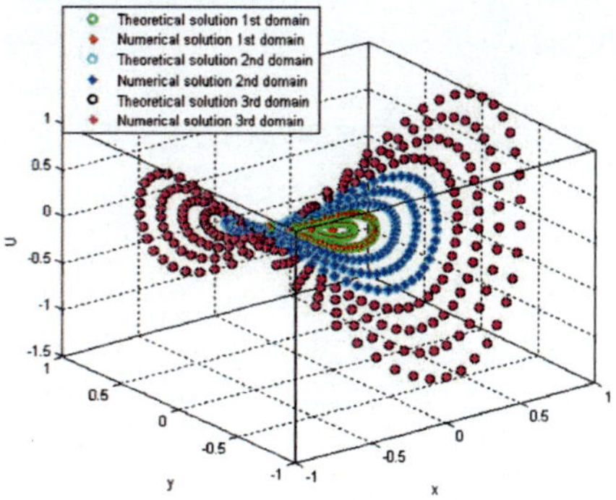

(b) Three sub-domains

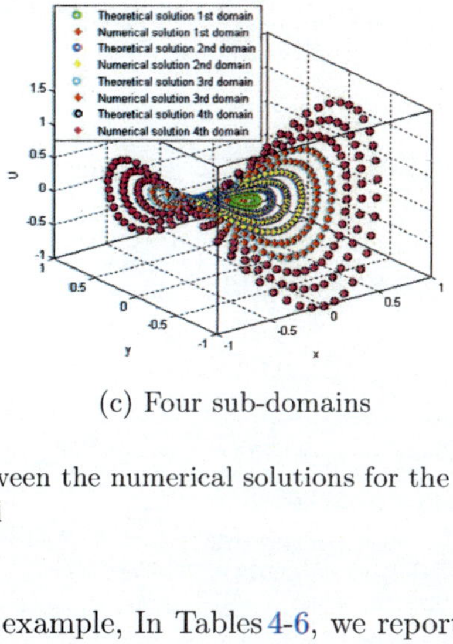

(c) Four sub-domains

Fig. 3. Comparison between the numerical solutions for the three cases of decomposition with that analytical

As in the previous example, In Tables 4-6, we report the mean square error and the conditioning (COND) for three cases of domain decomposition technique.

In Table 4, we notice that the error obtained by doing the domain decomposition technique is smaller than the one obtained by doing a single subdomain. However, the conditioning obtained by the domain decomposition technique is smaller than that obtained by a single sub-domain.

In Table 5, the more the domain is decomposed, the more the error decreases but the conditioning remains the same for the two methods.

In Table 6, the same remark is illustrated that in Table 5.

Table 4. Mean squared errors and conditioning for two subdomains

	RMSE and COND	Two sub-domains	One sub-domain
$g(x,y) = e^x sin(y)$	RMSE1	5.1920×10^{-14}	1.4346×10^{-13}
	RMSE2	4.5763×10^{-14}	
	COND	2.6962×10^{17}	1.1355×10^{18}
$g(x,y) = (x^2-1)-(y^2-1)$	RMSE1	9.3709×10^{-15}	9.8946×10^{-14}
	RMSE2	6.6020×10^{-15}	
	COND	2.6962×10^{17}	1.1355×10^{18}

Table 5. Mean squared errors and conditioning for three sub-domains

	RMSE and COND	Three sub-domains	One sub-domain
$g(x,y) = e^x sin(y)$	RMSE1	4.7355×10^{-14}	2.5609×10^{-13}
	RMSE2	3.5274×10^{-14}	
	RMSE3	7.9238×10^{-14}	
	COND	2.5609×10^{18}	1.1355×10^{18}
$g(x,y) = (x^2-1)-(y^2-1)$	RMSE1	1.0638×10^{-14}	9.8946×10^{-14}
	RMSE2	1.0447×10^{-14}	
	RMSE3	9.6199×10^{-15}	
	COND	2.5609×10^{18}	1.1355×10^{18}

Table 6. Mean squared errors and conditioning for four sub-domains

	RMSE and COND	Four sub-domains	One sub-domain
$g(x,y) = e^x sin(y)$	RMSE1	3.2372×10^{-14}	1.4346×10^{-13}
	RMSE2	4.3114×10^{-14}	
	RMSE3	2.9284×10^{-14}	
	RMSE4	6.9530×10^{-14}	
	COND	3.1461×10^{18}	1.1355×10^{18}
$g(x,y) = (x^2-1)-(y^2-1)$	RMSE1	7.5019×10^{-15}	9.8946×10^{-14}
	RMSE2	4.0911×10^{-15}	
	RMSE3	1.0969×10^{-14}	
	RMSE4	1.2987×10^{-14}	
	COND	3.1461×10^{18}	1.1355×10^{18}

The figures and tables of this second example confirm that the analytical and numerical solutions are in good agreement. Moreover, we notice that the RMSE errors decrease when the number of sub-domains increases. We also note that the conditioning stabilizes around 10^{17}. These results are very encouraging for a large number of subdomains, which will make the MFS more stable and therefore more attractive for solving complex problems.

4 Conclusion

In this work, we have combined the domain decomposition technique without covering with the Method of Fundamental Solutions for solving the Laplace equation. The domain decomposition technique is introduced to stabilise the system obtained by using the MFS approximation. The numerical examples showed the robustness of our algorithm such that the root mean square errors were very small for different types of geometries. We are underway to automate this technique in order to apply it to a very large number of subdomains.

References

1. Mathon, R., Johnston, R.L.: The approximate solution of elliptic boundary-value problems by fundamental solutions. SIAM J. Num. Ana. **14**, 638–650 (1977). https://doi.org/10.1137/0714043
2. Johnston, R.L., Fairweather, G.: The method of fundamental solutions for problems in potential flow. Appl. Math. Mod. **8**(4), 265–270 (1984). https://doi.org/10.1016/0307-904X(84)90161-6
3. Alves, C.J.S., Chen, C.S.: A new method of fundamental solutions applied to nonhomogeneous elliptic problems. Adv. Comput. Math. **23**(1), 125–142 (2005). https://doi.org/10.1007/s10444-004-1833-5
4. Li, X., Zhu, J.: The method of fundamental solutions for nonlinear elliptic problems. Eng. Anal. Boun. Elem. **33**(3), 322–329 (2009). https://doi.org/10.1016/j.enganabound.2008.07.004
5. Tri, A., Zahrouni, H., Potier-Ferry, M.: High order continuation algorithm and meshless procedures to solve nonlinear Poisson problems. Eng. Anal. Boun. Elem. **36**(11), 1705–1714 (2012). https://doi.org/10.1016/j.enganabound.2012.04.007
6. Askour, O., Tri, A., Braikat, B., Zahrouni, H., Potier-Ferry, M.: Method of Fundamental Solutions and high order algorithm to solve nonlinear elastic problems. Eng. Anal. Boun. Elem. **89**, 25–35 (2018). https://doi.org/10.1016/j.enganabound.2018.01.007
7. Askour, O., Tri, A., Braikat, B., Zahrouni, H., Potier-Ferry, M.: Bifurcation indicator for geometrically nonlinear elasticity using the method of fundamental solutions. Compt. Rend. Méc. **347**(2), 91–100 (2019). https://doi.org/10.1016/j.crme.2019.01.002
8. Tri, A., Askour, O., Braikat, B., Zahrouni, H., Potier-Ferry, M.: Fundamental solutions and asymptotic numerical methods for bifurcation analysis of nonlinear bi-harmonic problems. Num. Meth. Part. Diff. Equat. **35**(6), 2091–2102 (2019). https://doi.org/10.1002/num.22403
9. Smith, B.F., Bjorstad, P.E., Gropp, W.D., Pasciak, J.E.: Domain decomposition: parallel multilevel methods for elliptic partial differential equations. SIA. Rev. **40**(1), 169–170 (1998). https://lccn.loc.gov/95042362
10. Kupradze, V.D., Aleksidze, M.A.: The method of functional equations for the approximate solution of certain boundary value problems. USSR Comp. Math. Math. Phys. **4**(4), 82–126 (1964). https://doi.org/10.1016/0041-5553(64)90006-0

Modeling Elastoplastic Structures in Finite Transformation by a High-Order Algorithm

C. El Kihal[1,4](✉), A. Hamdaoui[1], B. Braikat[1], N. Damil[1,2], and M. Potier-Ferry[3]

[1] Laboratoire D'Ingénierie Et Matériaux LIMAT, Faculté des Sciences Ben M'Sik, Hassan II University of Casablanca, B.P 7955 Sidi Othman, Casablanca, Morocco
chafik.elkihal@gmail.com

[2] Centrale Casablanca, Centre de Recherche Systèmes Complexes et Interactions, Ville Verte, Bouskoura 27182, Morocco

[3] Université de Lorraine, CNRS, Arts et Métiers ParisTech, LEM3, 57000 Metz, France

[4] School of Civil Engineering, Wuhan University, 8 South Road of East Lake, Wuchang, Wuhan 430072, People's Republic of China

Abstract. A high-order algorithm to compute elastoplastic structures in large deformation. We will asses the robustness of this algorithm by studying four loading processes and and we will check the quality of the solution with the help of the residual curve.

Keywords: ANM algorithm · elastoplasticity · finite transformation

1 Introduction

Algorithms based on Newton-Raphson type prediction-correction methods for solving elastoplasticity problems in finite transformations are widely used in most of the industrial codes [1]. Recently, we have proposed a new algorithm [2,7,8], which generalizes the one used in [3,4], to solve problems in large elasto-plastic strain, based on techniques of the Asymptotic Numerical Method (ANM) [6]: computation of the solution branch by branch using Taylor series representations for each branch, discretization technique using FEM and use of a continuation technique. In order to make the problem analytical we proposed to replace the initial finite transformation elastoplastic constitutive law by a new regularized finite transformation elastoplastic constitutive law, using four regularization parameters.

In this work, we will asses the robustness of the proposed algorithm by studying four loading processes and we will check the quality of the solution with the help of the residual curve. Comparisons will be made with results obtained using commercial software.

S. Aniss et al. (Eds.): CMM 2022, LNME, pp. 46–54, 2024.
https://doi.org/10.1007/978-3-031-46973-2_5

2 Finite Transformation Elastoplasticity Problem and the Proposed Algorithm

We consider a continuum that occupies the domain Ω_0 at the initial time and the domain Ω at the current time. This body is subjected to an external loading $C(t)F^{ext}$ proportional to a time-dependent loading parameter $C(t)$. The Lagrangian formulation of the principle of virtual power, in the case where the body forces are zero, is written as:

$$\int_{\Omega_0} \boldsymbol{\tau}.{}^t\mathbf{f} : \boldsymbol{\delta}\mathbf{L}d\Omega_0 = C(t)\int_{\partial\Omega_0} \mathbf{F}_{ext}.\delta\mathbf{v}dS_0 \quad (1)$$

In (1), $\boldsymbol{\tau}$ denotes the Kirchhoff stress tensor, **L** the velocity gradient tensor, **F** the transformation gradient tensor, **f** its inverse and **v** the velocity vector. The elasto-plastic behavior in large deformation allows to link, on the one hand, the elastic part D_e of the strain rate tensor $D = 1/1(l + l^t) = D_e + D_p$, to the Jauman derivative of the stress tensor tau^j and, on the other hand, the plastic part of the strain rate tensor D_p to the gradient of the yield function fy by the following relations :

$$\begin{aligned} \boldsymbol{\tau}^J &= \mathbf{C}^{elas} : (\mathbf{D} - \mathbf{D}^{\mathbf{p}}) \\ \mathbf{D}^{\mathbf{p}} &= \quad \dot{\lambda}\frac{\partial f_y}{\partial \boldsymbol{\tau}} = \dot{\lambda}\mathbf{n} \end{aligned} \quad (2)$$

where $\mathbf{C}^{elas}$ is the fourth order tensor of elastic constants and, $\dot{\lambda}$ is the derivative with respect to time of the plastic multiplier, **n** is the flow direction tensor, in what follow the yield function is given by :

$$f_y = \frac{q - \tau_e}{\tau_e} \; ; \mathbf{n} = \frac{3\boldsymbol{\tau}^d}{2q} \quad (3)$$

with $\boldsymbol{\tau}^d$ is the deviatoric part of the Kirchhoff stress tensor, q is the von-Mises equivalent stress and τ_e is the effective stress. The constitutive law is completed by the Kuhn-Tucker's condition:

$$\dot{\lambda} \geq 0 \; ; f_y \leq 0 \; ; \dot{\lambda} f_y = 0 \quad (4)$$

In [2], we have proposed to regularize this non-analytical consistency relation, thus Eq. (4) has been written in the form:

$$\dot{\lambda} = G(f_y)H(\mathbf{D}) \quad (5)$$

The functions $G(f_y)$ and $H(D)$, regularize, respectively, the elastic-plastic transition and the elastic unloading:

$$G(f_y) = \frac{\eta_3}{\frac{f_y^2}{2\mu}\tau_e + \eta_3(\frac{3}{2} + \frac{h}{2\mu}(1+f_y))} \quad (6)$$

$$H(H - \mathbf{n} : \mathbf{D}) = \eta_2^2 \dot{\tau}_c^2 \quad (7)$$

where h is the hardening modulus and μ is the shear modulus. The von-Mises equivalent stress has been also regularized by:

$$q = \sqrt{\frac{3}{2}\boldsymbol{\tau}^d : \boldsymbol{\tau}^d + \eta_1^2 \tau_y^2} \tag{8}$$

Three regularization parameters η_1 in (8), η_2 in (7) and η_3 in (6) are introduced.

In this work, we will impose different loading processes in order to study the ability of the proposed algorithm to reproduce an elastic, plastic and elastic unloading process. For this purpose, we will assume that the loading parameter $C(t)$ is written in one of the following forms:

$$(C - \frac{Cm}{Tm}t)(C - C_m(2 - \frac{t}{T_m})) = \eta_4^2 C_m^2 \tag{9}$$

$$C(t) = -\frac{C_m}{T_m}t^2 + 2\frac{C_m}{T_m}t \tag{10}$$

Expression (9) corresponds to a regularized loading parameter $C(t)$, in fact we have replaced the two half-lines corresponding to loading phase and unloading phase, by a regular relationship by introducing a small parameter η_4. In formula (10) we do not need regularization, we have chosen a parabolic form to translate the two phases loading and unloading. These different types of loading (loading-unloading) are shown in Fig. 2, where C_m and T_m are two given parameters that determine the maximum of the loading $C(t)$ and the time interval. Three couples of values (C_m, T_m) will be tested.

We have extended the techniques of ANM to large strain elastoplasticity problems. First, the unknown $\mathbf{U}$ ($\mathbf{U}$ contains all the unknowns of the problem) and the loading parameter $C(t)$ have been represented using Taylor series truncated to order N:

$$\begin{cases} \mathbf{U}(a) = \mathbf{U}^j + \sum_{k=1}^{N} a^k \mathbf{U}_k \\ C(a) = C^j + \sum_{k=1}^{N} a^k C_k \end{cases} ; a \in [0, a_{max}] \tag{11}$$

where $(\mathbf{U}^j, C^j)$ is a known and regular solution and a_{max} is the validity range. A parametrization, based on the time "t", and the load parameter $C(t)$ is used to define the path parameter "a":

$$a = (t - t^j)t_1 + (C - C^j)C_1 \tag{12}$$

where (t_1, C_1) represents the tangent vector to the branch at point : (t^j, C^j).

In this paper, as in a recent work [2], we consider that the path parameter "a" depends on the time and we will assume in the following that the time t is written in the form: $t = t^j + at_1 + ...a^N t_N$ where t^j is the initial time of the branch "j".

The discretized solutions velocity vectors $\mathbf{v}_{k-1}$ at order k using a classical finite element method, are solutions of the following linear problems: [2,5,8]:

$$[K_t]\{\mathbf{v}_{k-1}\} = C_k\{F^{ext}\} + \{P_k^{nl}\}\ for\ 1 \le k \le N \tag{13}$$

where $\{P_k^{nl}\}$ is a vector that depends on the previous orders.

The whole solution is obtained by a continuation method. For this respect, we define a validity range of the parameter a_{max} by :

$$a_{max} = inf\left(\delta_s \frac{\|\{\mathbf{U}_1^i\}\|}{\|\{\mathbf{U}_N^i\}\|}\right)^{\frac{1}{N-1}} \tag{14}$$

where $\mathbf{U}_k^i(k = 1, N)$ is a component"i" of the vector $\mathbf{U_k}$ and δ_s is a tolerance parameter. In the case of elasto-plasticity, we calculate the domain of validity of the different components U^i of the unknown U and then take the smallest value in order to ensure the quality of the end of the solution branch during the calculation. The solution $\mathbf{U}^j = \mathbf{U}(a_{max})$, $C^j = C(a_{max})$ will be the starting solution for the following ANM step. The complete solution branch is obtained step by step in this manner.

3 Numerical Discussion

The numerical test is a 2D elastoplastic plate with length $L = 250$ mm, width $H = 40$ mm and two imperfections located at the mid-upper and mid-lower at the middle of the plate. The mechanical characteristics of the plate are: Young modulus $E = 190.10^3$ MPa, Poisson ratio $\mu = 0.29$, initial yield stress $\tau_y = 340$ MPa and a hardening modulus $h = 300$ MPa. The plate is subjected to an uniaxial tension by applying a time-dependent horizontal loading function at the left and right ends, see Fig. 1.

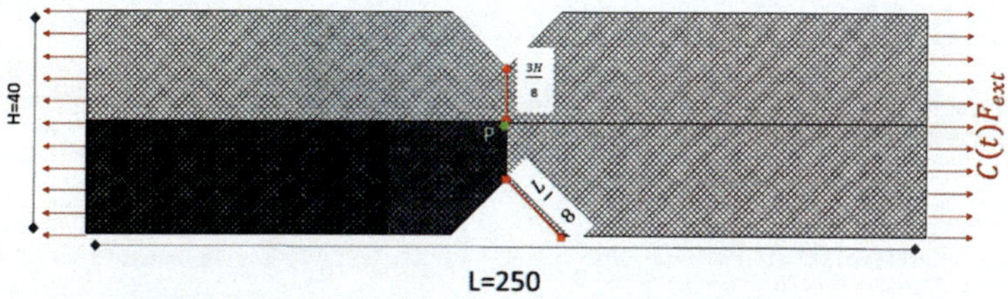

Fig. 1. Geometrical description and loading of the 2D plate. The imperfection has a triangular shape with a height of $\frac{3H}{8}$ and a hypotenuse of $\frac{L}{8}$.

For symmetry reasons, only the quarter of the plate will be considered. The chosen ANM order is $N = 15$ and the tolerance parameter is $\delta_s = 10^{-3}$. In Fig. 3, we have plotted the shapes of the two regularized functions $G(fy)$ and

$H(\mathbf{D})$ for different values of the regularization parameters η_2 and η_3. This figure shows that the regularized curves are close to the initial lines when the regularization parameters tend to zero. After several tests, the optimal regularization parameters used for this simulation are $\eta_1 = 10^{-2}, \eta_2 = 10^{-4}, \eta_3 = 10^{-4}$ and $\eta_4 = 5.10^{-4}$.

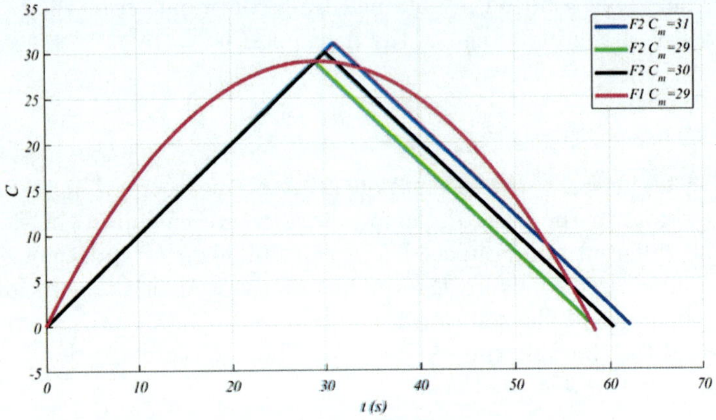

Fig. 2. Two loading functions $F1$ (10) and $F2$ (9). For $F1 : (C_m, T_m) = (29, 29)$ and for $F2 : (C_m, T_m) = (29, 29), (30, 30), (31, 31)$

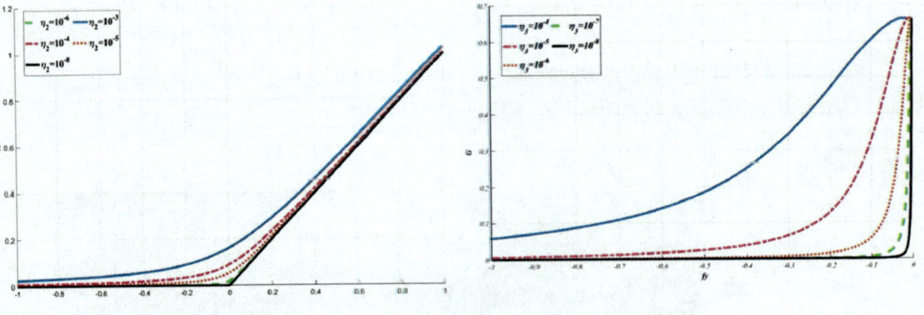

Fig. 3. Effect of regularization on the functions H and G defining the regularized constitutive law in Eqs. (5) and (6).

In Fig. 4, we represent the solution obtained by the proposed algorithm in the case of four loading processes, see Eqs. (9), (10) and Fig. 2: $(C_m = 29, T_m = 29)$, $(C_m = 30, T_m = 30)$ and $(C_m = 31, T_m = 31)$. In this figure, we represent the equivalent plastic strain versus the von-Mises equivalent stress during the simulation at the middle of the plate (point P in Fig. 1). For the four load tests, we can notice that the results obtained are in good agreement with the solution obtained by Abaqus in the three zones: elastic zone, plastic zone and elastic unloading zone, in Fig. 5, we have represented the Abaqus solution only in the case of loading $(C_m, T_m) = (31, 31)$. From Fig. 4, we can observe the great influence of a slight difference in the loading parameters (in the values of T_m and C_m) on the response curve and therefore on the residual deformation, this is due to the fact that we are in the presence of large elastoplastic deformation. At unloading, we obtain a deformation of about 25% in the third loading.

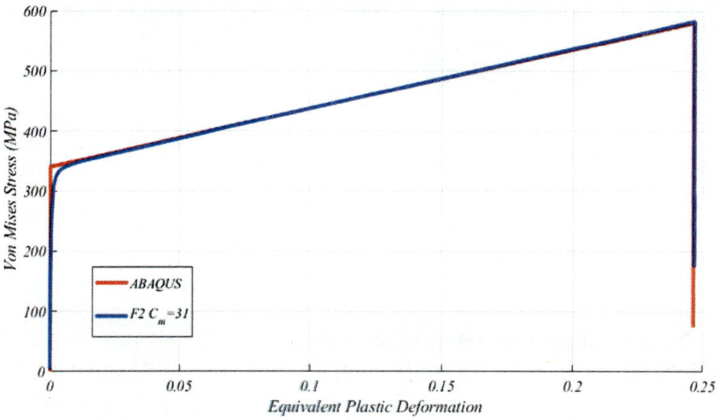

Fig. 4. Effect of the loading parameter C(t) on the response curve : q as function of equivalent strain, for four values of C_m and T_m. $F_1 : (C_m, T_m) = (29, 29)$ and $F_2 : (C_m, T_m) = (29, 29), (30, 30), (31, 31)$

In Fig. 6, we have represented the quality of the solutions obtained by the proposed algorithm for the four loading cases. We have represented the decimal logarithm of the norm of the residual as a function of time for the four loading cases. Note that the solution curve of Fig. 5 was obtained in 111 ANM steps which corresponds to 111 inversions of the tangent matrix, and Abaqus solution needed 280 steps (112 increments and 168 iterations). Additionally, we compare the computational efficiency of two algorithms: the proposed algorithm 277 seconds and ABAQUS 700 seconds. It is worth noting that we utilized identical degrees of freedom for both codes to ensure a fair and accurate comparison. We can observe on Figs. 4, 5 and 6, that for the four loading process we obtain a good quality of solutions in the elastic and plastic phases (until $t = T_m$) and that the quality degrades slightly during the elastic unloading phase for the chosen ANM parameters.

In Fig. 3, we have represented the curves of the G and H functions defining the regularized elastoplastic constitutive law for different values of the regularization parameters. We can thus observe that despite the fact that these curves are far from the initial non-regular curves, the proposed algorithm allows to account for the physics of the problem since the loaded and unloaded zones are automatically calculated. Indeed, on Fig. 7, **we can observe for different stages of loading the** elastic and the plastic zones for four values of time : $t = 1.45, t = 19.47, t = 30.99$ and $t = 61.94$.

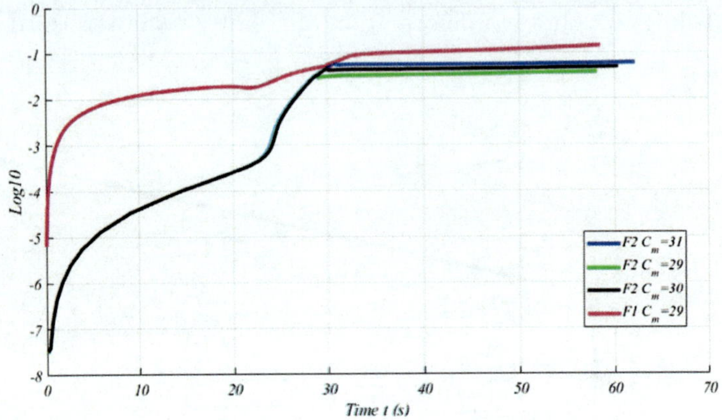

Fig. 5. Response curve : q as function of equivalent strain, for the regularized loading function with $C_m = 31$, $T_m = 31$ compared to the results obtained by ABAQUS code.

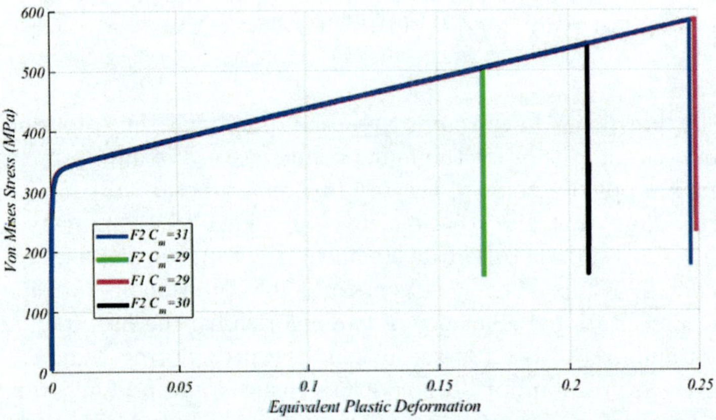

Fig. 6. Evolution of the residual as a function of time

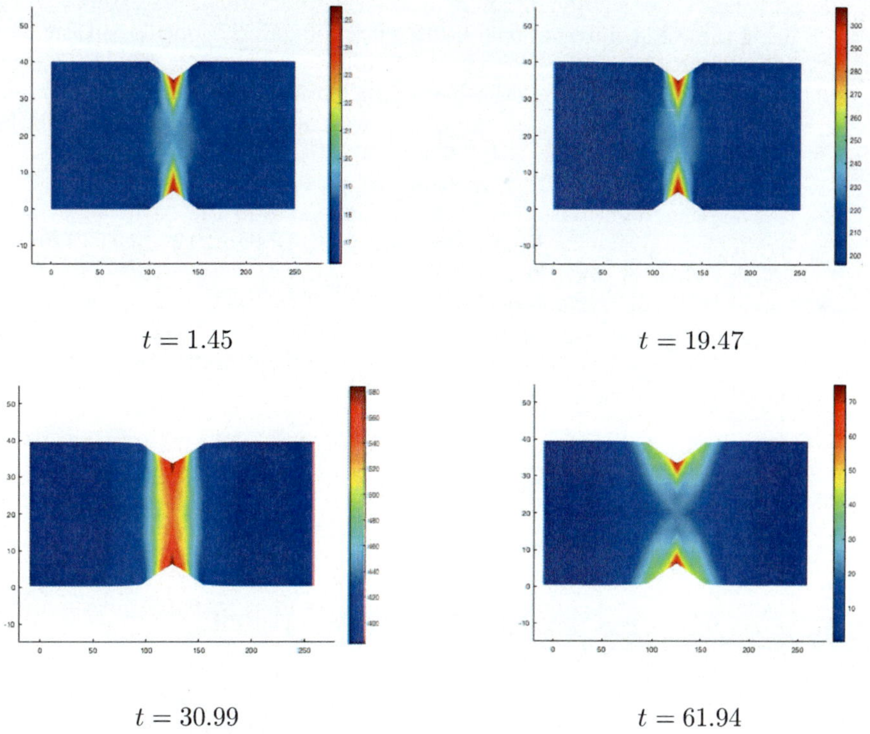

Fig. 7. Stress distribution at different stages during the process $t = 1.45, t = 19.47, t = 30.99$ and $t = 61.94$

4 Conclusion

A high order algorithm to solve elasto-plasticity problems in finite transformation has been presented. We have thus evaluated the robustness of the proposed algorithm by studying four loading processes. The quality of the solution, obtained by the proposed high order algorithm, was evaluated by examining the evolution of the residual.

References

1. Simo, J.C., Hughes, T.J.R. : Computational inelasticity. 7 (2006). ISBN 0-387-97520-9, SPIN 10760898
2. El Kihal, C., et al.: Asymptotic numerical method for finite plasticity. Fin. Ele. Analy. Des. **206**, 103759–103759 (2022). https://doi.org/10.1016/j.finel.2022.103759
3. Assidi, M., Zahrouni, H., Damil, N., Potier-Ferry, M.: Solving elastoplasticity problems by the asymptotic numerical method: influence of the parameterizations. Int. Jour. Mat. For. **2**(1), 1–14 (2009). https://doi.org/10.1007/s12289-008-0389-1
4. Hamdaoui, A., Braikat, B., Damil, N.: Regularization and perturbation technique to solve plasticity problems. Fin. Ele. Analy. Des. **115**, 33–42 (2016). https://doi.org/10.1016/j.finel.2016.03.001

5. El Kihal, C., et al.: Simulation of a 2D large transformation elastoplasticity problem using the ANM. International Conference Numerical Modelling in Engineering, pp. 185–193 (2022). https://doi.org/10.1007/978
6. Cochelin, B., Damil, N., Potier-Ferry, M.: The asymptotic-numerical method: an efficient perturbation technique for nonlinear structural mechanics. Revue Européenne des Éléments Finis **3**(2), 281–297 (1994)
7. El Kihal, C., Askour, O., Hamdaoui, A., Braikat, B., Damil, N., Potier-Ferry, M.: Convergence acceleration by Padé approximants in the ANM algorithm for large strain elasto-plasticity. Finite Elements Anal. Design **212**, 103853 (2022). https://doi.org/10.1016/j.finel.2022.103853
8. Chafik, E.K., et al.:Numerical computation of plasticity in large deformations using the asymptotic numerical method. International Conference Integrated Design and Production, pp. 195–202 (2022). https://doi.org/10.1007/978

Dynamic Modeling of the Multilayered Piezoelectric Beams Based on the Pseudo-Stroh Formalism

H. Balkissou[1,2(✉)], F. P. Ewolo Ngak[1], G. E. Ntamack[1], and L. Azrar[3,4]

[1] Groupe de Mécanique, Matériaux et Acoustique, B.P. 454, Département de Physique, Faculté des Sciences, Université de Ngaoundéré, Ngaoundéré, Cameroun
balkishamidou4@gmail.com

[2] National Committee for Development of Technologies, MINRESI, P. O. Box 1457, Yaoundé, Cameroon

[3] Research Center STIS, Team M2CS, Department of Applied Mathematics and Informatics, ENSAM, Mohammed V University in Rabat, Rabat, Morocco
l.azrar@um5r.ac.ma

[4] Department of Mechanical Engineering, Faculty of Engineering, King Abdulaziz University, Jeddah, Saudi Arabia

Abstract. In this paper, a semi-analytical solution for the dynamic behavior of the multilayered piezoelectric beams is derived. The beam is assumed to be simply supported and the interfaces between the layers are considered perfect. In each layer of the beams, the pseudo-Stroh formalism is developed and the set of partial differential equations is reduced to a system of second order ordinary equation of time. To solve that equation in time domain, the time interval is divided into several sub-intervals having equal length. In each sub-interval of time, the Lagrange polynomials have been used and a discrete eigenvalue problem is then obtained. The solutions are therefore constructed and interpolated in the time domain. Those solutions have been propagated throughout the thickness direction of beams using the propagator matrix method. In the computation process, the elastic material graphite epoxy (named E) and the piezoelectric material $BaTiO_3$ (named B) are used. The layering effect and the surfaces boundary conditions effect on the dynamic response of the three-layered piezoelectric beams are clearly illustrated. The obtained results can be helpful as benchmark.

Keywords: Multilayered beams · Pseudo-Stroh formalism · Lagrange polynomials · Propagator matrix

1 Introduction

The multifunctional multilayered system notably the piezoelectric beams are the smarts structures which know these last decades a considerable expansion in aeronautic, in aerospace, in shipbuilding industry, in medical ultrasonic imaging in building industry (smarts office building) etc. They contribute to the social evolution of many countries and

S. Aniss et al. (Eds.): CMM 2022, LNME, pp. 55–65, 2024.
https://doi.org/10.1007/978-3-031-46973-2_6

their cultivation. The aims of the research of new materials are: the great achievement profit, the drop of the cost of manufacturing product and the improvement of the reliability. In order to improve the lifetime of those structures in service, their static, dynamic, free vibration, the buckling and the post-buckling behaviors must be modeled accurately. Lezgy-Nazargah et al. (2013) studied the static and dynamic behaviors of functionally graded piezoelectric beams using an efficient finite element method. Using a sinus model based finite element method, (Beheshti-Aval and Lezgy-Nazargah 2010) investigated the static strains of piezoelectric composites beams. Based on the higher-order shear strain theory, (Doroushi et al. 2011) analyzed the vibrations and the transient response of functionally graded piezoelectric beams subjected to thermo-electro-mechanical loading. Ramirez et al. (2006) carried out the static and dynamic behaviors of laminates magneto-electro-elastic beams and plates using the discrete layer approach. Beheshti-Aval et al. (2011) proposed a refined sinus model based finite element method for the analysis of piezoelectric laminated beams. Based on the state-space approach and using the Kelvin Voigt viscoelastic model (Yan and Chen 2006) demonstrated the impact of the imperfect interfaces on the electromechanical properties of multilayered functionally graded piezoelectric beams.

Ajdour and Azrar (2013) derived an analytical solution based on the pseudo-Stroh formalism for the static behavior of the multilayered piezoelectric rectangular plates. Based on the same approach, (Pan 2001) studied the static strains of the multilayered magneto-electro-elastic plates. Heyliger and Pan (2004) investigated the cylindrical bending response of laminated magneto-electro-elastic plates using the discrete layer method. Coupling the pseudo-Stroh formalism with the Lagrange polynomials, (Ewolo et al. 2019) studied the dynamic behavior of the simply supported magneto-electro-elastic plates with imperfect interfaces. Yan et al. (2014) studied the effect of viscoelastic interfaces on the cylindrical bending response of angle-ply piezoelectric laminates. Pan and Han (2005) derived a benchmark solution based on the pseudo-Stroh formalism for the three-dimensional static behavior of the exponentially graded magneto-electro-elastic simply supported plates. The layering effect, the loading conditions effect and the material properties gradient index effect were clearly illustrated by them. Phoenix et al. (2009) investigated the static and dynamic behaviors of multi-layered magneto-electro-elastic plates using the Layerwise theory. Ajdour et al. (2016) studied the static behavior of composite plates reinforced of magneto-electro-elastic fibers. The effective properties of the fibers were obtained via the Mori-Tanaka homogenization scheme. The pseudo-Stroh formalism is thereby applied in order to obtain the macroscopic response of the reinforced plates.

After a close literature survey, we found that there are a few investigations on the dynamic behavior of the multilayered beams and plates with the pseudo-Stroh formalism in comparison with other theories. To the author's best knowledge, no literature on the dynamic behavior of the multilayered beams with the pseudo-Stroh formalism. We therefore present here the dynamic modeling of the multilayered piezoelectric beams coupling the pseudo-Stroh formalism with the Lagrange polynomials.

2 Mathematical Modeling

2.1 Geometry of Problem and Constitutive Equations

Let us consider a multilayered piezoelectric beam with a length L, a thickness h and a small width b. The kth layer of the beam is bounded by its lower interface z_{k-1} and its upper interface z_k with thickness $h_k = z_k - z_{k-1} (k = 1, \ldots, N)$; N being the total number of the layers of the beam. Cartesian coordinates system (O, x, y, z) is attached to the beam as shown on the Fig. 1.

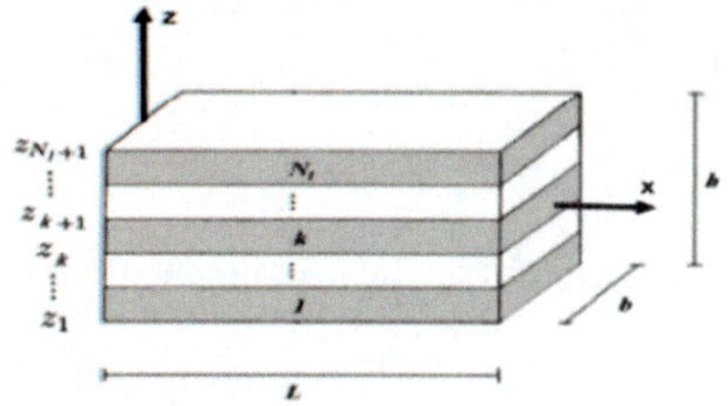

Fig. 1. Multilayered piezoelectric beam

The behavior laws of the piezoelectric materials are given by Beheshti-Aval et al. (2011):

$$\begin{cases} \sigma_{ij} = C_{ijkl}\varepsilon_{kl} - e_{lij}E_l \\ D_i = e_{ikl}\varepsilon_{kl} + \chi_{il}E_l \end{cases} \tag{1}$$

where D_i is the ith component of the electric displacement; E_i is the ith component of the electric field; σ_{ij} and ε_{kl} are respectively the components of the stress and strain tensors; C_{ijkl}, e_{lij} and χ_{il} being the components of the stiffness, piezoelectric and permittivity, respectively $(i, j, k, l = 1, 2, 3)$.

Since (ox) and (oz) are respectively the lengthwise and the transversal axis of the beam, $\sigma_{yy} = \tau_{yz} = \tau_{xy} = D_y = E_y = 0$. The behavior law (1) is then reduced to Pan and Han (2005):

$$\begin{cases} \vec{\sigma} = [\overline{C}]\vec{\varepsilon} - [\overline{e}]\vec{E} \\ \vec{D} = [\overline{e}]^T\vec{\varepsilon} + [\overline{\chi}]\vec{E} \end{cases}; \begin{cases} \vec{\sigma} = (\sigma_{xx}, \sigma_{zz}, \tau_{xz})^T; \vec{D} = (D_x, D_z)^T \\ \vec{\varepsilon} = (\varepsilon_{xx}, \varepsilon_{zz}, \gamma_{xz})^T; \vec{E} = (E_x, E_z)^T \end{cases} \tag{2}$$

$$[\overline{C}] = \begin{bmatrix} \overline{C}_{11} & \overline{C}_{12} & 0 \\ \overline{C}_{12} & \overline{C}_{22} & 0 \\ 0 & 0 & \overline{C}_{33} \end{bmatrix}; [\overline{e}] = \begin{bmatrix} 0 & \overline{e}_{31} \\ 0 & \overline{e}_{33} \\ \overline{e}_{15} & 0 \end{bmatrix}; [\overline{\chi}] = \begin{bmatrix} \overline{\chi}_{11} & 0 \\ 0 & \overline{\chi}_{33} \end{bmatrix}$$

The strains and the electric field are respectively linked to the elastic displacement u and electric potential ϕ by the following relationship (Beheshti-Aval et al. 2011):

$$\varepsilon_{ij} = \frac{1}{2}\left(u_{i,j} + u_{j,i}\right) \text{ and } E_i = -\phi_{,i} \tag{3}$$

The equation of motion of the mechanic of continuous milieu without body force and the Maxwell-Gauss equation without electric charge density are given for the piezoelectric beams by Pan and Han (2005):

$$\sigma_{ij,j} = \rho^{(k)} \frac{\partial^2 u_i}{\partial t^2} \text{ and } D_{i,i} = 0 \tag{4}$$

2.2 Pseudo-Stroh Formalism

The edge boundary conditions of the simply supported piezoelectric beams are given by Beheshti-Aval et al. (2011):

$$u_z = \phi = \sigma_{xx} = \sigma_{zz} = 0 \text{ at } x = 0 \text{ and } x = L \tag{5}$$

The Stroh solution satisfying the conditions (5) are given by:

$$\boldsymbol{u} = \begin{bmatrix} u_x \\ u_z \\ \phi \end{bmatrix} = e^{sz} \begin{bmatrix} a_1(t)\cos(px) \\ a_2(t)\sin(px) \\ a_3(t)\sin(px) \end{bmatrix}; \left\{ \begin{array}{c} p = \frac{m\pi}{L} \\ 0 \le x \le L \\ z_{k-1} \le z \le z_k \end{array} \right. \tag{6.1}$$

$$\boldsymbol{\Gamma} = \begin{bmatrix} \tau_{xz} \\ \sigma_{zz} \\ D_z \end{bmatrix} = e^{sz} \begin{bmatrix} b_1(t)\cos(px) \\ b_2(t)\sin(px) \\ b_3(t)\sin(px) \end{bmatrix} \tag{6.2}$$

$$\boldsymbol{\Psi} = \begin{bmatrix} \sigma_{xx} \\ D_x \end{bmatrix} = e^{sz} \begin{bmatrix} c_1(t)\sin(px) \\ c_2(t)\cos(px) \end{bmatrix} \tag{6.3}$$

Replacing Eqs. (6.1), (6.2), and (6.3) into Eqs. (2), (3) and (4) and using the method of direct elimination, the following equations are resulted:

$$\left[\boldsymbol{Q} + s\left(\boldsymbol{R} - \boldsymbol{R}^T\right) + s^2\boldsymbol{T}\right]\boldsymbol{a}(t) = \rho^{(k)}\boldsymbol{J}\frac{d^2\boldsymbol{a}(t)}{dt^2} \tag{7.1}$$

$$\boldsymbol{b}(t) = \left[s\boldsymbol{T} - \boldsymbol{R}^T\right]\boldsymbol{a}(t) \tag{7.2}$$

$$\boldsymbol{c}(t) = \left[\boldsymbol{M}_1 + \boldsymbol{M}_2\boldsymbol{T}^{-1}\boldsymbol{R}^T \;\; \boldsymbol{M}_2\boldsymbol{T}^{-1}\right]\begin{bmatrix} \boldsymbol{a}(t) \\ \boldsymbol{b}(t) \end{bmatrix} \tag{7.3}$$

where: $\boldsymbol{a}(t) = (a_1, a_2, a_3)^T$; $\boldsymbol{b}(t) = (b_1, b_2, b_3)^T$; $\boldsymbol{c}(t) = (c_1, c_2)^T$

$$\boldsymbol{Q} = \begin{bmatrix} -p^2\overline{C}_{11} & 0 & 0 \\ 0 & -p^2\overline{C}_{33} & -p^2\overline{e}_{15} \\ 0 & -p^2\overline{e}_{15} & p^2\overline{\chi}_{11} \end{bmatrix}; \boldsymbol{R} = \begin{bmatrix} 0 & p\overline{C}_{12} & p\overline{e}_{31} \\ -p\overline{C}_{33} & 0 & 0 \\ -p\overline{e}_{15} & 0 & 0 \end{bmatrix}; \boldsymbol{T} = \begin{bmatrix} \overline{C}_{33} & 0 & 0 \\ 0 & \overline{C}_{22} & \overline{e}_{33} \\ 0 & \overline{e}_{33} & -\overline{\chi}_{33} \end{bmatrix}$$

$$\boldsymbol{M}_1 = \begin{bmatrix} -p\overline{C}_{11} & 0 & 0 \\ 0 & p\overline{e}_{15} & -p\overline{\chi}_{11} \end{bmatrix}; \boldsymbol{M}_2 = \begin{bmatrix} 0 & s\overline{C}_{12} & s\overline{e}_{31} \\ s\overline{e}_{15} & 0 & 0 \end{bmatrix}; \boldsymbol{J} = \begin{bmatrix} 1\,0\,0 \\ 0\,1\,0 \\ 0\,0\,0 \end{bmatrix}$$

– **Stroh Fundamental Matrix**

Combining Eqs. (7.1) and (7.2), one gets:

$$\rho^{(k)}\boldsymbol{\Omega}\frac{d^2\boldsymbol{\xi}(t)}{dt^2} + (\mathbf{N} - s\mathbf{I}_{6\times 6})\boldsymbol{\xi}(t) = 0_{6\times 1};\ \boldsymbol{\xi}(t) = \begin{pmatrix} \boldsymbol{a}(t) \\ \boldsymbol{b}(t) \end{pmatrix} \tag{8.1}$$

$$\boldsymbol{\Omega} = \begin{bmatrix} 0_{3\times 3} & 0_{3\times 3} \\ \boldsymbol{J} & 0_{3\times 3} \end{bmatrix};\ \boldsymbol{N} = \begin{bmatrix} \boldsymbol{N}_1 & \boldsymbol{N}_2 \\ -\boldsymbol{N}_3 & -\boldsymbol{N}_1^T \end{bmatrix} \tag{8.2}$$

$$\boldsymbol{N}_1 = \boldsymbol{T}^{-1}\boldsymbol{R}^T;\ \boldsymbol{N}_2 = \boldsymbol{T}^{-1};\ \boldsymbol{N}_3 = \boldsymbol{Q} + \boldsymbol{R}\boldsymbol{T}^{-1}\boldsymbol{R}^T \tag{8.3}$$

where **N** is Stroh fundamental matrix, s and $\boldsymbol{\xi}(t)$ being respectively the eigenvalues and the eigenvector of Stroh.

– **Dynamic Analysis**

The solution of Eq. (8.1) is sought in polynomial form. Dividing the time domain into M intervals having equal length ΔT , we have in the rth $[(r-1)\Delta T, r\Delta T]$ as proposed by Ewolo et al. (2019):

$$\boldsymbol{\xi}^r(t) = \sum_{\alpha=0}^{N_p} \boldsymbol{\xi}\left(t_\alpha^r\right) L_\alpha(t);\ L_\alpha(t) = \prod_{\underbrace{\beta=0}_{\alpha\neq\beta}}^{N_p} \left(\frac{t - t_\beta^r}{t_\alpha^r - t_\beta^r}\right) \tag{9}$$

$$t_\alpha^r = (r-1)\Delta T + \frac{\Delta T}{2}\left(1 + \cos\left(\frac{(2\alpha+1)\pi}{2\left(N_p+1\right)}\right)\right) \tag{10}$$

where the functions $L_\alpha(t)$ are the Lagrange polynomials and the interpolation points $t_\alpha^r\left(\alpha = 0, 1, \ldots, N_p;\ r = 1, \ldots, M\right)$ are the Chebyshev points. Thus, considering this form of solution we have at the instant t_i of the rth interval the following relationships (Ewolo et al. 2019):

$$\begin{cases} \boldsymbol{\xi}^r(t_i) = \sum_{\alpha=0}^{N_p} \omega_{i\alpha}^{(0)} \boldsymbol{\xi}\left(t_\alpha^r\right) \\ \left[\frac{d^\delta \boldsymbol{\xi}^r(t)}{dt^\delta}\right]_{t_i} = \sum_{\alpha=0}^{N_p} \omega_{i\alpha}^{(\delta)} \boldsymbol{\xi}\left(t_\alpha^r\right) \end{cases};\ \begin{cases} \omega_{i\alpha}^{(0)} = L_\alpha(t_i) \\ \omega_{i\alpha}^{(\delta)} = \left[\frac{d^\delta L_\alpha(t)}{dt^\delta}\right]_{t_i} \end{cases} \tag{11}$$

Combining the Eqs. (8.1), (8.2), (8.3), (9), (10) and (11), we obtain the following discrete eigenvalues and eigenvectors equation:

$$\left(\boldsymbol{N} + \rho^{(k)}\omega_{\alpha\alpha}^{(2)}\boldsymbol{\Omega}\right)\boldsymbol{\xi}\left(t_\alpha^r\right) = s\boldsymbol{\xi}\left(t_\alpha^r\right) \tag{12}$$

After the resolution of Eq. (12), the solution in the kth layer of the beams at the instant t_α^r is given as follows:

$$\begin{bmatrix} \boldsymbol{u}\left(z, t_\alpha^r\right) \\ \boldsymbol{\Gamma}\left(z, t_\alpha^r\right) \end{bmatrix} = \begin{bmatrix} \boldsymbol{A}_1\left(t_\alpha^r\right) & \boldsymbol{A}_2\left(t_\alpha^r\right) \\ \boldsymbol{B}_1\left(t_\alpha^r\right) & \boldsymbol{B}_2\left(t_\alpha^r\right) \end{bmatrix} \langle e^{sz} \rangle \begin{bmatrix} \boldsymbol{K}_1\left(t_\alpha^r\right) \\ \boldsymbol{K}_2\left(t_\alpha^r\right) \end{bmatrix} \tag{13.1}$$

$\boldsymbol{K}_1\left(t_\alpha^r\right)$ and $\boldsymbol{K}_2\left(t_\alpha^r\right)$ being 3×1 matrices with:

$$\begin{cases} \boldsymbol{A}_1\left(t_\alpha^r\right) = \left(\boldsymbol{a}_1\left(t_\alpha^r\right)\ \boldsymbol{a}_2\left(t_\alpha^r\right)\ \boldsymbol{a}_3\left(t_\alpha^r\right)\right) \\ \boldsymbol{A}_2\left(t_\alpha^r\right) = \left(\boldsymbol{a}_4\left(t_\alpha^r\right)\ \boldsymbol{a}_5\left(t_\alpha^r\right)\ \boldsymbol{a}_6\left(t_\alpha^r\right)\right) \\ \boldsymbol{B}_1\left(t_\alpha^r\right) = \left(\boldsymbol{b}_1\left(t_\alpha^r\right)\ \boldsymbol{b}_2\left(t_\alpha^r\right)\ \boldsymbol{b}_3\left(t_\alpha^r\right)\right) \\ \boldsymbol{B}_2\left(t_\alpha^r\right) = \left(\boldsymbol{b}_4\left(t_\alpha^r\right)\ \boldsymbol{b}_5\left(t_\alpha^r\right)\ \boldsymbol{b}_6\left(t_\alpha^r\right)\right) \end{cases} \tag{13.2}$$

$$\begin{cases} \boldsymbol{a}_i(t_\alpha^r) = \left(a_{1i}(t_\alpha^r)\ a_{2i}(t_\alpha^r)\ a_{3i}(t_\alpha^r)\right)^T \\ \boldsymbol{b}_i(t_\alpha^r) = \left(b_{1i}(t_\alpha^r)\ b_{2i}(t_\alpha^r)\ b_{3i}(t_\alpha^r)\right)^T \\ \langle \boldsymbol{e}^{sz} \rangle = diag\left(e^{s_1 z}\ e^{s_2 z}\ e^{s_3 z}\ e^{s_4 z}\ e^{s_5 z}\ e^{s_6 z} \right) \end{cases} \tag{13.3}$$

The Eq. (13.1) leads to:

$$\begin{bmatrix} \boldsymbol{u}(z, t_\alpha^r) \\ \boldsymbol{\Gamma}(z, t_\alpha^r) \end{bmatrix} = \boldsymbol{P}_k\left(z - z_{k-1}, t_\alpha^r\right) \begin{bmatrix} \boldsymbol{u}(z_{k-1}, t_\alpha^r) \\ \boldsymbol{\Gamma}(z_{k-1}, t_\alpha^r) \end{bmatrix} \tag{14}$$

$$\boldsymbol{P}_k\left(z - z_{k-1}, t_\alpha^r\right) = \begin{bmatrix} \boldsymbol{A}_1\ \boldsymbol{A}_2 \\ \boldsymbol{B}_1\ \boldsymbol{B}_2 \end{bmatrix} \langle \boldsymbol{e}^{s(z-z_{k-1})} \rangle \begin{bmatrix} \boldsymbol{A}_1\ \boldsymbol{A}_2 \\ \boldsymbol{B}_1\ \boldsymbol{B}_2 \end{bmatrix}^{-1} \tag{15}$$

The matrices $\boldsymbol{P}_k\left(z - z_{k-1}, t_\alpha^r\right)$ are called the propagator matrices of the *kth* layer at the instant t_α^r.

The global transfer relationship of the multilayered beam which allows applying the surface boundary conditions is given at the instant t_α^r by:

$$\begin{bmatrix} \boldsymbol{u}(z_N, t_\alpha^r) \\ \boldsymbol{\Gamma}(z_N, t_\alpha^r) \end{bmatrix} = \prod\nolimits_{k=N}^{1} \boldsymbol{P}_k\left(h_k, t_\alpha^r\right) \begin{bmatrix} \boldsymbol{u}(z_0, t_\alpha^r) \\ \boldsymbol{\Gamma}(z_0, t_\alpha^r) \end{bmatrix} \tag{16}$$

The solution at any thickness and at any time of the *rth* interval of the piezoelectric beam is given by:

$$\begin{bmatrix} \boldsymbol{u}(z, t) \\ \boldsymbol{\Gamma}(z, t) \end{bmatrix} = \sum\nolimits_{\alpha=0}^{N_p} L_\alpha(t) \boldsymbol{P}_k\left(z - z_{k-1}, t_\alpha^r\right) \begin{bmatrix} \boldsymbol{u}(z_{k-1}, t_\alpha^r) \\ \boldsymbol{\Gamma}(z_{k-1}, t_\alpha^r) \end{bmatrix} \tag{17.1}$$

$$\begin{bmatrix} \boldsymbol{u}(z_{k-1}, t_\alpha^r) \\ \boldsymbol{\Gamma}(z_{k-1}, t_\alpha^r) \end{bmatrix} = \prod\nolimits_{\delta=k-1}^{1} \boldsymbol{P}_k\left(h_\delta, t_\alpha^r\right) \begin{bmatrix} \boldsymbol{u}(z_0, t_\alpha^r) \\ \boldsymbol{\Gamma}(z_0, t_\alpha^r) \end{bmatrix} \tag{17.2}$$

3 Numerical Results and Discussion

In this section, a three-layered piezoelectric beam made of piezoelectric material $BaTiO_3$ (named B) and elastic material of graphite epoxy (named E) is considered. The properties of these materials are given in Table 1. The dimensions of the beam are $L = 1$ m and $h = 0.3$ m. Two staking sequences namely $E/B/E$ and $B/E/B$ are studied and we assume that the thickness of each layer of the beam is identical $h_E = h_B = h/3$.

3.1 Example 1: Mechanical Loading

The Fig. 2a and b present the dynamic response of the elastic displacements u_x and u_z on the top and bottom surface of the beam induced by the normal stress $\sigma_{zz}(x, h, t) = \sigma_0 \sin(2\pi f t)\sin(\pi x/L)$, $\sigma_0 = 1\ (N/m^2)$. The solutions have been calculated in the four first intervals ($r = 1, 2, 3, 4$), for $x = 0.5\,$L, $\Delta T = 1/f = 0.01$ s and $N_p = 6$.

From those figures, it appears that the elastic displacements u_x and u_z have the same appearance, only their amplitudes are different. Moreover, those variables are the

Table 1. Properties of the materials used (Ewolo et al. 2019)

Properties	$BaTiO_3$	G. Epoxy	Properties	$BaTiO_3$	G. Epoxy
C_{11}(GPa)	166	183.43	$e_{15}\left(C/m^2\right)$	11.6	0
C_{22}(GPa)	166	11.662	$e_{24}\left(C/m^2\right)$	11.6	0
C_{33}(GPa)	162	11.662	$e_{31}\left(C/m^2\right)$	−4.4	0
C_{12}(GPa)	77	4.363	$e_{32}\left(C/m^2\right)$	−4.4	0
C_{13}(GPa)	78	4.363	$e_{33}\left(C/m^2\right)$	18.6	0
C_{23}(GPa)	78	3.918	$\chi_{11}\left(10^{-9}C^2/Nm^2\right)$	11.2	15.3
C_{44}(GPa)	43	2.870	$\chi_{22}\left(10^{-9}C^2/Nm^2\right)$	11.2	15.3
C_{55}(GPa)	43	7.170	$\chi_{33}\left(10^{-9}C^2/Nm^2\right)$	12.6	15.3
C_{66}(GPa)	44.5	7.170	ρ (Kg/m^3)	5800	1590

sinusoidal function of the time. Furthermore, it can be clearly observed that the staking sequence has a strong influence on the dynamic behavior of the multilayered piezoelectric beam. In addition, Fig. 3a shows that the electric potential ϕ at the bottom surface in the configuration E/B/E and B/E/B are in phase contrast. While at the top surface it is phase in those configurations (Fig. 2b).

3.2 Example 2: Electric Loading

The Fig. 4a and b present the dynamic response of the elastic displacements u_x and u_z at the top and bottom surfaces of the beam induced by the electric displacement loading $D_z(x, h, t) = D_0\sin(2\pi ft)\sin(\pi x/L), D_0 = 1\,C/m^2$.

From Fig. 4a, it is clearly displayed that the elastic displacements u_x and u_z at the lower surface of the beam are in phase contrast. Whereas, at the upper surface (Fig. 4b) those variables are in phase. In addition, from Fig. 4b it is shown that the staking sequence has not effect on the transverse elastic displacement u_z.

Figure 5a and b show the dynamic evolution of electric potential induced by the electric potential loading $\phi(x, h, t) = \phi_0\sin(2\pi ft)\sin(\pi x/L), \phi_0 = 1V$. From these figures, it is demonstrated that the staking sequences do not influence the electric potential.

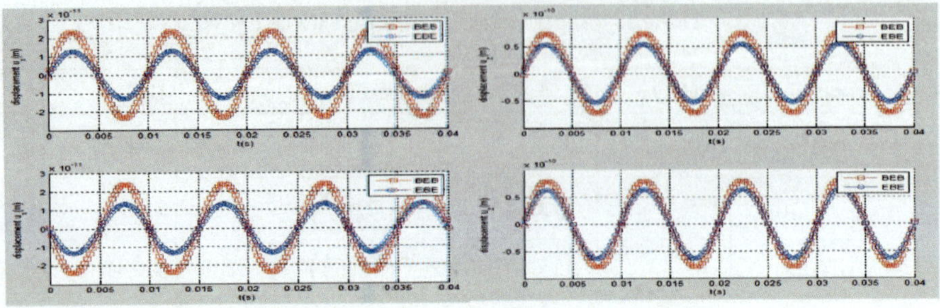

(a) (b)

Fig. 2. Displacements u_x (m) (a) and u_z (m) (b) on the bottom and top surface of the beam with the configurations E/B/E and B/E/B induced by the mechanical loading

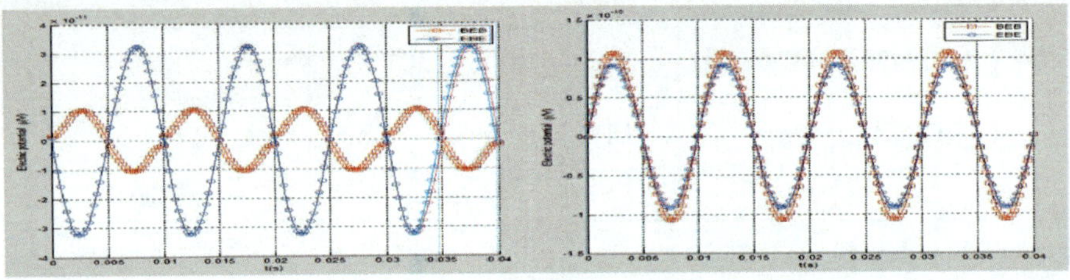

(a) (b)

Fig. 3. Electric potential *ϕ (V)* on the bottom (a) and top (b) surface of the beam with the configurations *E/B/E* and *B/E/B* induced by the mechanical loading

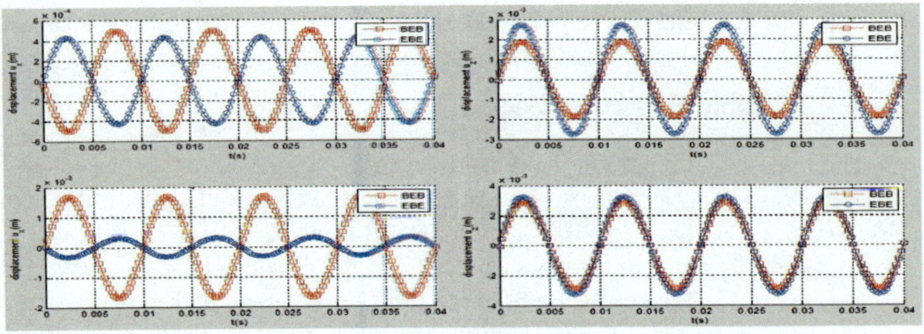

(a) (b)

Fig. 4. Displacements u_x (m) (a) and u_z (m) (b) on the bottom and top surface of the beam with the configurations E/B/E and B/E/B induced by the electric displacement loading

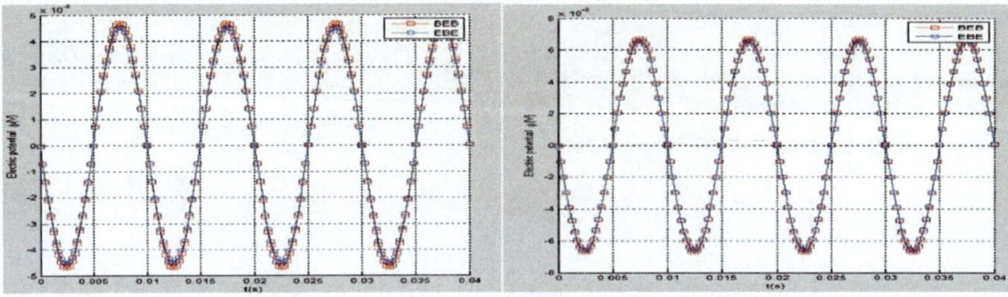

(a) (b)

Fig. 5. Electric potential ϕ (V) on the bottom (a) and top (b) surface of the beam with the configurations E/B/E and B/E/B induced by the electric potential loading

3.3 Example 3: Dynamic Response Along the Thickness

The dynamic through-thickness distribution of the three-layered beam (E/B/E and B/E/B) is presented on the Fig. 6 and Fig. 7 for electric potential loading. The solutions have been calculated at the instants t = 1.25 ms, t = 3.75 ms, t = 6.25 ms and t = 8.75 ms of the first interval. It emerges that the potential and the electrical displacement are the same in the two configurations (E/B/E and B/E/B). The stresses (in the two configurations) are symmetrical with respect to the thickness. The layering effect significantly influences the displacements (u_x, u_z).

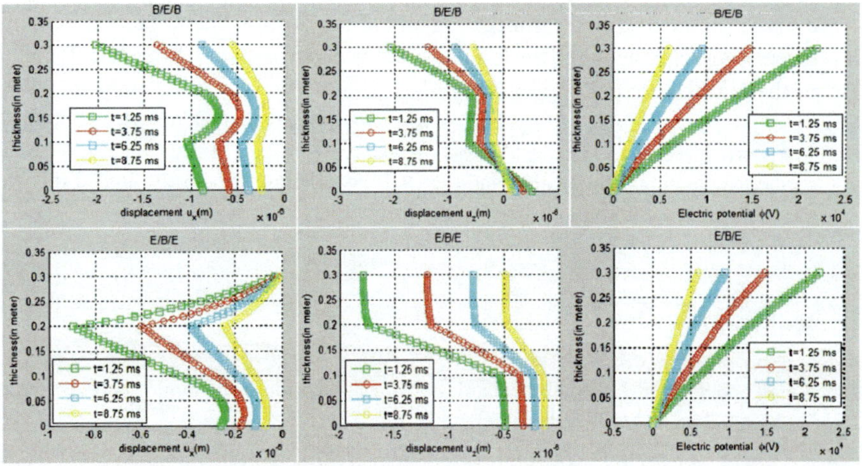

Fig. 6. Dynamic through-thickness distribution of the elastic displacements u_x (m) and u_z (m) and electric potential ϕ (V) in the beam with the configurations E/B/E and B/E/B at the instants t = 1.25 ms, t = 3.75 ms, t = 6.25 ms and t = 8.75 ms induced by the electric potential loading

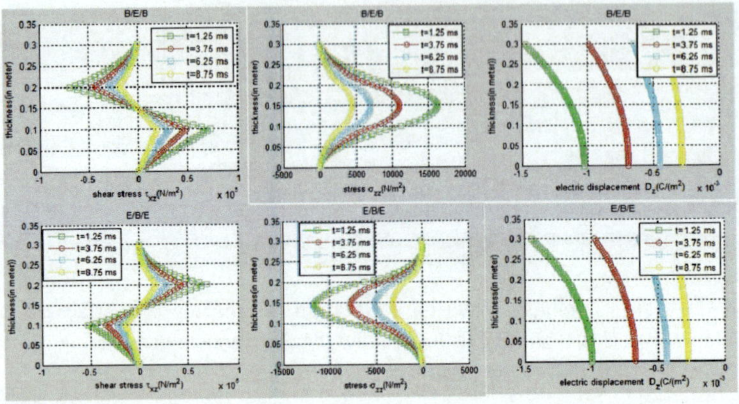

Fig. 7. Dynamic through-thickness distribution of the stresses τ_{xz} (N/m^2) and σ_{zz} (N/m^2) and electric displacement D_z (C/m^2) in the beam with the configurations E/B/E and B/E/B at the instants t = 1.25 ms, t = 3.75 ms, t = 6.25 ms and t = 8.75 ms induced by the electric potential loading

4 Conclusion

The aim of this present work was to propose a semi-analytical approach based on the pseudo-Stroh formalism coupled by the Lagrange polynomials for the dynamic behavior of multilayered piezoelectric beams. The elaborated methodological procedure has been derived for an arbitrary number of layers of the beams. The numerical results show that the staking sequences and the loading conditions strongly influence the dynamic response of the multilayered structures.

Acknowledgements. We would like to thank the Moroccan Ministry of Higher Education, Scientific Research and Innovation and the OCP Foundation who partly funded this work through the APRD research program.

References

Ajdour, M., Azrar, L.: Analytical solutions of loaded multifunctional multilayered plates. Adv. Mater. Res. **682**, 127–134 (2013)

Ajdour, M., Bakkali, A., Azrar, L., El Omri, A.: Modeling and analysis of loaded multilayered magnetoelectroelastic structures composite materials: applications. Adv. Electromagnet. **5**(3), 91–97 (2016)

Beheshti-Aval, S.B., Lezgy-Nazargah, M., Vidal, P., Polit, O.: A refined sinus finite element model for the analysis of piezoelectric laminated beams. J. Intell. (2011)

Beheshti-Aval, S.B., Lezgy-Nazargah, M.: A finite element model for composite beams with piezoelectric layers using a sinus model. J. Mech. **26**(3), 335–344 (2010)

Doroushi, A., Eslami, M.R., Komeili, A.: Vibration analysis and transient response of an FGPM Beam under thermo-electro-mechanical loads using higher-order shear strain theory. J. Intell. Mater. Syst. Struct. **22**(3), 231–243 (2011)

Ewolo Ngak, F.P., Ntamack, G.E., Azrar, L.: Dynamic analysis of multilayered magnetoelectroelastic plates based on a pseudo-Stroh formalism and lagrange polynomials. J. Intell. Mater. Syst. Struct. **30**(6), 939–962 (2019)

Heyliger, P.R., Pan, E.: Static fields in magnetoelectroelastic laminates. AIAA J. **42**(7), 1435–1443 (2004)

Lezgy-Nazargah, M., Vidal, P., Polit, O.: An efficient finite element model for static and dynamic analyses of functionally graded piezoelectric beams. Compos. Struct. **104**, 71–84 (2013)

Pan, E.: Exact solution for simply supported and multilayered magneto-electro-elastic plates. J. Appl. Mech. **68**(4), 608–618 (2001)

Pan, E., Han, F.: Exact solution for functionally graded and layered magneto-electro-elastic-plates. Int. J. Eng. Sci. **43**, 321–339 (2005)

Phoenix, S.S., Satsangi, S.K., Singh, B.N.: Layer-wise modelling of magneto-electro-elastic-plates. J. Sound Vib. **324**, 798–815 (2009)

Ramirez, F., Heyliger, P.R., Pan, E.: Free vibration response of two-dimensional magneto-electro-elastic laminated plates. J. Sound Vib. **292**(3–5), 626–644 (2006)

Yan, W., Chen, W.Q.: Electro-mechanical response of functionally graded beams with imperfectly integrated surface piezoelectric layers. Sci. China Ser. G Phys. Mech. Astron. **49**, 513–525 (2006)

Yan, W., Wang, J., Chen, W.Q.: Cylindrical bending response of angle-ply piezoelectric laminates with viscoelastic interfaces. Appl. Math. Model. **38**, 6018–6603 (2014)

Integrated Design and Production, Manufacturing Processes

Detailed Study of Endurance and Takeoff Distance Effect on Pre-sizing of Unmanned Aerial Vehicle

A. Kottat(✉) and M. E. Ait Ali

Université Mohammed V de Rabat, Ecole Mohammadia d'Ingénieurs, ERG2 (ME), Av. Ibn Sina, Rabat, Morocco
amina.kottat@research.emi.ac.ma, aitali@emi.ac.ma

Abstract. The objective of this study is to investigate the effect of endurance and takeoff distance on the result of an adapted pre-sizing of an unmanned aerial vehicle (uav). Pre-sizing methods aim to define an initial geometry of the uav that necessitates little modifications in the next phase of uav design: detailed dimensioning. Endurance and takeoff distance are two important parameters, among others, that the end user wishes to control. The non-linear formulas and equations used in pre-sizing do not give a direct relation between these two parameters and the initial geometry. In this work, a summary presentation of an adapted pre-sizing method is given. A parametric study is then conducted of the effect of endurance and takeoff distance in two cases, an uav with one engine and one with two engines. The results will help to quantify and control the effect of the required parameters on the geometry at early stages of an uav design project and assist designers when they must choose one or two motors to better meet the end user's expectations.

Keywords: Unmanned Aerial Vehicle · Pre-sizing of uav · Aircraft Sizing · Endurance Effect on uav · Takeoff Effect on uav

1 Introduction

The importance of pre-sizing is reflected in the difficulty in determining the initial geometry of an unmanned aerial vehicle (uav) [2]. The uav design is a recursive process divided into three essential phases: imposing requirements, pre-sizing, and detailed dimensioning [17]. We start with requirements by imposing at least ten parameters (endurance, payload weight, range, ceiling, etc…) in concertation with the end user. Then we choose a configuration for the uav, that the designer uses with semi-empirical formulas extracted from statistical surveys of existing uavs to dimension the initial geometry of the uav. Loftin [8] and Raymer [10] describe a detailed method for initial sizing of airplanes that is easily adaptable to uav. It gives more than twenty-nine coefficients extracted from these statistical surveys. The output of this method is an initial geometry that is supposed to satisfy the imposed requirements. One of the wishes of the end user is to have a better performance [3], for example bigger endurance, smallest takeoff and landing distances. The effect of changing these parameters on the initial geometry is difficult to

S. Aniss et al. (Eds.): CMM 2022, LNME, pp. 69–77, 2024.
https://doi.org/10.1007/978-3-031-46973-2_7

deduce directly from the used formulas. Quantifying these effects will help the designer in making a good compromise between having a better performing uav and controlling the parameters of the initial geometry. In this work we focus on studying the effect of two parameters from the requirements: endurance and takeoff distance on the initial geometry of the uav. In the first part, we start with a summary description of the adapted pre-sizing method used for our case. And we give more details about relevant formulas in relation with the two parameters of interest. After defining our requirement, we choose an uav configuration, define a reference initial geometry and we conduct a parametrical analysis where we change endurance and takeoff distance of the uav and keep the other requirements fixed. We conduct this parametrical analysis in the case of an uav with one engine and another with two engines. The results give a quantification of the effect of each parameter and identifies limits of each parameter where it is more suitable to use one engine or two engines to better meet requirements. Our study helps the designer in guiding the decision between having better performing uav and controlling the uav's initial geometry. Our study is a method that should be followed by every designer to give a clear link between the requirements and initial geometry resulting from the adapted pre-sizing method.

2 Summary Presentation of the Adapted Pre-sizing Method

After specifying the requirements, a configuration is selected as first step, followed by power-to-weight ratio, wing loading, needed engine power and takeoff weight estimation. After these steps, a first estimation is carried out of the uav layout that includes the wing, the fuselage, and the tail's dimensions. Several authors include these steps in the design process of the uav using different designations. Loftin [8] and Raymer [10] include them in conceptual design and sizing. Howe [5] Includes them in the feasibility study as synthesis/analysis step. In our study we combine these three approaches and adopt the pre-sizing method described in Fig. 1.

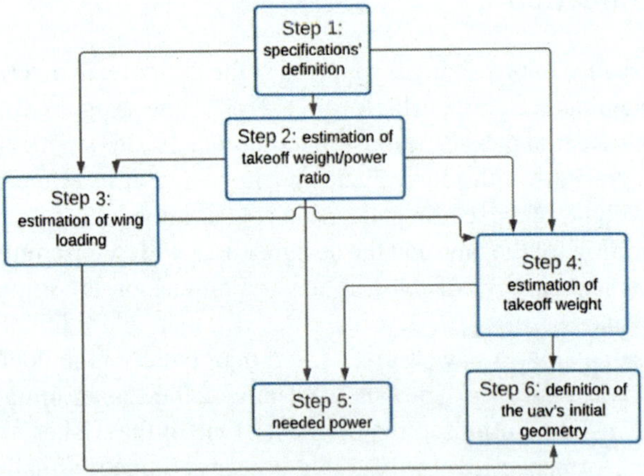

Fig. 1. Steps of the adopted pre-sizing method of an uav

The first step consists in - specifying the input data of this pre-sizing method (takeoff distance, landing distance, endurance…); - selecting the configuration and number and type of engines, as well as the flight mission's segments.

In the second step, we estimate the takeoff weight to needed power ratio, (W_{TO}/P), according to the formula recommended by Raymer [10].

In the third step, we estimate wing loading for the uav: weight to wing area ratio (W/S), in every flight segment and we chose the smallest value. In this step we used the proposed values for the needed coefficients given by Roskam [11]: maximum lift coefficient, maximum lift coefficient at takeoff (C_{LTO}) and at landing, aspect ratio and zero-lift drag coefficient.

In the fourth step, to estimate the takeoff weight W_{TO}, we start by estimating the fuel mass fraction FM_{ca} and the fuel mass fraction $FM_{ca,i}$ for every flight segment using expressions given by Raymer [10]. In our study, we chose a single cruise flight mission with five segments as the uav's mission (takeoff, climbing, landing, cruise and float segment) and propeller engine. Then we estimate the empty mass fraction designated by FM_{vide} by using the coefficients' value proposed by Gundlach [4] and Raymer [10], these coefficients are determined from the database of aircraft given by Taylor [16].

In the end of this step, we obtained the total takeoff weight W_{TO} by solving Eq. (1) with Newton Raphson algorithm.

$$W_{T0} = \frac{W_{PL}}{1 - (FM_{ca} + FM_{vide})} \tag{1}$$

where W_{PL} is the payload weight. In the fifth step, we calculate the needed power (P) using the results of both step two and step four. In the sixth step, the goal is to define the initial geometry of the uav. As a start, we calculated the six following parameters [13]: wing area (S), span (b), chord at the root level (C_0), chord at the wing tip (C_e), mean aerodynamic chord (CAM) and its position noted PCAM. After defining the wing's parameter required, we calculate the fuselage length of the uav using the expression given by Gundlach [4]. Then, we determine the parameters related to the horizontal and vertical tail using the expressions and coefficients' value given by Raymer [10] and Sadraey [13].

3 Effect of Endurance and Take-Off Distance on Pre-sizing of an uav (with One and Two Engines)

In this section we carry out a study of the effect of endurance and takeoff distance on the uav's initial geometry for the chosen mission called "simple cruise" (Fig. 2). We use the following conditions: payload 8.5 kg, maximum cruise speed 22.2 m/s, range 20 km, stall speed 7 m/s, rate of climb 3 m/s, takeoff distance and landing distances 10 m, endurance 4 h and ceiling 200 m.

We applied our pre-sizing method in the case of an uav with one engine, noted case "One engine", and one with two engines, noted "Two engines". To study the effect of endurance we vary its value from 30 min to 31 h with a step of 15 min, and we fix all the other parameters. Similarly, we vary the takeoff distance and fix the other parameters, to study its effect on the initial geometry of the uav. We quantify for each study the variation

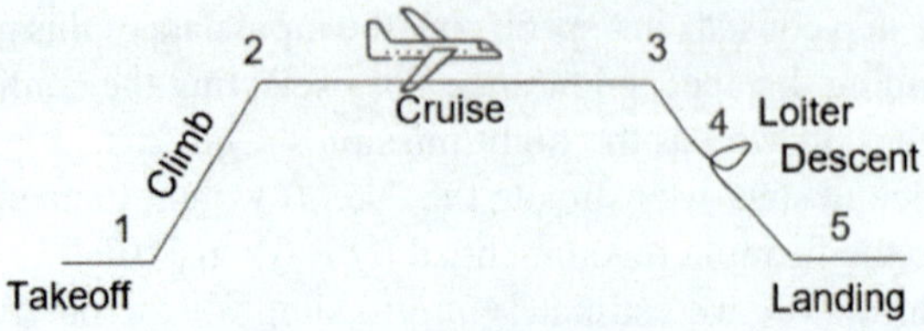

Fig. 2. Flight mission segments used in our study

of takeoff weight, the needed power, and the parameters of the uav's initial geometry. A Matlab script was developed to solve the non-linear equations with Newton-Raphson algorithm.

3.1 Effect of Endurance

Figure 3 summarizes the effect of endurance on the takeoff weight in the case "One engine" and "Two engines". We notice that the takeoff weight increases more rapidly in the case "One engine" compared with the case "Two engines". We deduce that it is more favorable to use two engines than one if we consider only the takeoff weight.

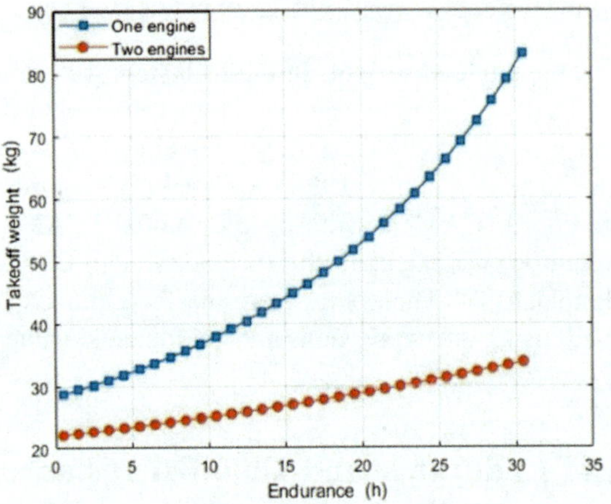

Fig. 3. The takeoff weight variation with endurance of the uav with one and two engines

Figure 4 summarizes the effect of endurance on the needed power in the cases "One engine" and "Two engines". Similarly, to the effect on takeoff weight, the necessary power increases more rapidly in the case "One engine" than the case "Two engines". The individual power of one engine (of the case "Two engines") is smaller than the power of engine of the case "One engine". When endurance exceeds 25h, it is more favorable to use two engines than one to meet the requirements.

Figure 5 presents the variation of the ten linear dimensions of the initial geometry of the uav (with one engine and with two engines) when the endurance varies between

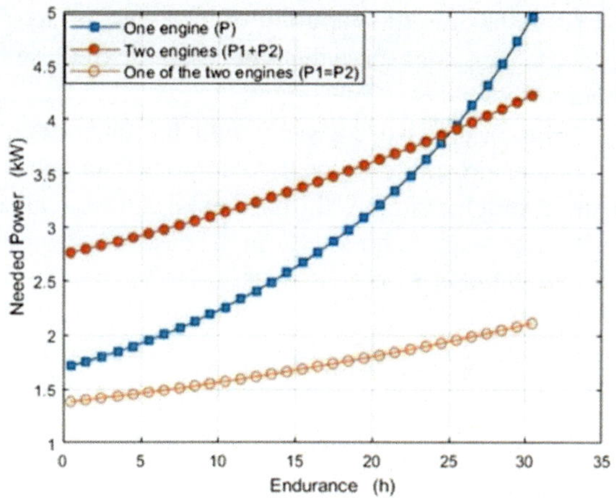

Fig. 4. The needed power variation with endurance of the uav with one and two engines

30 min and 31 h. We observe for wingspan an increase of 72% in the case. "One engine" and of 24.1% in the case "Two engines" when the endurance varies between 30 min and 31 h. We also notice when the endurance exceeds 15.5 h the wingspan in the case "One engine" becomes bigger than the case "Two engines".

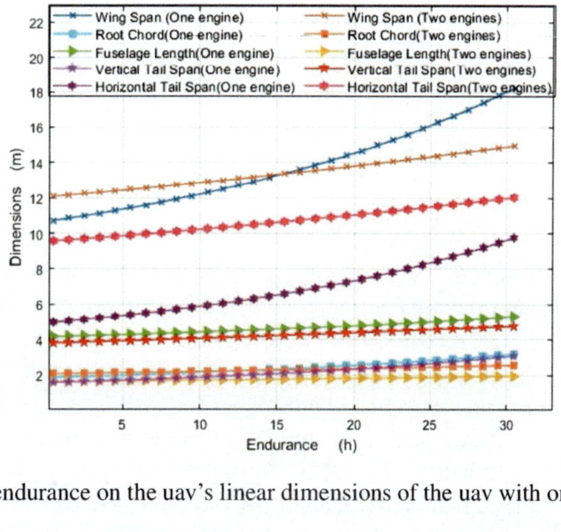

Fig. 5. Effect of endurance on the uav's linear dimensions of the uav with one and two engines

Moreover, for the root chord we observe an increase of 72% and 24.1% respectively for the case "One engine" and "Two engines". We also notice an intersection between the root chord curves at 12.75 h. Above this limit the wing's root chord of the case "One engine" becomes bigger than that of the case "Two engines". We observe an increase of the fuselage's length with a percentage of 28.3% in the case. "One engine" and 19.9% in the case "Two engines". For the horizontal and vertical tail span, we observe an increase in the case "One engine" of 99.1% and 26.3% in the case "Two engines".

From Fig. 6 synthesizes the variation of different surfaces' area of the uav's initial geometry surfaces. We can observe the rapid increase of these surfaces in the case "One engine" compared to the case "Two engines". The wing area in the case. "One engine" is smaller than the case "Two engines" until the endurance surpasses 14.5 h. After this value, it is more favorable to use two engines than one to meet the requirements. Moreover, in the case "One engine" the horizontal and vertical tail's surface increase with a percentage of 296.5% and with 59.4% in the case "Two engines" when the endurance varies between 30 min and 31 h.

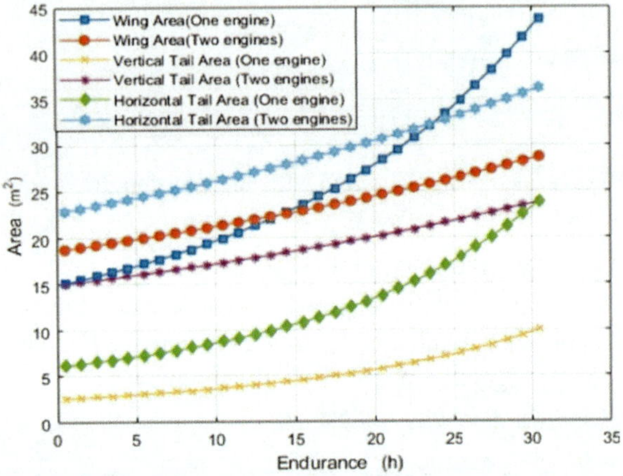

Fig. 6. Effect of endurance on area of wing, horizontal tail and vertical tail of the uav with one and two engines

3.2 Effect of Takeoff Distance

The takeoff distance of the uav's have an effect on the initial geometry of the uav only when the minimum wing loading, in the third step of the pre-sizing method, is the takeoff wing loading, see formula (2) given by Roskam [11]. Using pre-sizing formulas, we find that this condition is satisfied only when takeoff distance d_{to} as given by formula (3) according to Loftin [8], is below a limit, in our case 13 m in the case "One engine" and less than 3.1 m in the case "Two engines". Over these limits, the takeoff distance has no effect on the parameters of the uav's initial geometry. The second limit (3.1 m) is too small to conduct a parametrical analysis, therefore, we vary the takeoff distance from 1 to 13 m, with a step of 0.5 m, in the case "One engine".

$$\left(\frac{W}{S}\right)_{dto} = TOP \times \sigma \times C_{LTO} \times \left(\frac{P}{W_{TO}}\right) \tag{2}$$

$$d_{TO} = 8.134 \times TOP + 0.01494 \times TOP^2 \tag{3}$$

where TOP is a characteristic parameter of takeoff [8–11] and σ is the ratio of air density at the takeoff altitude by the density at sea level.

Figure 7 shows respectively the takeoff weight and the needed power variation. We observe a decrease of takeoff weight, needed power by a percentage of respectively. 3.3%, 4% when the takeoff distance increase from 1 and 13 m.

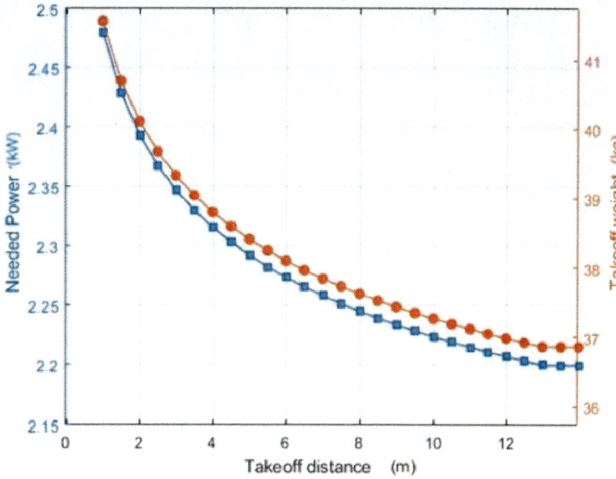

Fig. 7. Effect of takeoff distance on needed power and takeoff weight of an uav with one engine

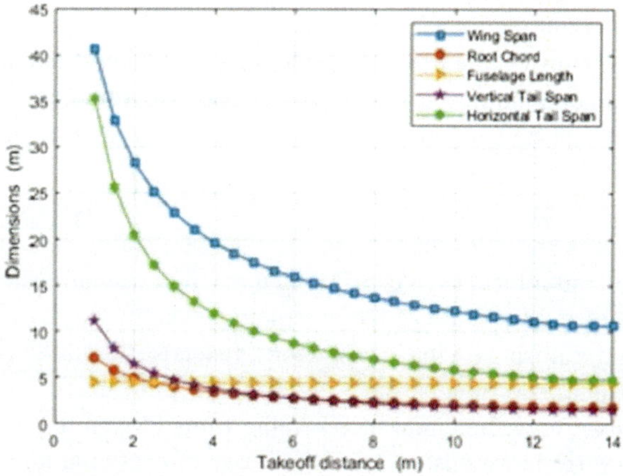

Fig. 8. Effect of takeoff distance on the uav's linear dimensions of an uav with one engine

Figure 8 shows the linear dimensions variation of the uav's initial geometry. We observe a decrease of the wingspan, root chord, vertical and horizontal tails by a percentage of respectively 38.6%, 38.6%, 51.7% and 51.7% when the takeoff distance increases from 1 to 13 m, The fuselage's length stays constant (variation of 0.6%). We deduce that the value of these dimensions decreases when the distance of takeoff increases (the minimum value in our study is 13 m).

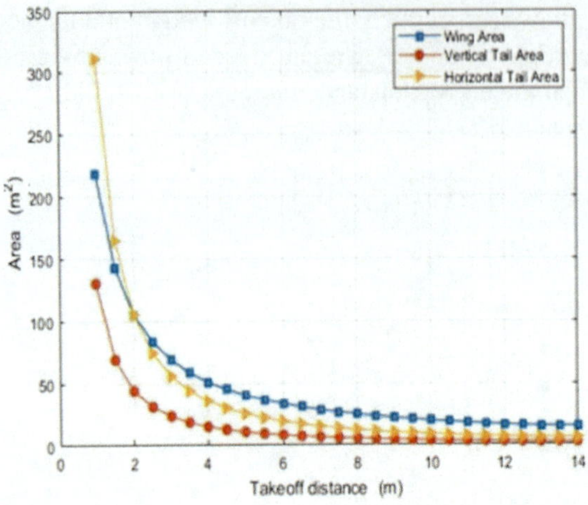

Fig. 9. Effect of takeoff distance on the surfaces' area of an uav with one engine

Figure 9 synthesizes the variation of different surfaces' area of the uav's initial geometry surfaces. We observe an important decrease of the area of the wings, the area of the horizontal and vertical tail respectively of 62.3%, 76.7% and 51.7%, when the takeoff distance increases from 1 to 13 m.

4 Conclusion

This study quantifies the effect of endurance and takeoff distance to help the designer in making a good compromise between longer flight time and smaller takeoff distance in one hand and the uav's initial geometry found by the pre-sizing method in the other hand.

We gave a summary presentation of the adapted pre-sizing method by giving the important steps and relevant formulas to the two parameters in this method. We studied the effect of endurance, in two cases: an uav with one engine and one with two engines. We quantified the variation of ten parameters: takeoff weight, needed power, five linear dimensions (wingspan, wing's root chord, fuselage's length, vertical tail span and horizontal tail span) and three surface's area (wing area, area of the horizontal and vertical tail), when we vary the endurance's value from 15 min to 31 h and keeping the other requirement fixed. We observe rapid increase of every parameter in the case of the uav with one engine compared with the one with two engines. We also identified intersection points, above which it is more favorable to use two engines than one in the case of the specifications used.

To study the effect of takeoff distance, we identified a limit (13 m for an uav with one engine) above which it has no effect on the uav's initial geometry. Under this limit, decreasing the takeoff distance increases rapidly each of the ten parameters.

These quantifications give a better insight into how to meet the needs of the end user and how to control the uav's initial geometry in the early stages of uav design. As a perspective, this study should be extended to the rest of the uav's requirements.

References

1. Abbott, I.H., Von Doenhoff, A.E.: Theory of Wing Sections, 1st edn. Dover Publications, New York (1959)
2. Basset, P.M., Tremolet, A., Lefebvre, T.: Rotary Wing UAV pre-sizing - Past and Present Methodological Approaches at Onera. Aerial Robotics **8**, 1–12 (2014). https://doi.org/10.12762/2014.AL08-10
3. Bronz, M.: A contribution to the design of long endurance mini unmanned aerial vehicles. Dissertation PhD thesis, Institut Supérieur de l'Aéronautique et de l'Espace-ISAE (2012)
4. Gundlach, J.: Designing Unmanned Aircraft Systems, 2nd edn. American Institute of Aeronautics & Astronautics, Reston (2014)
5. Howe, D.: Aircraft conceptual design synthesis, 1st edn. Professional Engineering Publishing, UK (2000)
6. Madireddy, J.: Aircraft design and weight estimation nomenclature. Glob. J. Res. Eng. **14**, 35–39 (2014)
7. Leng, Y.: Aerodynamic design of long endurance convertible UAV. Disseration PhD thesis, University of Toulouse (2020)
8. Loftin, L.K.: Subsonic aircraft: Evolution and the matching of size to performance, NASA (1980)
9. Raymer, D.P.: Enhancing Aircraft Conceptual Design using Multidisciplinary Optimization. Disseration PhD thesis, Institutionen för flygteknik (2002)
10. Raymer, D.P.: Aircraft Design a Conceptual Approach, 6th edn. American Institute of Aeronautics and Astronautics, USA (2018)
11. Roskam, J.: Airplane Design: Part 1 Preliminary Sizing of Airplanes. DARcorporation, Kansas (1997)
12. Sadraey, M.H.: Aircraft Design A Systems Engineering Approach, 1st edn. John Wiley, USA (2013)
13. Sadraey, M.H.: Unmanned aircraft design: A review of fundamentals. Morgan & Claypool Publishers, USA (2017)
14. Sharma, H., Suraj, C.S., Antony, R., Ramesh, G., Ahmed, S., Narayan, P.: Design of a High-Altitude Fixed Wing Mini UAV – Aerodynamic Challenges. In: Proceedings of International Conference on Intelligent Unmanned Systems, India (2013)
15. Struett, R.C.: Empennage sizing and aircraft stability using Matlab. https://api.semanticscholar.org/CorpusID:16770742 (2012), last accessed 11/10/2023
16. Taylor, J.W.R.: Jane's All the World's Aircraft. Jane's London, England, UK (1976)
17. Verstraete, D., Palmer, J.L., Hornung, M.: Preliminary sizing correlations for fixed wing unmanned aerial vehicle characteristics. J. Aircr. **55**, 715–726 (2018). https://doi.org/10.2514/1.C034199

Porous Media and Granular Media

Infiltration in Unsaturated Porous Media: Phenomenology and Modeling (Overview)

R. Ababou(✉)

IMFT - Institut de Mécanique des Fluides de Toulouse, CNRS-INPT, Toulouse, France
ababou@imft.fr, rachidababou@gmail.com

Abstract. We review several infiltration phenomena and modeling approaches for unsaturated porous media. First, we present a recent extension of the piston flow model of Green & Ampt, called Moving Multi-Front (MMF), which propagates vertically a discrete number of "fronts" or pressure iso-values. Secondly, based on the general Richards PDE governing variably saturated flow, we analyze several infiltration phenomena like capillary barrier phenomena due to heterogeneities in 1D (vertically discontinuous water contents) and in 2D (perched water over a clay lens in vertical cross-section). Thirdly, we present previously unpublished modeling and analyses of the 2D/3D dynamics of moisture plumes and ponded zones during localized infiltration from point sources and line sources (drip irrigation).

Keywords: Porous media · Unsaturated flow · Infiltration · Moving Multi-Front (MMF) · Capillary barrier · Perched water · Drip irrigation · Surface ponding

1 Introduction

This paper presents an overview of several infiltration phenomena and modeling approaches for unsaturated or partially saturated porous media; the paper includes reviews of recent developments in the literature such as the 1D MMF method, but also, several previously unpublished modeling approaches and results by the author on capillary barrier effects, and on localized infiltration and ponding. These different topics are presented in three distinct Sects. 2, 3, 4 which are independent in terms of physical phenomena and modeling methods.

First, in Sect. 2, we briefly review a recent extension of the piston flow model of Green & Ampt (1911), called Moving Multi-Front (MMF), proposed by Alastal & Ababou (2019). The MMF method propagates vertically a discrete number of unsaturated "fronts", or pressure iso-values. This method reduces the 1D Richards PDE to a quasi-analytical system of ODE's governing the fronts positions; the model is Lagrangian and gridless (moving mesh). The MMF model is compared to a finite volume model (BIGFLOW3D) and a finite element model (HYDRUS1D) for various unsaturated flow problems like infiltration to a water table, and capillary rise from a water table, with encouraging results in terms of CPU time and precision. Comparisons indicate that MMF is very advantageous in terms of CPU time and can be quite precise, with only a

S. Aniss et al. (Eds.): CMM 2022, LNME, pp. 81–98, 2024.
https://doi.org/10.1007/978-3-031-46973-2_8

small number of discrete moving fronts compared to the fine grids required with Eulerian codes.

Secondly, we present in Sect. 3 several analyses of variably saturated infiltration problems with capillary barrier effects due to heterogeneities, in both 1D (vertically discontinuous water contents in a layered soil) and 2D (perched water over a clay lens in vertical cross-section). These investigations are based on the general Richards PDE governing variably saturated flow in heterogeneous media. The simulations are carried out with the Finite Volume BIGFLOW 3D code (Ababou and Bagtzoglou 1993), a numerical solver of the generalized Richards PDE for variably saturated/unsaturated porous media, which can accept partially saturated zones, and accommodates fully 3D heterogeneous nonlinear water retention and hydraulic conductivity curves. We present some remarkable flow phenomena caused by material interfaces in multilayered or composite subsurface media, such as 1D capillary barrier effects in a two-layered loam/sand soil, and 2D perched water due to a clay lens.

Finally, and thirdly, we present in Sect. 4 some simulation results and theoretical analyses of the 2D/3D dynamics of the wetted volume, and of surface ponding, during localized infiltration, as occurs in drip irrigation systems with either an isolated dripper (point source), or a continuous line of drippers (line source). The simulations are based on a custom-made Finite Difference model of unsaturated flow based on the Kirchhoff transform, which has a special provision for controlling the extent of the surface ponded zone based on a specific integral boundary condition at soil surface satisfying mass balance (Ababou 1981). This code was used for 3D point source infiltration (axial symmetry) as well as 2D line source infiltration (plane symmetry). Simulation results based on this custom-made Kirchhoff-Richards code are analyzed, and a theoretical analyzis of the dynamics of the surface ponded zone is indicated based on previously unpublished analytical arguments by Ababou 1981 (which are corrected here).

2 The MMF Infiltration Model: Moving Multi-front

2.1 Summary

We briefly review here the recent work of Alastal & Ababou (2019) who proposed an extension of the piston flow model of Green & Ampt (1911), called Moving Multi-Front (MMF). The MMF method propagates vertically a discrete number of unsaturated "fronts", or pressure iso-values, in a quasi-analytical fashion. It reduces the 1D Richards PDE to a system of ODE's governing the fronts positions. The MMF model is *Lagrangian* and *gridless* (with moving mesh). It has been compared to a finite volume model (BIGFLOW3D) and a finite element model (HYDRUS1D) for typical flow problems like infiltration to a water table, and capillary rise from a water table, with encouraging results in terms of CPU time and precision. In particular, the MMF model requires only a small number of discrete moving fronts compared to the fine grids required with Eulerian codes. The comparisons in Alastal & Ababou (2019) indicate that the MMF model is advantageous in terms of CPU time and can be quite precise, with only a small number of discrete moving fronts, as can be seen below in this brief overview of their work.

2.2 Presentation of the MMF Model

The Moving Multi-Front model consists in solving the Richards PDE for a vertical flow by discretizing the pressure head variable «h» into a number of iso-values or "fronts".

Let us first present a reminder on the 1D Richards equation in a homogeneous unsaturated porous medium. The vertical flow equation of Richards (1931) is a Darcy-based mass conservation equation for incompressible water, which can be expressed as follows:

$$\frac{\partial \theta(h)}{\partial t} = \frac{\partial}{\partial z}\left(K(h)\frac{\partial h}{\partial z} - gK(h)\right) \tag{1}$$

Note that "g" here is not the acceleration of gravity, but a symbol to reflect the orientation of axis "z" with respect to gravity: take $g = +1$ if axis "z" is depth, directed downwards, otherwise take $g = -1$ if axis "z" is elevation, directed upwards.

This 1D Richards PDE, Eq. (1), expresses mass conservation of incompressible water in the presence of air, per cubic meter of porous medium. In fact, it can be expressed as $\partial\theta/\partial t = -\partial q_Z/\partial z$ (minus flux divergence), where q_Z [m/s] is the areal water flux density (sometimes called "Darcy flux" for short). Darcy's law expresses the quasi-linear relation between q_Z [m/s] and pressure gradient in the presence of gravity, as follows: $q_Z = -\{K(h).\partial h/\partial z - g.K(h)\}$.

The variable h [m] is the pore pressure head relative to atmospheric pressure, expressed in meters. Note that $h < 0$ in the unsaturated zone. For this reason, it is convenient to define a "water suction" variable $\psi = -h$ [m], such that $\psi > 0$ in the unsaturated zone. The variable θ [m^3/m^3] is the volumetric water content. Coefficient K [m/s] is the unsaturated hydraulic conductivity of the soil, which depends on pressure (h). The relation $\theta(h)$ is the water retention curve of the soil, and K(h) is the unsaturated hydraulic conductivity curve of the soil. These two hydrodynamic constitutive relations $\{\theta(h), K(h)\}$, taken together, characterize the unsaturated porous medium. They can also be expressed as $\{\theta(\psi), K(\psi)\}$.

The MMF model consists in:

a. discretizing the vertical suction profile $\psi(z,t)$ into a discrete number of moving iso-values called "fronts" (ψ_{Fi}), characterized by their moving positions $Z_{Fi}(t)$;
b. discretizing the corresponding water content profile $\theta(z,t)$ into moving moisture packets $\theta_{i+1/2}$, corresponding to the moisture comprised between the two moving fronts "i" and "i + 1";
c. applying Darcy's flux law on the moving grid by expressing the vertical flux density $q_{i-1/2}(t)$ through the zone comprised between fronts "i" and i-1″ (see below);
d. finally, expressing the quasi-analytical system of ODE's for the unknown moving positions $Z_{Fi}(t)$, and solving this ODE system for the positions of the M discrete moving "fronts" (see below).

Figure 1 illustrates graphically the discretization of the vertical unsaturated soil column into discrete pressure iso-values or «fronts».

The above procedure (i) to (iv) is now developed mathematically, leading to an algebraic differential system of equations (see Alastal & Ababou 2019 for algorithmic details and comparisons with other moving grid methods in the literature).

First, as explained earlier, the suction profile in the porous medium is discretized into M mobile “fronts” (i = 1, 2, …, M). Then Darcy’s flux law is applied to each discrete mobile zone [i-1, i], also denoted [i-1/2], as follows:

$$q_{i-1/2}(t) = -K_{i-1/2}\left[\frac{\partial h}{\partial z} - g\right] = K_{i-1/2}\left[\frac{\psi_{F_i} - \psi_{F_{i-1}}}{Z_{F_i}(t) - Z_{F_{i-1}}(t)} + 1\right] \tag{2}$$

…where ψ represents suction head ($\psi = -h$, opposite of pressure head). Note that the mid-front fluxes $q_{i-1/2}(t)$ are defined at the intermediate positions between the moving fronts $Z_{F_{i-1}}(t)$ and $Z_{F_i}(t)$. Each mid-front flux $q_{i-1/2}(t)$ can be considered as an approximation of the flux through the mobile zone $\left[Z_{F_{i-1}}(t), Z_{F_i}(t)\right]$.

Secondly, water mass conservation is applied around each mobile front $Z_{F_i}(t)$:

$$q_{i+1/2}(t) - q_{i-1/2}(t) = \left(\theta_{i+1/2} - \theta_{i-1/2}\right)\frac{dZ_i}{dt} \tag{3}$$

Finally, a differential system of size M, governing the unknown front positions $Z_i(t)$, is obtained by inserting the mid-frontal fluxes $q_{i-1/2}(t)$ of Eq. (2) into the mass conservation Eq. (3). The result is a nonlinear differential system of size M, to be solved for front positions $Z_{F_i}(t)$, subject to an initial suction profile at t = 0. The ODE system is of the form $\underline{C}\ d\underline{Z}/dt = \underline{F}(\underline{Z})$ with a nonlinear right-hand side vector function. The solution $\underline{Z}(t) = [Z_{F1}(t), \ldots, Z_{Fi}(t), \ldots, Z_{FM}(t)]$ can be obtained either quasi-analytically (in some cases), or else numerically with an ODE solver such as the explicit stiff MATLAB™ ODE solver function “ODE15S”.

Note that, for M = 1, the MMF model has a single “front” and reduces to the classical piston flow model of Green and Ampt (1911). On the other hand, for M → ∞, the “z” coordinate becomes continuous, and it is expected that the MMF model converges to the Richards PDE, with the correct nonlinear coefficients K(h) and θ(h) characterizing the porous medium. This has been confirmed by the various tests implemented in Alastal & Ababou (2019), some of which are summarized in the next section.

2.3 Review of Applications and Tests of the MMF Model

We summarize below the comparison benchmarks and tests that were developed for assessing the performance of the “gridless” Moving Multi-Front infiltration model (MMF) compared to classical Eulerian methods with fixed mesh, like Finite Volumes (BIGFLOW 3D) or Finite Elements (HYDRUS 1D). The flow problems tested in Alastal and Ababou (2019) include a diversity of vertical flows, such as (i) infiltration in a dry soil towards a deep water table; (ii) infiltration towards a shallow water table initially at equilibrium; (iii) capillary rise from a fixed water table; and (iv) capillary rise due to the gradual ascent of the water table.

Only one of these tests is discussed in this brief review. Thus, Fig. 2 shows a comparison of the MMF method (Alastal and Ababou 2019) with the HYDRUS 1D code (Simunek et al. 2013) for the case of infiltration in a dry soil towards a deep water table. The comparison results were excellent, with only 30 discrete moving fronts in the MMF model (M = 30) compared to 1001 fixed nodes in the HYDRUS 1D code.

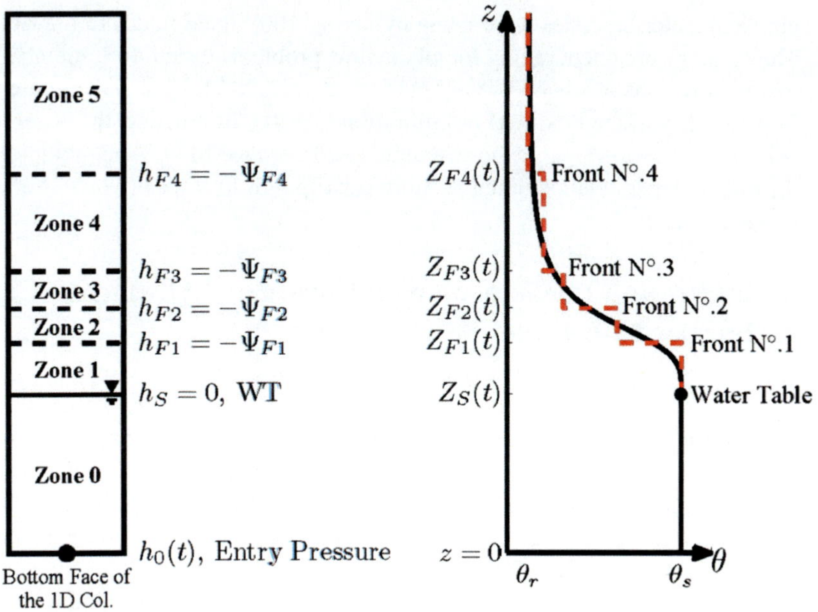

Fig. 1. Graphical illustration of the discretization of a vertical profile of pore pressures "h" or suctions "Ψ" into discrete iso-values (or "fronts")

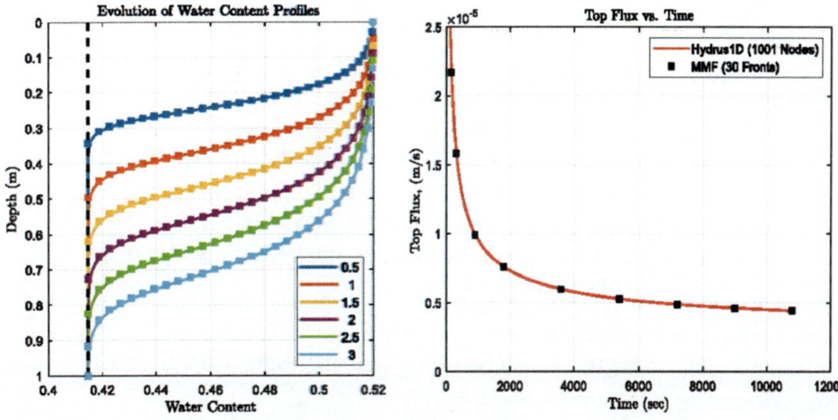

Fig. 2. Test of the MMF model for downward infiltration in a dry soil towards an infinitely deep water table, for the Guelph Loam soil (GL). The simulation with the HYDRUS1D Finite Elements code with 1001 fixed nodes (continuous curves), is compared to the MMF solution with M = 30 moving fronts (square symbols). Left: water content profiles at times t = 0.5, 1.0, 1.5, 2.0, 2.5, 3.0 h. Right: temporal evolution of infiltration flux rate at soil surface, vs. time t in seconds (up to 11000 s)

Note: all the tests implemented in Alastal & Ababou (2019) were also analyzed in terms of CPU time performance and precision (order of accuracy) with respect to the number of fronts in the MMF model (e.g. M = 1, 2, 4, 8, 16, 31, 64). Comparisons with classical Eulerian codes were made by taking 1000 fixed nodes in the Eulerian codes. The results were encouraging for all the flow problems tested. In summary, it was found that the order of precision of the MMF model is $1/M^2$ for the water content $\theta(z,t)$, and $1/M$ for the surface flux $q_z(t)$ or infiltration rate $i(t)$. In practice, the observed precision was excellent for $M \geq 32$, and still quite good even for $M = 8$ (but not less). Recall that the piston flow model of Green & Ampt corresponds to $M = 1$ (which is of course much less accurate).

3 Infiltration Phenomena with Layers and Obstacles: Capillary Barrier Effects and Perched Water

3.1 Summary

We investigate in this section the case of variably saturated flow problems with "capillary barrier" effects due to heterogeneities. Two cases are considered for illustration. First, 1D infiltration is simulated in a two-layered loam/sand soil with vertically discontinuous water content, demonstrating a "blocked moisture" effect to be discussed below. Secondly, 2D flow is simulated in vertical cross-section, demonstrating perched groundwater formation due to a clay lens inclusion in a sandy soil. The simulations are carried out with the implicit Finite Volume BIGFLOW 3D code (Ababou and Bagtzoglou 1993), which is based on the general Richards PDE governing variably saturated flow in 3D heterogeneous porous media, and which can accept the occurrence of partially saturated zones.

3.2 Capillary Barrier Effect: Vertical Infiltration in a Two-Layer Soil

Figure 3 illustrates the case of transient vertical infiltration through a two-layered "GL/FS" soil, with a fixed water table at 5 m depth. The top layer is a loamy soil (Guelph Loam "GL"), and the bottom layer is a coarser soil (Fine Sand "FS"). The material interface between the two layers is at depth 2.5 m. The initial pressure and moisture profiles are hydrostatic, at equilibrium with the water table.

Table 1. Hydrodynamic parameters in the Mualem - Van Genuchten curves $\{\theta(h),K(h)\}$ for the porous media used in this study (Fine Sand, Guelph Loam)

Parameters	Fine sand (FS)	Guelph Loam (GL)
K_S (m/s)	1.5×10^{-4}	3.66×10^{-6}
θ_S (m³/m³)	0.38	0.520
θ_r (m³/m³)	0.02	0.218
α (m⁻¹)	4.6	1.15
n	5.0	2.03

The unsaturated properties of the Guelph Loam and the Fine Sand are modeled with the Mualem – van Genuchten model (Mualem 1976, van Genuchten 1980), with the parameters given in Table 1. The two-layer simulation was conducted by the author with the BIGFLOW 3D code. Infiltration lasts for 3.5 days in this simulation, under top boundary condition of total saturation, or equivalently zero pressure h = 0 at soil surface (i.e., h = 0 at the top surface of the top « GL» layer).

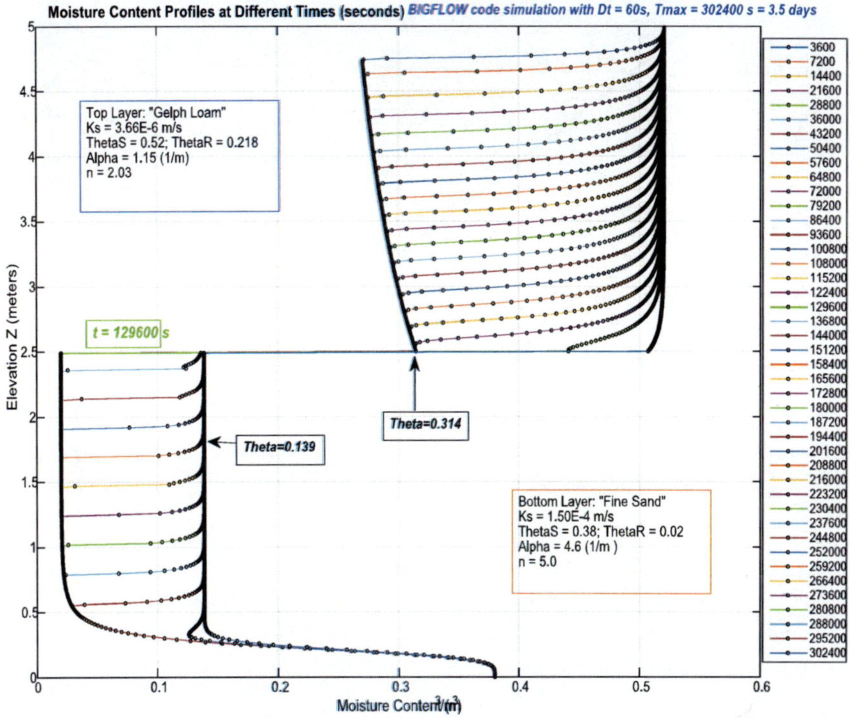

Fig. 3. Transient vertical infiltration in a two-layer soil GL/FS: Guelph Loam on top / Fine sand at bottom. Water content profiles θ(z,t) are shown up to time t = 302400 s (or 84 h, or 3.5 days). The boxes inside the figure contain the parameter values already listed in Table 1 for the GL (top) and the FS (bottom). The time legend at right reproduces the color codes for the moisture profiles every 3600 s (t = 3600 s, 7200 s, …, 302400 s), i.e., every hour from t = 1 h to t = 84 h (3.5 days). Time t = 129600 s (36 h) is the time of wetting front penetration in the bottom layer

From the water content profiles θ(z,t) of Fig. 3, one can observe that:

- The downwards moving water content "front" reaches the GL/FS interface after about 1.5 days of infiltration (t = 129600 s = 1.5 days).
- At all times ($\forall$ t $\geq$ 0), the water content profiles θ(z,t) are discontinuous with respect to "z" at the interface between the two layers.
- At large times, a gravitational flow regime ($\partial\theta/\partial z \approx 0$) is established in the bottom Fine Sand layer, with a constant moisture profile θ(z) $\approx$ 0.139 within a large zone, between elevations z = 0.25 m and z = 2.50 m.

Finally, compared to moisture profiles, note that the pressure profiles h(z,t) exhibit a different pattern, as shown in Fig. 4. In particular, unlike moisture content θ(z), water pressure h(z) is continuous at the loam/sand interface.

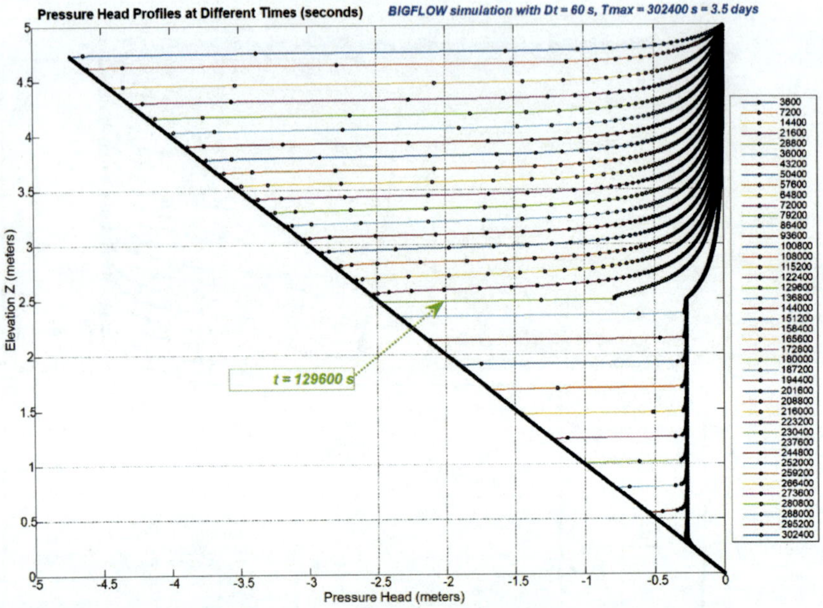

Fig. 4. Transient vertical infiltration in the two-layer soil GL/FS (Guelph Loam on top / Fine sand at bottom). Water pressure profiles h(z,t) are shown up to time t = 302400 s (or 84 h, or 3.5 days). The time legend at right reproduces the color codes for the pressure profiles every 3600 s, and the profile indicated by an arrow at t = 129600 s (36 h) corresponds to wetting front penetration in the bottom layer

In conclusion, in this two-layer infiltration test, with the loamy soil on top and the sandy soil at bottom, we see that the bottom layer, which is coarser than the top layer, plays the role of a «capillary barrier» impeding the descending soil moisture profiles θ(z,t).

3.3 Formation of a Perched Water Table Due to a Clay Lens

We now turn to another phenomenon, the formation of perched groundwater over a clay lens inclusion in a sandy aquifer. The goal here is to observe, via simulation, the genesis of perched water starting from a relatively dry unsaturated porous domain. Note: the case presented below is one of several similar cases investigated over the years by the author and collaborators (G. Trégarot, A. Larabi, M. Bouzelboudjen, et al.); see for instance Ababou & Trégarot (2002).

Figure 5 presents a 2D simulation illustrating the genesis of perched water table over a clay lens. Infiltration is due here to submersion or flooding of the soil surface, which is constantly saturated, submitted to a positive water depth at all times. The clay lens blocks temporarily the passage of water during infiltration. At any given time, the multiple water tables are calculated from the pressure field h(x,z) as the iso-surfaces of atmospheric pressure (iso-surfaces $h = 0$).

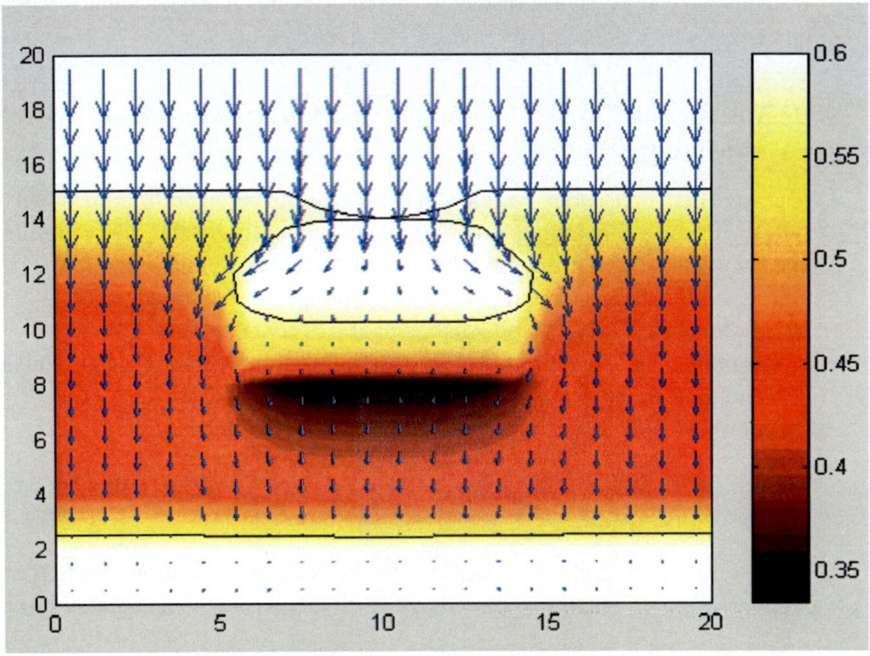

Fig. 5. Snapshot from a transient simulation of variably saturated flow in a sandy soil containing a rectangular clay lens. The figure shows, at a given time, a color map of water contents, as well as Darcy flux vectors, and multiple water table contours (which are calculated as iso-surfaces of atmospheric pressure). The fully saturated zones are in white and the dryer zones are in black

This simulation was performed using the BIGFLOW 3D code (Ababou & Bagtzoglou 1993), which solves the generalized Richards PDE, particularly here with variably saturated/partially saturated conditions (Ababou and Trégarot 2002).

4 Drip Irrigation: Dynamics of Moisture Plume and Ponded Zone During Point and Line Source Infiltration

4.1 Summary

We present in this section previously unpublished simulation results and theoretical analyses of the 2D/3D dynamics of the wetted volume and of surface ponding during localized infiltration, as occurs in drip irrigation systems with either an isolated dripper (point source), or a continuous line of drippers (line source). The simulations are based on a custom-made Finite Difference model of unsaturated flow derived from the Richards equation based on the Kirchhoff transform. This code has a special provision for controlling the extent of the ponded zone based on a specific integral boundary condition at soil surface satisfying mass balance (Ababou, 1981). The simulation code was used for 3D point source infiltration (axial symmetry) as well as 2D line source infiltration (plane symmetry). Some of the simulation results based on this custom-made Kirchhoff transform code are analyzed here, and a theoretical analyzis of the dynamics of the surface ponded zone is also indicated, based on Ababou's 1981 unpublished analytical arguments (which are partially revised and corrected here).

4.2 The Kirchhoff-Based Infiltration Model in Axial and Plane Symmetry with Surface Ponding

We present here first a brief overview of the Kirchhoff-based infiltration model of Ababou (1981) with surface ponding, assuming plane symmetry for line source infiltration, and axial symmetry for point source infiltration. (*Note*: this simulation tool was restricted to homogeneous soils; concerning 3D moisture plumes in heterogeneous soils, see Ababou 1991, Polmann et al. 1991, and Ababou et al., 1992).

The numerical solver used for localized infiltration in homogeneous soils was based on a Kirchhoff transform of the Richards PDE, expressed in planar (x,z) or cylindrical (r,z) coordinates. Briefly, the Kirchhoff transform is defined by:

$$U(h) = \int_{ho}^{h} K(h')dh' \tag{4}$$

Applying this Kirchhoff transform to the Richards equation yields the following Kirchhoff-Richards PDE (expressed here with the axis "z" oriented downwards):

$$f(U)\frac{\partial U}{\partial t} = \nabla^2 \mathrm{U} - \mathrm{g}(U)\frac{\partial U}{\partial z} \tag{5}$$

...where "z" is depth below soil surface, and the symbol ∇^2 represents the Laplacian operator div(grad), to be expressed in either planar coordinates (x,z) or axial coordinates (r,z). Expressing both cases in a single equation yields:

$$f(U)\frac{\partial U}{\partial t} = \frac{\partial^2 U}{\partial r^2} + \frac{\in}{r}\frac{\partial U}{\partial r} + \frac{\partial^2 U}{\partial z^2} - \mathrm{g}(U)\frac{\partial U}{\partial z} \tag{6}$$

...where we let either $\varepsilon = 0$ and $\mathrm{r} = \mathrm{x}$ for plane symmetry, or else $\varepsilon = 1$ and $r = \sqrt{x^2 + y^2}$ for axial symmetry, where (x,y) are the horizontal coordinates and "z" is depth. The functions f(U) and g(U) are nonlinear coefficients deduced from application of the Kirchhoff transform to the Richards PDE; they are defined as:

$$f(U) = C(h)/K(h) \ with \ C(h) = d\theta/dh; \ and \ g(U) = \frac{1}{K}\frac{dK}{dh} \tag{7}$$

where the relation $\mathrm{h} = \mathrm{h(U)}$ should be used. For more details on the Kirchhoff transform and the nonlinear coefficients of unsaturated flow, see Ababou (2018,Chap. 5).

The Kirchhoff-Richards PDE (6) is a nonlinear parabolic PDE, to be solved numerically for either line source conditions ($\varepsilon = 0$: plane symmetry) or point source conditions ($\varepsilon = 1$: axial symmetry). In Ababou (1981), a stable implicit 2D ADI (*Alternate Directions Implicit*) scheme was implemented (Peaceman & Rachford 1955). A "direct" or "explicit" linearization was performed at each time step, that is, the nonlinear coefficients at each new step (n + 1) were estimated from the known values obtained at the previous step (n).

A special global mass balance scheme was devised to track, at each time step, the extent of the ponded surface of radius $R_S(t)$ around the point/line source. The ponded area inside [0, $R_S(t)$] was represented as a zone under small positive water depth (Dirichlet condition), while the outer area was submitted to zero flux (Neuman condition). More details on this numerical procedure can be found in the unpublished thesis of Ababou (1981) which is available on line.

4.3 Simulated Moisture Plumes and Ponded Zones

Figure 6 below illustrates the extension of the wet bulb (wetted zone), and the effect of the discharge rate Q [l/h/m], for a line source infiltration, as would occur for instance in the case of irrigation from a line of drippers. Note that there exists a ponded zone that extends at soil surface, in addition to the wetted volume in depth. Obviously, the surface ponding is more pronounced for the higher discharge rate Q = 3.0 l/h/m compared to Q = 0.5 l/h/m, all other things kept equal (with same total supply of water 72 L/m in both cases).

Recall that the results of Fig. 6 were obtained by numerical simulation using a custom made «*Kirchhoff*» code designed by the author. Briefly, the Richards PDE was transformed based on the *Kirchhoff* transform. The resulting *Kirchhoff-Richards* PDE was discretized in space using standard Finite Differences, and then solved in time using the 2D ADI method (*Alternate Directions Implicit*), either with planar coordinates (x,z) [*as done in Fig. 6 for the case of a line source of drippers*], or else with cylindrical coordinates (r,z) [*for the case of an individual dripper represented as a point source*]. A special procedure was implemented at each time step in order to compute the extension of the ponded zone due to excess water beneath the dripper at soil surface. For more details, see Ababou (1981).

Figure 7 presents in more details the pattern of Darcy flux density vector in (r,z) coordinates for the case of a point-source. Note in particular that the vertical fluxes in the ponded surface are largest at the borders of the ponded zone.

Furthermore, Ababou (1981, Part 2, pp.109–140) also developed semi-empirical laws for the horizontal extent (R(t)) and the vertical extent (Z(t)) of the moisture plume, i.e., the wetted volume. The latter was defined as the volume enclosed by a relatively dry moisture iso-value (or "wetting front"). The horizontal extent R(t) was defined as the maximum radius or half-width of the moisture plume (maximum over all depths). The vertical extent Z(t) was defined as the depth of the moisture plume on the central vertical axis.

The results were expressed in terms of an integer parameter, M = 2 or M = 3, representing the number of degrees of freedom of the problem: M = 2 for a line source (plane symmetry), or M = 3 for a point source (axial symmetry). Thus, M can be viewed as a "geometric parameter". With these definitions, the findings of Ababou (1981, Part 2, pp.109–140) are as follows.

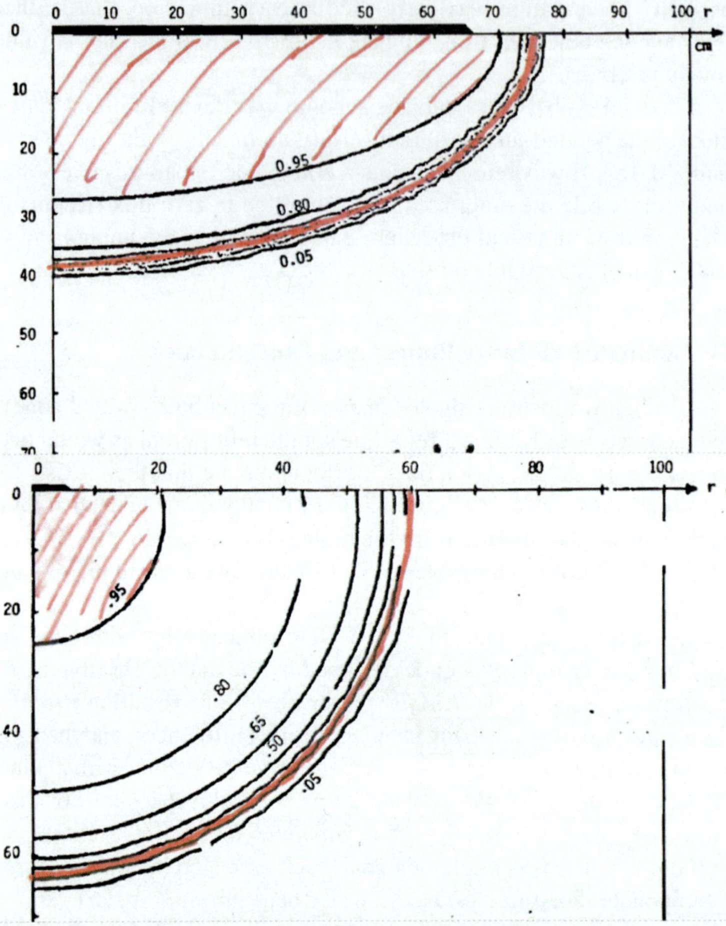

Fig. 6. Line source infiltration in a homogenous soil (Loess MC4): effect of discharge rate Q [l/h/m] for the same water supply volume [72 L/m]. Top: high flow rate Q = 3.0 l/h/m (t = 24 h). Bottom: low flow rate Q = 0.5 l/h/m (t = 144 h). [after Ababou 1981]

- The depth Z(t) of the moisture plume on the central axis (x = 0 or r = 0) is governed by:

$$Z(\tilde{t}) = Z_1(Q) \times (\tilde{t})^{1/M} + Z_2(Q) \times \tilde{t} \tag{8}$$

- The horizontal extent R(t) of the moisture plume is governed by:

$$R(\tilde{t}) = R_1(Q) \times (\tilde{t})^{1/M} - R_2(Q) \times \tilde{t} \tag{9}$$

In these expressions, $\tilde{t}$ represents a rescaled "time" having the same units as an infiltrated volume ([m^3] for M = 3, or [m^3/m] for M = 2):

$$\tilde{t} = Q \times t / (\theta_S - \theta_0) \tag{10}$$

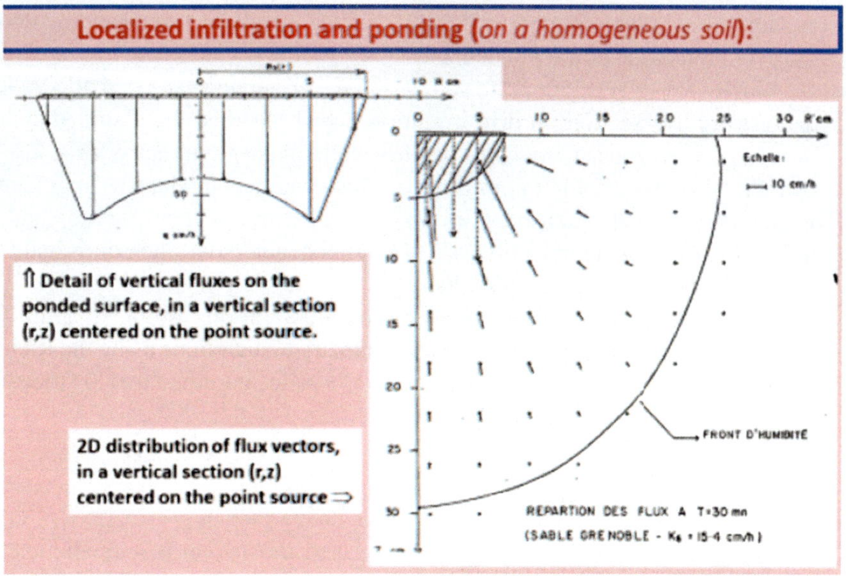

Fig. 7. Axi-symmetric point source infiltration shown at time t = 30 mn in the homogenous Grenoble Sand. Top left: detailed pattern of vertical fluxes in the ponded zone (its horizontal extent is $R_S \approx 7.5$ cm). Bottom right: 2D pattern of flux density vectors in the wet bulb in the (r,z) plane (the extend of the wet bulb is $R(t) \approx 25$ cm and $Z(t) \approx 30$ cm). [after Ababou 1981]

The empirical coefficients $Z_1(Q)$, $Z_2(Q)$, $R_1(Q)$, $R_2(Q)$, depend on discharge rate Q, but also on the geometry of the problem (i.e., they also depend on the geometric parameter M = 2 or M = 3). Note that the units of the discharge rate Q also depend on M: $[m^3/s]$ for M = 3, and $[m^2/s]$ for M = 2.

The transient evolution laws of the wet bulb (Eqs. 8, 9, 10) are valid only within a restricted time range; they cannot hold for very large times (the corresponding time range t_{LIM} was also analyzed in Ababou 1981, Part 2, pp. 109–140). Within these limits, however, the above expressions for Z(t) and R(t) were successfully fitted with numerically simulated wetting fronts, showing good fits for both point & line sources, for both silty & sandy soils, and this for several discharge rates Q.

Equations (8, 9, 10) can be interpreted physically in terms of diffusive and convective effects, as follows:

- The terms $Z_1(Q)$, $R_1(Q)$, which are proportional to $\tilde{t}^{1/M}$, are interpreted as "diffusive" terms driven by capillary diffusion.
- The terms $Z_2(Q)$, $R_2(Q)$, which are proportional to $\tilde{t}$, are interpreted as "convective" terms driven by gravity.

It was also observed, in Ababou (1981, p. 126), that coefficients Z1(Q), Z2(Q), R1(Q), R2(Q), were all roughly on the same order…except in the case of strong ponding effects. Therefore, at short times, when gravitational terms are negligible, we have roughly $R(t) \approx Z(t)$: the horizontal extent of the plume (radius or half-width) is on the same order as plume depth. On the other hand, if and when the extent of surface ponding $R_S(t)$

becomes large, the horizontal extent R(t) of the wet plume may become dominated by the evolution $R_S(t)$ of the ponded zone.

Finally, let us recall that the soil was assumed homogeneous in all simulations and analyses in this section on drip irrigation, which justified the assumed symmetries in the multidimensional *Kirchhoff-Richards* code: planar symmetry for a line source, or cylindrical symmetry for a point source. This numerical model was upgraded under the name "IRLOC" ("Irrigation Localisée") to deal with direct evaporation and root water uptake by plants under irrigation cycles in homogenous soils, including again the surface ponding process (Ababou, 1981, Part 3).

On the other hand, for *heterogeneous* unsaturated soils, problems of fully 3D localized infiltration can be tackled without symmetry assumptions using the Richards-based Finite Volumes BIGFLOW 3D code: see for instance the simulated 3D moisture plumes in Ababou (1991), Polmann et al. (1991), Ababou et al. (1992).

4.4 Theoretical Analyses of Surface Ponding Dynamics

We now present some previously unpublished theoretical results about the ponding dynamics for point sources & line sources. Ababou (1981, *Part 2, pp. 115–125)*, focusing on the phenomenon of surface ponding, developed theoretical laws, combined with heuristic arguments, for the evolution of the horizontal extent "$R_S(t)$" of the local ponded area, in two cases: point source infiltration ("$R_S(t)$" being the *radius* of the saturated/ponded zone at soil surface), and line source infiltration ("$R_S(t)$" being the *half-width* of the saturated/ponded zone).

The main physical basis of these analyses was the following global mass balance relation at soil surface, where Darcy's law is applied to express vertical flux on the saturated ponded surface, and a zero vertical flux is assumed outside the ponded area:

$$\textit{Point source} \text{ rate Q [m}^3\text{/s]:} \quad Q = 2\pi \int_0^{R_S(t)} K_S\left(-\frac{\partial h}{\partial z} + 1\right) r dr \tag{11}$$

$$\textit{Line source} \text{ rate Q [m}^2\text{/s]:} \quad Q = 2 \int_0^{R_S(t)} K_S\left(-\frac{\partial h}{\partial z} + 1\right) dx \tag{12}$$

In this mass balance relation, the transient ponded radius "$R_S(t)$" is *a priori* unknown. However, a typical "maximum" (or rather "upper bound") gravitational radius of the ponded zone "$R_{S\,MAX}$" can be inferred by inserting zero vertical pressure gradients on the saturated/ponded surface. This procedure yields a gravitational estimate of the upper bound ponded radius (called here $R_{S\,MAX}$):

$$\textit{Point source, discharge rate } Q\,[m3/s]: \quad R_{S\boldsymbol{max}} = \sqrt{\frac{Q}{\pi K_S}} \tag{13}$$

$$\textit{Line source, discharge rate } Q\,[m2/s]: \quad R_{S\boldsymbol{max}} = \frac{Q}{2K_S} \tag{14}$$

Dynamics of the Ponding Radius $R_S(t)$ Due to a Point Source

For point source infiltration, a theoretical law was obtained by Ababou (1981) for the radius $R_S(t)$ of the saturated/ponded area. The result was obtained by considering Turner & Parlange (1974)'s analyzis of the lateral movement at the periphery of a one-dimensional flow, and adapting their analyzis to the axi-symmetric flow at the periphery of the ponded zone around the point source. The final result is as follows, reproduced from Ababou (1981, pp. 119, Eq. 33):

$$R_S(t) = \left[\left(\frac{0.24S^2}{(\theta_S - \theta_0)}\right)^2 \times \left(\frac{1}{S.t^{-\frac{1}{2}} + \lambda.K_S}\right)^2 + \frac{Q}{\pi\left(S.t^{-\frac{1}{2}} + \lambda.K_S\right)}\right]^{\frac{1}{2}} - \frac{0.24S^2}{(\theta_S - \theta_0)} \times \frac{1}{S.t^{-1/2} + \lambda.K_S} \quad (15)$$

...where S [$m/s^{1/2}$] represents Philip's sorptivity coefficient (Ababou 2018, Ch. 5, Sect. 5.5.3), K_S [m/s] represents the saturated hydraulic conductivity, and λ is the dimensionless ratio A/K_S where A is Philip's gravitational coefficient. The ratio $\lambda = A/K_S$ is usually between 1/3 and 2/3; Ababou (1981) chose $\lambda = 1/3$ for transient ponding calculations of $R_S(t)$ ($0 \le t < \infty$), and $\lambda = 1$ for the asymptotic value denoted R_{SLIM} or $R_S(\infty)$ for $t \to \infty$ (*as explained further below*).

At short times, the flow regime is dominated by capillarity (sorptivity): the sorptivity term S dominates over the gravitational term K_S. The previous law then simplifies into the following result, reproduced from Ababou (1981, pp. 120, Eq. 35 - with a correction of a typographical error in that equation):

$$\textit{Point source(short times)}: \; R_S(t) \approx \sqrt{\frac{Q}{\pi S}} \times t^{1/4} \quad (16)$$

At large times, asymptotically as $t \to \infty$ (quasi-steady gravitational regime), the sorptivity term can be neglected with respect to the gravitational term K_S, and the coefficient $\lambda = 1/3$ must be replaced by $\lambda = 1$ (taking $\lambda = 1$ is equivalent to replacing "A" by K_S in the Philip formalism when $t \to \infty$). This yields the following asymptotic steady-state radius of the ponded zone, R_{SLIM} or $R_S(\infty)$ as $t \to \infty$, for a point-source (Ababou 1981, p.120, Eq. 36):

$$\textit{Point Source}(t \to \infty): \; R_S(\infty) = \left[\left(\frac{0.24S^2}{K_S(\theta_S - \theta_0)}\right)^2 + \left(R_{Smax}^{PointS}\right)^2\right]^{1/2} - \frac{0.24S^2}{K_S(\theta_S - \theta_0)} \quad (17)$$

As can be seen, the latter result on the asymptotic steady state regime ($t \to \infty$) provides a tighter bound on $R_S(t)$ than the gravitational upper bound R_{SMAX} defined earlier. Indeed, we have the following inequality (*for a point source here, but also for a line source to be discussed further below*):

$$0 \le R_S(t) < R_S(\infty) < R_{SMAX}. \quad (18)$$

Let us end up this sub-section by illustrating the extent of the point source ponded zone for a given case which we have simulated, taking the case of point source infiltration with discharge rate Q = 3.0 L/hour, on a silty soil "MC4 Montfavet Loess" (*note: this soil was used in a lysimeter experiment which was modeled in* Ababou 1981 *with evapotranspiration under line-source irrigation cycles, although we use it here to illustrate the case of a point source on bare soil*):

- Simulated radius of ponded zone at time t $\approx$ 10 h: $R_S(t_{10h}) \approx$ 50cm.
- Theoretical steady-state radius as t $\to \infty$: $R_S(\infty) \approx$ 104 cm.
- Maximum gravitational radius (upper bound): $R_{SMAX} \approx$ 173 cm.

It should be noted that the ponded area depends strongly on both discharge rate and soil properties. For instance, keeping the same point source discharge rate Q = 3.0 L/h, the asymptotic steady-state ponded radius for the Dek Sand (documented in Ababou 1981) is only $R_S(\infty) \approx$ 7 cm, compared to $R_S(\infty) \approx$ 104 cm for the MC4 Loess soil. Finally, doubling the point source discharge rate to Q = 6.0 L/h (instead of 3.0 L/h) increases these values to $R_S(\infty) \approx$ 15 cm for the Dek Sand and $R_S(\infty) \approx$ 115 cm for the MC4 Loess soil.

Dynamics of the Ponding Half-Width R_S(t) Due to a Line Source (Summary)
Finally, we summarize here in less detail some of the unpublished results from Ababou (1981) for the case of *line source* ponding.

Without going into details, (Ababou 1981, p. 122) used other analytical solutions by Parlange in the 1970's to infer, heuristically, an analytical law for Rs(t), the temporal evolution of the half-width of the ponded zone around a line source of specific discharge rate Q [m^2/s]. The latter can be interpreted as the discharge rate per unit length of line source: Q [(m^3/s)/m]. The asymptotic steady-state value R_{SLIM} or Rs(∞) of Rs(t) was also obtained. These analytical results are given below, after Ababou (1981, p. 122, Eqs. 37, 38, 39), provided a correction of these equations, with the factor 0.24 corrected here as 2 $\times$ 0.24):

Transient ponding half-width (line source) :

$$R_S(t) = \frac{1}{2}\left(Q - \frac{2 \times 0.24S^2}{\theta_S - \theta_0}\right) \times \frac{1}{S.t^{-\frac{1}{2}} + \lambda.K_S} \tag{19}$$

Asymptotic steady-state ponding (line source) :

$$RS(\infty) = R_{Slim} = R_{Smax} - \frac{1}{2K_S}\left(\frac{2 \times 0.24S^2}{\theta_S - \theta_0}\right) \tag{20}$$

As before, the value λ = 1/3 was adopted for the transient law for Rs(t). On the other hand, The steady asymptotic limit Rs(∞) was obtained by taking t $\to \infty$ and then inserting λ = 1 instead of λ = 1/3, as explained earlier.

A numerical illustration of the half-width extent of ponding for a specific line source infiltration problem can be seen in the previous Fig. 6 commented earlier. Looking at the largest discharge rate Q = 3.0 L/h/m, we see that surface ponding can be quite significant for a silty soil like the Loess MC4. The significant lateral extent of the wet "bulb" below soil surface was probably driven by the strong surface ponding phenomenon for this high flow rate of Q = 3.0 L/h/m.

To sum up, in this section, the dynamics of the wet plume, and of the ponded zone under point source & line source drip irrigation, have been analyzed numerically and analytically, through Figs. 6 and 7 and Eqs. (4) to (20).

End Note: for a global view of the contents of this text, the reader is invited to examine the introduction in Sect. 1, as well as the "Summary" sub-sections located at the start of each section: Sect. 2 (Moving Multi-Front method), Sect. 3 (capillary barrier effects), and Sect. 4 (drip irrigation and surface ponding).

Acknowledgements. This overview paper covers a broad range of research on unsaturated porous media flow modeling over the years. I would like to acknowledge previous collaborations with my former Dr-Ing. Advisors G. Vachaud and M. Vauclin (Grenoble, France) in the 1980's, and with former Ph.D. students G. Trégarot, N. Mansouri, and K. Alastal (Toulouse, France) in the 2000's and 2010's. I also thank Pr. A. Maslouhi and one anonymous reviewer for their helpful comments. Finally, I thank Pr. S. Aniss and the co-organizers of the CMM 2022 congress in Casablanca, Morocco, where this paper was initiated.

References

Ababou, R.: Capillary Flows in Heterogeneous and Random Porous Media 1, vol. 1, 371 p. ISTE Ltd and Wiley, Hoboken (2018). ISBN 978-1-84821-528-3, First Print, November 2018

Ababou, R.: Modélisation des Transferts Hydriques dans le Sol en Irrigation Localisée. Thèse de Dr-Ing., Institut de Mécanique de Grenoble (IMG), USMG & Institut National Polytechnique de Grenoble (INPG), France, 218 p. (1981). [Available online]

Ababou, R.: Three-dimensional flow in heterogeneous geologic media: high-resolution simulations. In: Proceedings, International Hydrology and Water Resources Symposium, Perth, Western Australia, vol.3, pp. 725–731. The Institution of Engineers, Australia, National Conference Publication No.92/22 (1991)

Alastal, K., Ababou, R.: Moving Multi-Front (MMF): a generalized Green-Ampt approach for vertical unsaturated flows. J. Hydrol. **579**, 124184 (17 pp.) (2019). https://doi.org/10.1016/j.jhydrol.2019.124184

Ababou, R., Bagtzoglou, A.C.: BIGFLOW: A Numerical Code for Simulating Flow in Variably Saturated, Heterogeneous Geologic Media (Theory and User's Manual, Version 1.1). Report NUREG/CR-6028, U.S. Nuclear Regul. Comm., Gov. Printing Office, Washington D.C., USA, 139p. (1993). [Available online]

Ababou, R., Trégarot, G.: Coupled modeling of partially saturated flows: macro-porous media, interfaces, and variability. In: Hassanizadeh, S.M., Schotting, R.J., Gray, W.G., Pinder, G.F. (eds.) Proceedings of CMWR2002: Computational Methods in Water Resources (23–28 June 2002, Delft, The Netherlands), vol. 47, no. C, pp. 49–56. Developments in Water Science (2002). Elsevier Science, Amsterdam. ISSN 01675648; https://doi.org/10.1016/S0167-5648(02)80044-2

Ababou, R., Sagar, B., Wittmeyer, G.: Testing procedures for spatially distributed flow models. Adv. Water Resour. **15**, 181–198 (1992)

Green, W.H., Ampt, G.A.: Studies on soil physics. J. Agric. Sci. **4**(1), 1–24 (1911). https://doi.org/10.1017/S0021859600001441

Mualem, Y.: A new model for predicting the hydraulic conductivity of unsaturated porous media. Water Resour. Res.Resour. Res. **12**(3), 513–522 (1976)

Peaceman, D.W., Rachford, H.H.: The numerical solution of parabolic and elliptic differential equation. J. SIAM **3**, 1 (28–41) (1955)

Polmann, D.J., McLaughlin, D., Gelhar, L.W., Ababou, R.: Stochastic modeling of large-scale flow in heterogeneous unsaturated soils. Water Resour. Res.Resour. Res. **27**(7), 1447–1458 (1991)

Richards, L.A.: Capillary conduction of liquids through porous mediums. Physics **1**(5), 318–333 (1931). https://doi.org/10.1063/1.1745010

Simunek, J., Sejna, M., Saito, H., Sakai, M., Van Genuchten, M.T.: The HYDRUS-1D Software Package for Simulating the One-Dimensional Movement of Water, Heat, and Multiple Solutes in Variably-Saturated Media, V.4.17. June 2013. Univ. California, Riverside, CA. Report, 308 p. (2013)

Turner, N.C., Parlange, J.-Y.: Lateral movement at the periphery of a one-dimensional flow of water. Soil Sci. **118**(2), 70–77 (1974)

Van Genuchten, M.T.: A closed-form equation for predicting the hydraulic conductivity of unsaturated soils. Soil Sci. Soc. Am. J. **44**, 892–898 (1980)

Fluids Mechanics, Rheology, Hydrodynamics Instabilities and Transition

Bifurcation Analysis of Incompressible Fluid Flow by a High Order Mesh-Free Indicator with Padé Approximants

M. Rammane[1(✉)], S. Mesmoudi[2], O. Askour[3], A. Tri[4,5], B. Braikat[6], and N. Damil[6,7]

[1] Laboratory of Materials, Signals, Systems and Physical Modeling, Department of Physics, Faculty of Sciences Agadir, University Ibn Zohr, BP 8106, 80000 Agadir, Morocco

mohamed.rammane.mc@gmail.com

[2] Hassan First University of Settat, National School of applied Sciences, LISA Laboratory, 26100 Berrechid, Morocco

[3] Mathematics, Computer Science and Communication Systems Laboratory, National School of Applied Sciences of Safi, Cadi Ayyad University, BP 63, 46000 Sidi Bouzid, Safi, Morocco

[4] Laboratory of Mechanics, Faculty of Sciences Ain Chock, Hassan II University of Casablanca, Casablanca, Morocco

[5] Higher Institute of Maritime Studies (ISEM), Km 7 Route d'El Jadida, Casablanca, Morocco

[6] Laboratory of Engineering and Materials, Faculty of Sciences Ben M'sik, Hassan II University of Casablanca, B.P 7955, Sidi Othman, Casablanca, Morocco

[7] Central Casablanca, Complex Systems and Interactions Research Center, Green City, 27182 Bouskoura, Morocco

Abstract. In this work, a mesh-free indicator is proposed to detect the bifurcation points of incompressible viscous fluid flow in a 2D sudden expansion. This indicator is mainly based on the high order mesh-free algorithm (HO-MFA) to solve the nonlinear problems obtained from the linear stability analysis. The Padé approximants are then used to improve the validity range of the HO-MFA and to determinate the roots of the indicator. The combination of these mathematical tools allows us to determine the zero of the indicator corresponding to the critical point. The Lyapunov-Schmidt reduction is used to search all the bifurcated branches from the critical point. To show the accuracy of our model, comparisons of the obtained results with those of the literature are made.

Keywords: Bifurcation point · mesh-free method · indicator · Padé approximants · Lyapunov-Schmidt reduction

1 Introduction

Many experimental and numerical studies have been carried out to study the bifurcation phenomenon of laminar flows in a sudden expansion (Drikakis 1997),

S. Aniss et al. (Eds.): CMM 2022, LNME, pp. 101–110, 2024.
https://doi.org/10.1007/978-3-031-46973-2_9

(Cherdron et al. 1978), (Alleborn et al. 1997), (Guevel et al. 2014), (Medal and Cochelin 2015), (Rammane et al. 2020), (Rammane et al. 2021a). It is well known that these flows can provide symmetrical or non-symmetrical profiles, for example, a Pitchfork bifurcation "symmetry breaking" is encountered at the critical Reynolds number. Indeed, the stability of the system changes; the flow is initially symmetrical and becomes asymmetric beyond the critical value (Drikakis 1997) ,(Cherdron et al. 1978), (Alleborn et al. 1997), (Guevel et al. 2014). Several numerical models were proposed to numerically detect these bifurcation points. We can find the direct simulation model (Drikakis 1997), the linear stability analysis using either the Eigenvalue model (Alleborn et al. 1997) or the bifurcation indicator model (Guevel et al. 2014). In all these models the finite element method is the most used, due to its large panel of application in mechanical engineering. However several difficulties are encountered in these models, such as numerical integration, mesh generation, computational cost due to determination of the Eigenvalues or steps used in the classical indicator. A Geometric Progression model (Medale and Cochelin 2015), (Rammane et al. 2021a) was proposed to reduce the time of computation consumed in the classical indicator models. We propose in this concept, a high order mesh-free indicator that uses the Padé approximants (HO-MFI-PA) to overcome the difficulties encountered in the classical models. In this approach, a linear perturbation is used for the stability analysis. The HO-MFA (Rammane et al. 2020), (Rammane et al. 2021a) is used to solve the nonlinear problem, to avoid the numerical integration, the mesh-generation and the computational cost consumed by the iterative algorithms (Drikakis 1997), (Alleborn et al. 1997). Then, the Padé approximants (Elhage-Hussein et al. 2000) are used to increase the validity range of the HO-MFA. The Padé approximants allow to compute directly the null values of the indicator, which for the bifurcation point is detected. This is practically achieved by analyzing the roots of its numerator, this is why this approximation is chosen in this model. By this way, we are able to determinate the critical point without recourse to the follow-up procedures of the indicator curve as in (Cadou et al. 2006), (Rammane et al. 2022) or the Eigenvalues sign change as in (Alleborn et al. 1997). To compute all post bifurcated branches that cross at the critical point, the Lyapunov-Schmidt reduction (Medale and Cochelin 2015), (Rammane et al.2021a) is used in this model. The obtained results will be compared and verified by those presented in some works of the literature (Cherdron et al. 1978), (Drikakis 1997), (Alleborn et al. 1997), (Guevel et al. 2014), (Medale and Cochelin 2015).

The rest of this paper is organized as follows. First, the steady state Navier-Stokes equations for incompressible viscous isothermal flows are recalled in Sect. 2. Then, the proposed indicator for solving the stationary incompressible Navier-Stokes equations and for detecting the critical point with their corresponding post-bifurcated branches, is described in Sect. 3. After that, the numerical results obtained for classical flow in a sudden expansion are reported in Sect. 4. Finally, a conclusion of the obtained results and a list of references is given in the end of this paper.

2 Governing Equations

The numerical analysis of stationary bifurcation phenomena encountered in incompressible, isothermal Newtonian fluid flows involves solving the following Navier-Stokes equations (Guevel et al. 2014), (Medale and Cochelin 2015), (Alleborn et al. 1997), (Rammane et al. 2021a) :

$$\begin{cases} \rho\,(\mathbf{V}\cdot\boldsymbol{\nabla}\mathbf{V}) = \mu\boldsymbol{\Delta}\mathbf{V} - \boldsymbol{\nabla}p & in\ \Omega \\ \boldsymbol{\nabla}\cdot\mathbf{V} = 0 & in\ \Omega \\ \mathbf{V} = \lambda\mathbf{V}_d & on\ \partial\Omega_{\mathbf{V}} \end{cases} \tag{1}$$

where $\mathbf{V}$ is the field velocity, ρ is the mass density, μ is the dynamic viscosity, p is the pressure and $\mathbf{V}_d$ is the imposed velocity. λ is a dimensionless control parameter, used here to search numerically the path solution of velocity variation as function of Reynolds number defined by $Re = (\lambda v_d \rho D))/\mu$. In the constraint (1)b ($\boldsymbol{\nabla}\cdot\mathbf{V} = 0$), the penalty technique (Cadou et al. 2006), (Guevel et al. 2014), (Rammane et al. 2020), (Rammane et al. 2021a) is used to ensure the incompressibility condition by introducing the penalty parameter ξ and to express the the pressure as function of the velocity ($p = \xi\boldsymbol{\nabla}\cdot\mathbf{V}$). The previous equations can be rewritten in the following compact form:

$$L(\mathbf{V}) + Q(\mathbf{V},\mathbf{V}) = \lambda F \tag{2}$$

where $L(\mathbf{V}) ={}^t\!< \mu\boldsymbol{\Delta}\mathbf{V} - \boldsymbol{\nabla}(\xi\boldsymbol{\nabla}\cdot\mathbf{V}), \mathbf{V} >$ and $Q(\mathbf{V},\mathbf{V}) = -^t < \rho\,(\mathbf{V}\cdot\boldsymbol{\nabla}\mathbf{V})\,, 0 >$ are the linear and nonlinear operators respectively, $F ={}^t\!< 0, \lambda\mathbf{V}_d >$ is the second member obtained from the boundary on $\partial\Omega_{\mathbf{V}}$.

3 Description of the Mesh-Free Indicator

In the linear stability analysis, the solution is sought as linear combination of a perturbation $\boldsymbol{\delta}\mathbf{V}$ and a stationary solution $\mathbf{V}^{\boldsymbol{\lambda}}$ (such as $\mathbf{V} = \mathbf{V}^{\boldsymbol{\lambda}} + \delta\mathbf{V}$). By introducing this combination in (2) and by considering only the responses to the 1st order of perturbation (since $\boldsymbol{\delta}\mathbf{V}^2 << \boldsymbol{\delta}\mathbf{V}$), we obtain:

$$\begin{cases} L(\mathbf{V}^{\boldsymbol{\lambda}}) + Q(\mathbf{V}^{\boldsymbol{\lambda}},\mathbf{V}^{\boldsymbol{\lambda}}) = \lambda F \\ L(\boldsymbol{\delta}\mathbf{V}) + \tilde{Q}(\mathbf{V}^{\boldsymbol{\lambda}},\boldsymbol{\delta}\mathbf{V}) = \eta^s f \end{cases} \tag{3}$$

where $\tilde{Q}(\mathbf{V}^{\boldsymbol{\lambda}},\boldsymbol{\delta}\mathbf{V}) = Q(\mathbf{V}^{\boldsymbol{\lambda}},\boldsymbol{\delta}\mathbf{V}) + Q(\boldsymbol{\delta}\mathbf{V},\mathbf{V}^{\boldsymbol{\lambda}})$. We have introduced in (3)-b a perturbation denoted by $\eta^s f$, where the cancellation of the scalar η^s define the proposed bifurcation indicator and f is a random function (Cadou et al. 2006), (Guevel et al. 2014), (Askour et al. 2019).

3.1 Solving the Nonlinear Problem

The computation of a bifurcation requires solving the equation (3) by using the HO-MFA presented in (Rammane et al. 2020), where the unknowns of the problem are sought into a Taylor series expansion, truncated at the order k, with respect to a parameter a as follows:

$$\mathcal{U} = \mathcal{U}_0 + \sum_{i=1}^{k} a^i \mathcal{U}_i \tag{4}$$

where $\mathcal{U}^t =< \mathbf{V}^\lambda, \lambda, \delta\mathbf{V}, \eta^s >$ and the parameter a is defined by the arc-length parametrization (Guevel et al. 2014), (Rammane et al. 2020) : $a =< \mathbf{V}^\lambda, \mathbf{V}_1^\lambda >$ $+\lambda\lambda_1$. By using the HO-MFA techniques of (Rammane et al. 2020) and an orthogonality condition to the solution (chosen as $< \delta\mathbf{V} - \delta\mathbf{V}_0, \delta\mathbf{V}_0 >= 0$), we obtain the following linear systems:

$$\begin{cases} \mathbf{K_t}(\mathbf{V}_0^\lambda)\mathbf{V}_i^\lambda & = \lambda_i \mathbf{F} - \sum_{r=1}^{i-1} \mathbf{Q}(\mathbf{V}_r^\lambda, \mathbf{V}_{i-r}^\lambda) \\ < \mathbf{V}_i^\lambda, \mathbf{V}_1^\lambda > + \lambda_i\lambda_1 & = \delta_{i1} \\ \mathbf{K_t}(\mathbf{V}_0^\lambda)\delta\mathbf{V}_i & = \eta_i^s \mathbf{f} - \sum_{r=1}^{i-1} \widetilde{\mathbf{Q}}(\mathbf{V}_r^\lambda, \delta\mathbf{V}_{i-r}) \\ < \delta\mathbf{V}_i, \delta\mathbf{V}_0 > & = 0 \end{cases} \tag{5}$$

The tangent matrix $\mathbf{K_t}(\mathbf{V}_0^\lambda)$ and the second member vectors $\mathbf{F}$, $\mathbf{f}$, $\mathbf{Q}$ and $\widetilde{\mathbf{Q}}$ are obtained by approximating the spatial derivatives of each field unknowns using the mesh-free MLS approximation, as described in (Rammane et al. 2020). It should be noted that the previous linear systems (5) have the same tangent operator. First, the unknowns of (5) are calculated which gives $(\mathbf{V}_i^\lambda, \lambda_i)$. Then, the other unknowns are computed $(\delta\mathbf{V}_i, \eta_i^s)$.

3.2 Continuation Techniques and Singular Point Detection

Classically, the continuation procedure is used by considering the end of each step $\mathcal{U}(a_{max})$ as the new starting point for the next branch, where a_{max} is defined by (Cadou et al. 2006), (Medale and Cochelin 2015), (Rammane et al. 2020), (Rammane et al. 2021a), (Rammane et al. 2021b) :

$$a_{max} = \left(\epsilon \frac{||\mathcal{U}_1||}{||\mathcal{U}_k||} \right)^{\frac{1}{k-1}} \tag{6}$$

In the works of (Cadou et al. 2006), (Askour et al. 2019), (Rammane et al. 2022), the detection of a singular point is achieved by observing the indicator curve according to λ and the critical point λ_c is determined by the point $(\eta^s = 0, \lambda_c)$ of the curve. Then, the critical Reynolds is calculated by $Re_c = (\lambda_c v_d \rho D))/\mu$. Here, we propose an alternative continuation scheme to indicate analytically the singular points. The asymptotic expansions of (4) are replaced by equivalent rational approximations called Padé approximants:

$$\mathcal{U}_{Padé,k} = \mathcal{U}_0 + \sum_{i=1}^{k-1} \frac{\mathcal{D}_{k-1-i}(a)}{\mathcal{D}_{k-1}(a)} a^i \mathcal{U}_i \tag{7}$$

where $\mathcal{D}_i$ is a polynomial of degree k defined by :

$$\begin{cases} \mathcal{D}_i & = 1 + \sum_{n=1}^{i} a^n d_n \\ d_i & = -\frac{\beta_{k(k-i)}}{\beta_{(k-i)(k-i)}} - \sum_{m=1}^{i-1} \frac{\beta_{(k-m)(k-i)}}{\beta_{(k-i)(k-i)}} d_m \\ \beta_{ii} & = \sqrt{<\boldsymbol{\mathcal{U}}_i^p, \boldsymbol{\mathcal{U}}_i^p>} \\ \beta_{im} & = <\boldsymbol{\mathcal{U}}_i, \boldsymbol{\mathcal{U}}_m^*> \\ \boldsymbol{\mathcal{U}}_i^p & = \boldsymbol{\mathcal{U}}_i - \sum_{l=1}^{i-1} \boldsymbol{\mathcal{U}}_l^* \\ \boldsymbol{\mathcal{U}}_i^* & = \boldsymbol{\mathcal{U}}_i^p / \beta_{ii} \end{cases} \tag{8}$$

In general, the representation (7) has a validity range "a_{max}^p" greater than the polynomial approximation (4). The validity range of (7) is determined by (Guevel et al. 2014), (Elhage-Hussein et al. 2000):

$$\varepsilon_d = \frac{||\boldsymbol{\mathcal{U}}_{Padé,k}(a_{max}^p) - \boldsymbol{\mathcal{U}}_{Padé,k-1}(a_{max}^p)||}{||\boldsymbol{\mathcal{U}}_{Padé,k}(a_{max}^p) - \boldsymbol{\mathcal{U}}_0||} \tag{9}$$

Once the parameter a_{max}^p is calculated, a new starting point can be defined by introducing the value a_{max}^p in the expression (7) allowing to calculate the solution $\boldsymbol{\mathcal{U}}$ for any Reynolds number. Near a singularity, a unique real root a_c of the dominator $\mathcal{D}_{k-1}(a)$ of (7) appears, where a_c corresponds here to the root of the indicator. Finally, a stationary bifurcation point is determined by $\eta^s(a_c) = 0$ and the singular solution is calculated by $\mathcal{U}_c = \mathcal{U}(a_c)$.

3.3 Post-critical Branches Determination

Once, the point of bifurcation is obtained $\boldsymbol{\mathcal{U}}_c$, we propose to compute the bifurcated solutions by using $\boldsymbol{\mathcal{U}}_c$ as a new starting point of (4). In theory, the tangent operator evaluated at the bifurcation point is singular, for that the Lyapunov-Schmidt reduction (Medale and Cochelin 2015) is combined with the HO-MFA to compute all the post-bifurcated branches as in (Rammane et al. 2021a). Each branch, noted bj, is searched using the HO-MFA as follows :

$$\begin{cases} \mathbf{K_t}(\mathbf{V}_c^{\boldsymbol{\lambda}})\mathbf{V}_i^{\boldsymbol{\lambda}^{bj}} & = \lambda_i^{bj}\mathbf{F} - \sum_{r=1}^{i-1} \mathbf{Q}(\mathbf{V}_r^{\boldsymbol{\lambda}^{bj}}, \mathbf{V}_{i-r}^{\boldsymbol{\lambda}^{bj}}) \\ < \mathbf{V}_i^{\boldsymbol{\lambda}^{bj}}, \mathbf{V}_1^{\boldsymbol{\lambda}^{bj}} > + \lambda_i^{bj}\lambda_1^{bj} & = \delta_{i1} \end{cases} \tag{10}$$

The use of Lyapunov-Schmidt reduction makes it possible to rewrite the unknowns $\mathbf{V}_i^{\boldsymbol{\lambda}^{bj}}$ in the following linear combination :

$$\mathbf{V}_i^{\boldsymbol{\lambda}^{bj}} = \lambda_i^{bj}\mathbf{W} + \varsigma_i^{bj}\boldsymbol{\delta}\mathbf{V}_c + \mathbf{V}_i^Q \tag{11}$$

$\boldsymbol{\delta}\mathbf{V}_c = \boldsymbol{\delta}\mathbf{V}(a_c)$ is the bifurcation mode, ς_i^{bj} its unknown scalar of order i, $\mathbf{W}$ and $\mathbf{V}_i^Q$ are the complementary solutions computed by :

$$\mathbf{K_t}(\mathbf{V}_c^{\boldsymbol{\lambda}})\mathbf{W} = \mathbf{F} \;\; ; \;\; \mathbf{K_t}(\mathbf{V}_c^{\boldsymbol{\lambda}})\mathbf{V}_i^Q = -\sum_{r=1}^{i-1} \mathbf{Q}(\mathbf{V}_r^{\boldsymbol{\lambda}^{bj}}, \mathbf{V}_{i-r}^{\boldsymbol{\lambda}^{bj}}) \tag{12}$$

The unknown scalars λ_i^{bj} and ς_i^{bj} are determined by projecting the eigenvector $\boldsymbol{\vartheta}$ (defined by ${}^t\boldsymbol{\vartheta}\mathbf{K_t}(\mathbf{V}_c^{\boldsymbol{\lambda}}) = 0$ with $< \boldsymbol{\vartheta}, \boldsymbol{\delta}\mathbf{V}_c >= 0$ and $< \boldsymbol{\vartheta}, \boldsymbol{\delta}\mathbf{V}_c >= 1$) on the algebraic system of (10) at the order $i+1$. For example, to obtain λ_1^{bj} and ς_1^{bj}, the projection at the second order $i=2$ gives $< \boldsymbol{\vartheta}, \mathbf{Q}(\mathbf{V}_1^{\lambda^{bj}}, \mathbf{V}_1^{\lambda^{bj}}) >= 0$. This leads to the following bifurcation equation :

$$(\lambda_1^{bj})^2\alpha + \lambda_1^{bj}\varsigma_1^{bj}\beta + (\varsigma_1^{bj})^2\gamma = 0 \qquad (13)$$

where $\alpha =< \boldsymbol{\vartheta}, \mathbf{Q}(\delta\mathbf{V}_c, \delta\mathbf{V}_c) >$, $\beta =< \boldsymbol{\vartheta}, \tilde{\mathbf{Q}}(\mathbf{W}, \delta\mathbf{V}_c) >$ and $\gamma =< \boldsymbol{\vartheta}, \mathbf{Q}(\mathbf{W}, \mathbf{W}) >$. This equation allows to know if the singular point correspond to a limit point or bifurcation point and to compute the post-critical branches by using the discriminant $\Delta = \sqrt{\beta^2 - 4\alpha\gamma}$. After the computation of all terms of the series for each branch bj, the continuation technique via Padé approximants is used.

4 Numerical Results

The classical flow in a sudden expansion is considered to illustrate the accuracy of the proposed model. The domain, the boundary conditions and the properties of the fluid are presented in Fig. 1. The inflow maximum velocity value is chosen as $\mathbf{V}_d =^t < 1, 0 >$. This problem is often used in order to investigate stationary bifurcation phenomenon in fluid flows (Cherdron et al. 1978), (Drikakis 1997), (Alleborn et al. 1997), (Guevel et al. 2014), (Medale and Cochelin 2015). According to the cited references, an aspect ratio between the outflow and the inflow heights $A_r = H/D = 3$ is used to test the validity of the numerical scheme to determinate the critical Reynolds number Re_c compared to the experimental studies of (Cherdron et al. 1978).

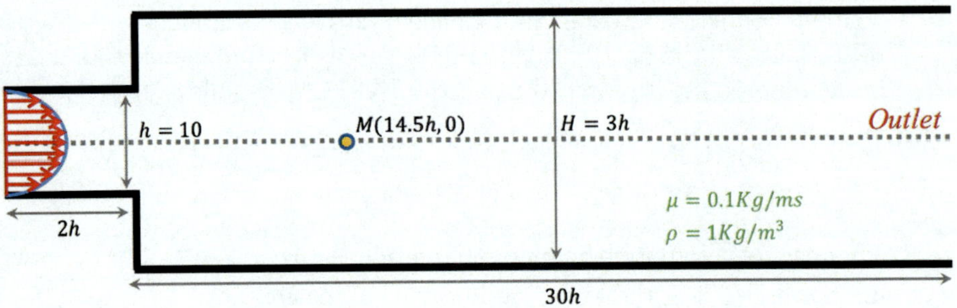

Fig. 1. Geometry and boundary conditions

In Fig. 2, we present the evolution of the indicator as a function of the Reynolds number. Let note that the second system of equation (3) corresponds to a simple eigenvalue problem when $\eta^h = 0$. Consequently, the determination of a singular point consists on searching the value of λ when $\eta^h = 0$, instead of solving an eigenvalue problem for each value of λ. The obtained results show that the indicator reaches a null value at $Re_c = 82.87$ ($\lambda_c = 8.287$). In the literature investigations, the value of Re_c are: $Re_c > 81$ in the experimental studies of (Cherdron et al. 1978), $Re_c = 79.95$ in (Drikakis 1997) using direct simulation model, $Re_c = 82.97$ in (Alleborn et al. 1997) using the Eigenvalue model, $Re_c = 80.46$ in (Guevel et al. 2014) using the indicator model with FEM, $Re_c = 82.08$ in (Medale and Cochelin 2015) using 3D investigation by FEM with Geometric Progression and $Re_c = 82.87$ in (Rammane et al. 2021a) using mesh-free approach with Geometric Progression. The obtained results are very similar to those obtained in the literature and the closest to the experimental results of (Cherdron et al. 1978) using 2D investigations.

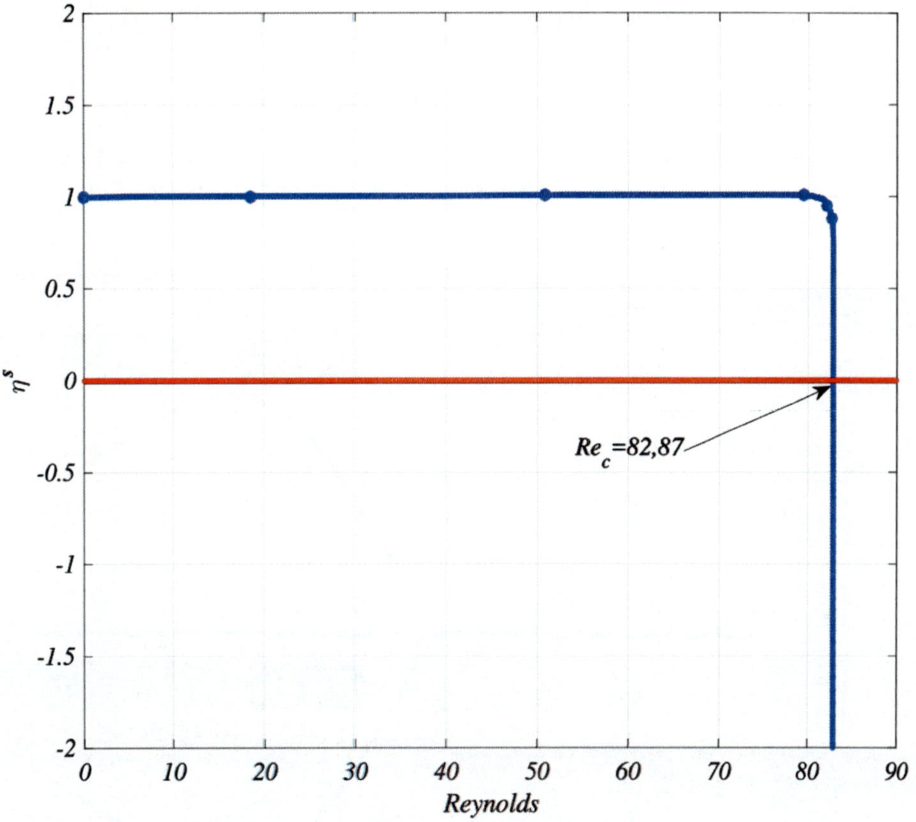

Fig. 2. Curve of the bifurcation indicator

In Table 1, we present the truncation order influence on the critical point detected using the classical series continuation (5), the Padé approximants (7) and the indicator model of (Guevel et al. 2014). The results show that the Padé approximants have a precise detection and by using lower truncation order, continuation steps and time of computation.

Table 1. Influence of k on Re_c

Order k		15	20	25	30
Padé (7)	Re_c	82.85	82.86	82.87	82.87
	Step	7	7	6	6
	CPU	55 s	63 s	42 s	49 s
Series (4)	Re_c	–	–	–	82.94
	Step	–	–	–	10
	CPU	–	–	–	101 s
FEM with Padé (7) (Guevel et al. 2014)	Re_c	–	–	–	82.79
	Step	–	–	–	11
	CPU	–	–	–	3200 s

Figure 3 presents the diagram of bifurcation obtained by the proposed model with respect to the streamlines solution corresponding to each bifurcated branch

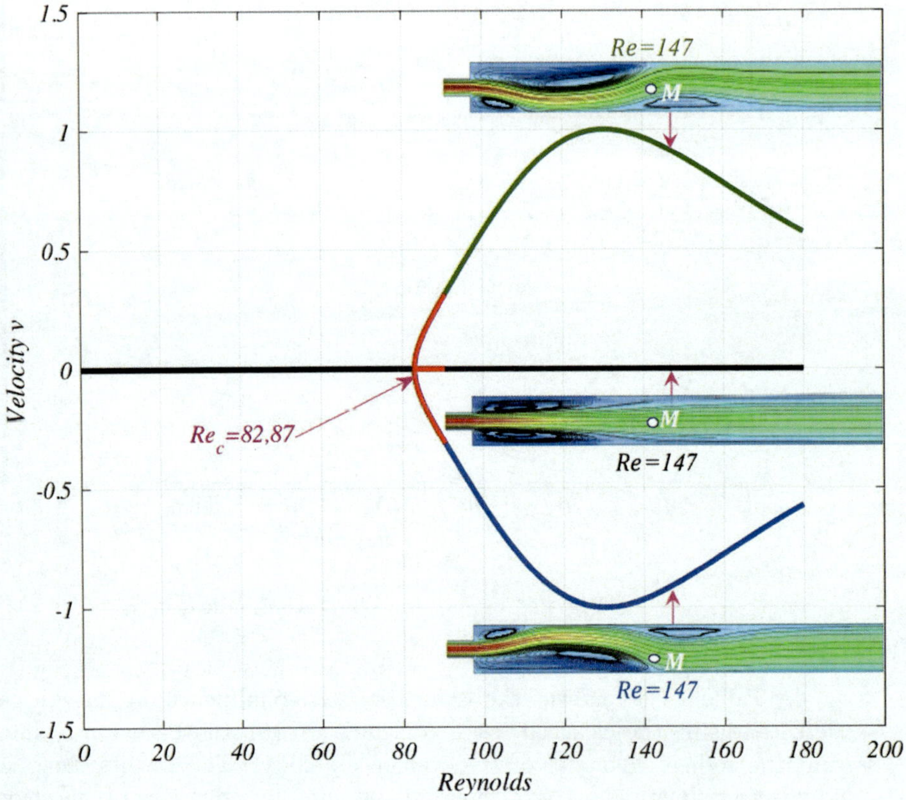

Fig. 3. Bifurcation diagram at the registration point $M(145, 0)$

solution. The diagram represents a Pitchfork bifurcation characterized by the asymmetrical flow (where one being the mirror image of the other) and a second fundamental solution where the flow remains symmetrical. The post-critical branches represented in this diagram by red color, are obtained using the critical solution $(\mathbf{V}_c, \lambda_c)$ as a starting point in the Lyapunov-Schmidt with HO-MFA procedure, which allows to determinate all the flows possibilities encountered at a stationary bifurcation. We note that these results are similar to those of the literature (Cherdron et al. 1978), (Drikakis 1997), (Alleborn et al. 1997), (Guevel et al. 2014), (Medale and Cochelin 2015).

5 Conclusion

We have proposed in this paper an algorithm to compute stationary bifurcation points in fluid mechanics. This model gave the results similar to those of literature. It is an efficiency approach to detect precisely the critical Reynolds points and to visualize the bifurcating solutions, consuming a less matrix inversions and less expensive computation time compared to the classical investigations. The obtained results show a pitchfork bifurcation, where the bifurcated branches (non values of speed) describe the non-symmetrical flows beyond the Reynolds critical value detected by the proposed indicator. It should be noted that literature studies have shown that the critical Reynolds value decreases as the channel aspect ratio increases. For this reason, we aim to use the proposed algorithm in future work to enable detection of these critical points, taking into account the variation in aspect ratio and channel length.

References

Cherdron, W., Durst, F., Whitelaw, J.H.: Asymmetric flows and instabilities in symmetric ducts with sudden expansions. J. Fluid Mech. **84**(1), 13–31 (1978)

Drikakis, D.: Bifurcation phenomena in incompressible sudden expansion flows. Phys. Fluids **9**(1), 76–87 (1997)

Alleborn, N., Nandakumar, K., Raszillier, H., Durst, F.: Further contributions on the two-dimensional flow in a sudden expansion. J. Fluid Mech. **330**, 169–188 (1997)

Elhage-Hussein, A., Potier-Ferry, M., Damil, N.: A numerical continuation method based on Padé approximants. Int. J. Solids Struct. **37**(46–47), 6981–7001 (2000)

Cadou, J.M., Potier-Ferry, M., Cochelin, B.: A numerical method for the computation of bifurcation points in fluid mechanics. Europ. J. Mech.-B/Fluids **25**(2), 234–254 (2006)

Guevel, Y., Girault, G., Cadou, J.M.: Parametric analysis of steady bifurcations in 2D incompressible viscous flow with high order algorithm. Comput. Fluids **100**, 185–195 (2014)

Medale, M., Cochelin, B.: High performance computations of steady-state bifurcations in 3D incompressible fluid flows by Asymptotic Numerical Method. J. Comput. Phys. **299**, 581–596 (2015)

Askour, O., Tri, A., Braikat, B., Zahrouni, H., Potier-Ferry, M.: Bifurcation indicator for geometrically nonlinear elasticity using the Method of Fundamental Solutions. Comptes Rendus Mécanique **347**(2), 91–100 (2019)

Rammane, M., Mesmoudi, S., Tri, A., Braikat, B., Damil, N.: Solving the incompressible fluid flows by a high-order mesh-free approach. Int. J. Numer. Meth. Fluids **92**(5), 422–435 (2020)

Rammane, M., Mesmoudi, S., Tri, A., Braikat, B., Damil, N.: Bifurcation points and bifurcated branches in fluids mechanics by high order mesh-free geometric progression algorithms. Int. J. Numer. Meth. Fluids **93**(3), 834–852 (2021)

Rammane, M., Mesmoudi, S., Tri, A., Braikat, B., Damil, N.: A dimensionless numerical mesh-free model for the compressible fluid flows. Comput. Fluids **221**, 104845 (2021)

Rammane, M., Mesmoudi, S., Tri, A., Braikat, B., Damil, N.: Mesh-free model for Hopf's bifurcation points in incompressible fluid flows problems. Int. J. Numer. Meth. Fluids (2022). https://doi.org/10.1002/fld.5101

Three–Dimensional Numerical Investigation of a New Drag Reduction Device Applied to Ahmed Body

A. Agriss(✉), M. Agouzoul, A. Ettaouil, and A. Mehdari

Mechanical and Energetic Engineering Research Team: Modelling and Experimentation ERG2(ME), Mohammed V University in Rabat, Mohammadia School of Engineers (EMI), Avenue Ibn Sina, P. O. Box 765 – Agdal, Rabat, Morocco
amineagriss@research.emi.ac.ma

Abstract. In this work, a novel drag reduction device applied to the Ahmed body is presented. This consists of perforating a rectangular conduit in the body. The purpose of this conduit is leading part of the airflow from the front and inject it in the recirculation region behind the body. Three–dimensional simulations have been carried out, at a Reynolds number of $Re = 2.85 \times 10^6$, with the software of computational fluid dynamics ANSYS FLUENT. The influence of changing the conduit dimensions and its position from the lower slant edge on drag coefficient is presented. This study is a continuation of the two–dimensional numerical study of this drag reduction technique.

Keywords: drag reduction · Ahmed body · conduit · three–dimensional · ANSYS FLUENT

1 Introduction

For the diminishment of fuel consumption and harmful gas emissions, the automotive industry tries to reduce drag by designing more aerodynamically efficient vehicles. In this context, many flow control strategies have been proposed in order to reduce drag. These techniques are tested on simplified car models (i.e. The Ahmed body) before using them in reality.

The flow control techniques can be passive or active. The passive ones don't require external energy supplies. They can be a modification of geometrical shapes (Agriss et al. 2019) or the use of additional external devices (oruç 2017). While the active ones require external energy supplies in order to operate.

Among the passive techniques of flow control applied to the Ahmed body, placing vertical splitter plates at the rear of the body could reduce drag (Gilliéron and kourta 2010). The use of flaps on the body slant edges reduces drag (tian et al. 2017). By rounding the Ahmed body roof/backlight intersection, drag could be reduced (Rossitto et al. 2016). By modifying the flow topology using large–scale coherent stripes forced on the Ahmed body roof, drag could be decreased (Pujals et al. 2010).

S. Aniss et al. (Eds.): CMM 2022, LNME, pp. 111–119, 2024.
https://doi.org/10.1007/978-3-031-46973-2_10

Using a tapered cavity and asymmetric flaps on a yaw square tail vehicle could reduce drag (Uquhart et al. 2020).

Among active flow control techniques, using a synthetic jet to control the wake flow could reduce drag (Kourta and leclerc 2013). The use of a continuous jet at the top of the slant part of the body reduces drag (Rouméas et al. 2009). Using different types of actuators (steady and periodic) simultaneously could reduce drag (Brunn et al. 2008). By mixing the use of periodic jets and deflectors acting on the wake of the Ahmed body, drag could be reduced (Eulalie 2014). The flow control technique by the wake symmetrization based on sliding mode control could reduce drag of the Ahmed body (Mariette et al. 2020). All these flow control methods are described in detail in scientific literature.

In this work, a 3D numerical study of the drag reduction device treated in the 2D case (Agriss et al. 2021) is presented. Numerical simulations are carried out with the CFD software ANSYS FLUENT 17.0. A rectangular conduit is perforated into the Ahmed body. It leads part of airflow from the body front to be inject in the recirculation zone behind the body. Various conduit configurations are tested. For each position of the conduit from the lower edge of the slant, various dimensions of the conduit are tested. Finally, the conduit is narrowed at the exit in order to accelerate the flow. Various configurations are tested based on the reduction ratio of the conduit dimensions. Effects of all these modifications on drag coefficient are evaluated.

2 Numerical Modeling

In this study, an incompressible and viscous flow is considered. For turbulent flows, the Reynolds–averaged Navier–Stokes (RANS) equations are defined by:

$$(\mathbf{U}.\nabla)\mathbf{U} = \nu\nabla^2\mathbf{U} - \frac{1}{\rho}\nabla p + \nabla\tau \tag{1}$$

where **U** (u, v, w) is the mean velocity vector, ν is the kinematic viscosity, ρ is the density, p is the pressure and τ is the Reynolds stresses.

Turbulence models are used to capture turbulence impacts and determine Reynolds stresses.

In this study, the SST k–ω turbulence model is used. It is a two–equation model with advanced near–wall treatment, transition modeling capacity, and accurate prediction of shear stress transport. This model gives good drag predictions of flows around bluff bodies (Hirschel 2003). More information about this model, developed by Menter, can be found in literature (Menter 1994).

The Reynolds number is defined by:

$$\mathrm{Re} = \frac{UL_c}{\nu} \tag{2}$$

where U is the freestream velocity and L_c s the characteristic length taking as the length of the body.

The drag coefficient is given by:

$$C_d = \frac{F_d}{\frac{1}{2}\rho U^2 S} \tag{3}$$

where F_d is the drag force and S is the body frontal projected area.

F_d is computed in the numerical calculations by integrating the pressure and viscous forces acting in the totality surface of the Ahmed body including the conduit.

3 Test Configuration

The idea in this work is reducing drag of the Ahmed body by using a conduit perforated into the body. This conduit allows leading part of airflow from the body front and inject it behind the body, in the recirculation zone. This is a continuation of a previous study treating the same idea in the 2D case.

The Ahmed body is a simplified vehicle model first introduced by Ahmed in 1984 (Ahmed et al. 1984). It becomes a benchmark for aerodynamic simulations since it allows capturing the most important flow features around a vehicle.

The Ahmed body is 1044 mm in length, 288 mm in height and 389 mm in width. It is mounted on four 30 mm stilts. These 50 mm high stilts allow fixing the body in experimental investigations. They have a low drag contribution and they are not taken into account. In this study, the slant angle φ considered is 25°. The Ahmed body geometry is presented in the Fig. 1. In Figs. 1 and 2, dimensions are in mm.

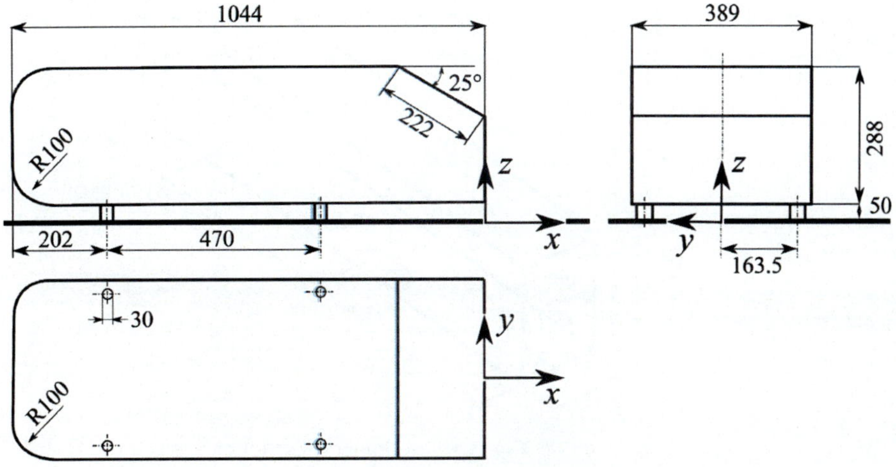

Fig. 1. Ahmed body geometry

A rectangular conduit is perforated along the body to lead part of airflow from the body front to be injected in the recirculation region behind the body. Figure 2 shows the conduit dimensions, L is its length and l is its width. For various distances D from the slant lower edge, different tests are performed for various conduit lengths and widths.

The conduit is narrowed in order to accelerate the flow at the exit. L and l are reduced by the same ratio R. For the suitable distance D, various conduit configurations depending on the ratio R are tested.

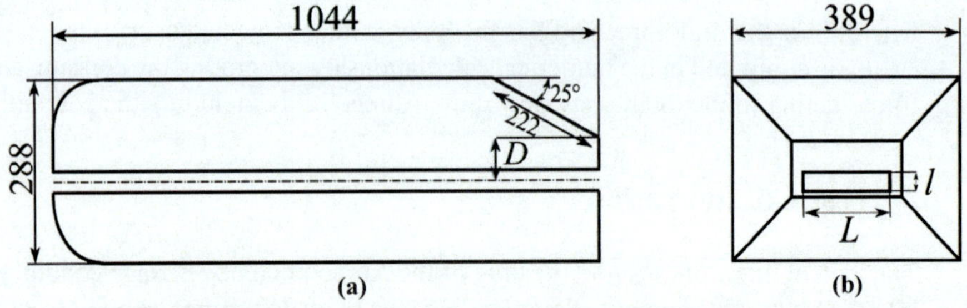

Fig. 2. Conduit description (a) Conduit location (b) Conduit characteristics

The present numerical simulations are performed using the computational domain described in the Fig. 3. The blockage ratio, defined by the ratio of the area occupied by the Ahmed body to the total cross-sectional area of the flow passage, is equal to 1.71%. This ensures suitable air growth around the body.

In this study, we consider a viscous and incompressible fluid (air) with a velocity of 40 m/s, a density of 1.225 kg/m^3 and a viscosity of 1.7894 $\times$ 10^{-5} kg/m.s.

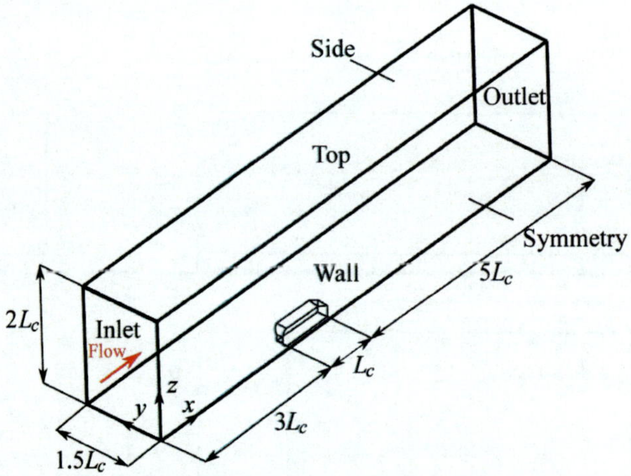

Fig. 3. Flow domain description

Boundary conditions are detailed in the Fig. 4. The boundary conditions velocity inlet and pressure outlet are used. The Ahmed body itself, including the conduit, is considered as a stationary wall with the no-slip boundary condition. The ground is considered to move relative to the body. The symmetry condition is used to cut the model in half and reduce the computational time.

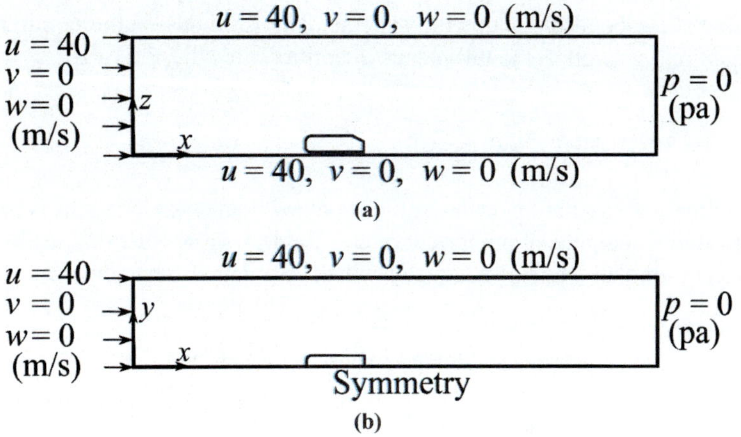

Fig. 4. Boundary conditions description (a) Front view (b) Top view

The inlet turbulent intensity (= 1%) and the turbulent viscosity ratio (= 10) are used to define the inlet boundary conditions for the turbulent kinetic energy k and the specific dissipation rate ω.

Figure 5(a) shows the unstructured tetrahedral mesh of 1 502 900 elements used in this study. For the treatment of the boundary layer, an inflation of 5 layers is used such as the non–dimensional grid spacing $y+$ is between 30 and 300 (Fig. 5(b)).

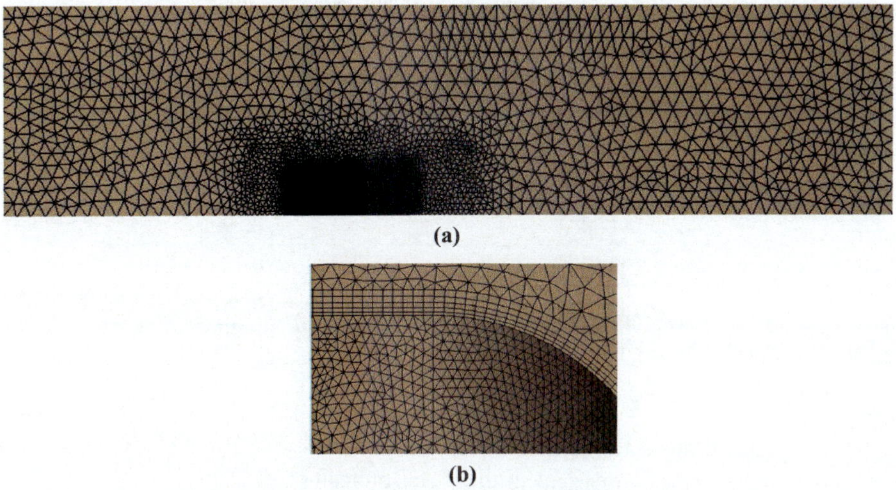

Fig. 5. Mesh description (a) General view (b) Zoom

Numerical simulations are performed in the steady state using the SST k–ω turbulence model. The coupled scheme for the pressure–velocity coupling method is used. Gradients are solved with the Green–Gauss node–based discretization. The pressure is solved using the second order scheme. Momentum and turbulence equations are solved using the second order upwind scheme. The residuals of the continuity and velocity equations, as well as the turbulence quantities, are adjusted to 10^{-7}.

4 Results and Discussions

A numerical simulation on the Ahmed body without considering the conduit is performed to validate the present numerical model. Table 1 shows that the results obtained by the current simulations are in adequacy with the previous experimental and numerical results.

Table 1. Drag coefficient validation

	Present study	Experimental result (Ahmed et al. 1984)	Numerical result (Guilmineau 2008)
C_d	0.287	0.285	0.3074

The study of the mesh influence on numerical results is carried out. Drag coefficients are calculated for various mesh densities. The grid independence study is summarized in Table 2. Therefore, the mesh of 1 502 900 elements is used in the next simulations in order to ensure simulations accuracy.

Table 2. Grid independence study

Number of mesh elements	C_d
356651	0.302
508887	0.299
762646	0.294
905083	0.290
1167258	0.288
1319008	0.287
1502900	0.287

Figure 6 shows the velocity vectors obtained by the current study. This is in adequacy with the previous numerical studies (Guilmineau et al. 2018).

In the front corners of the Ahmed body, the narrowing space leads to a compression effect, causing the flow acceleration and reaching higher local velocities than the freestream velocity. This is normal according to the principle of continuity, when the flow area decreases, the flow velocity increases.

In the case of uniform conduits placed at different distances D with respect to the lower edge of the slant, different conduit configurations characterized by a length L and width l are tested. Figure 7 shows the evolution of drag coefficient with respect to the distance D of the conduit from the lower edge of the slant.

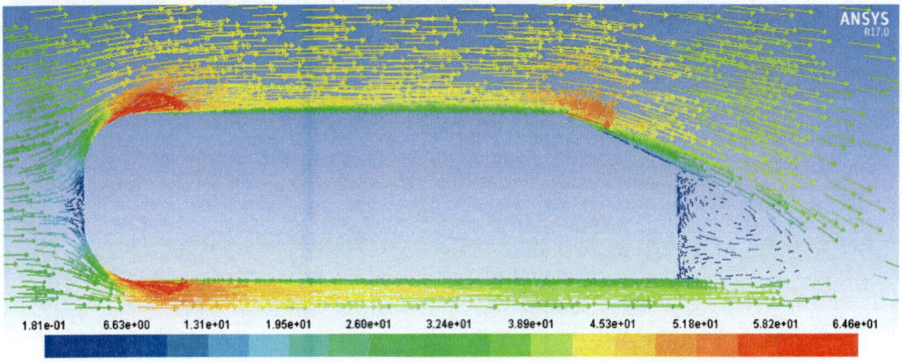

Fig. 6. Velocity vectors for the current study [m/s]

Almost all configurations increase the drag coefficient of the Ahmed body. The more the dimensions of the conduits increase, the more the drag coefficients increase for all distances. The distance $D = 0.03$ m is the favorable distance. It allows having drag coefficients lower than the other distances for all configurations of the conduit.

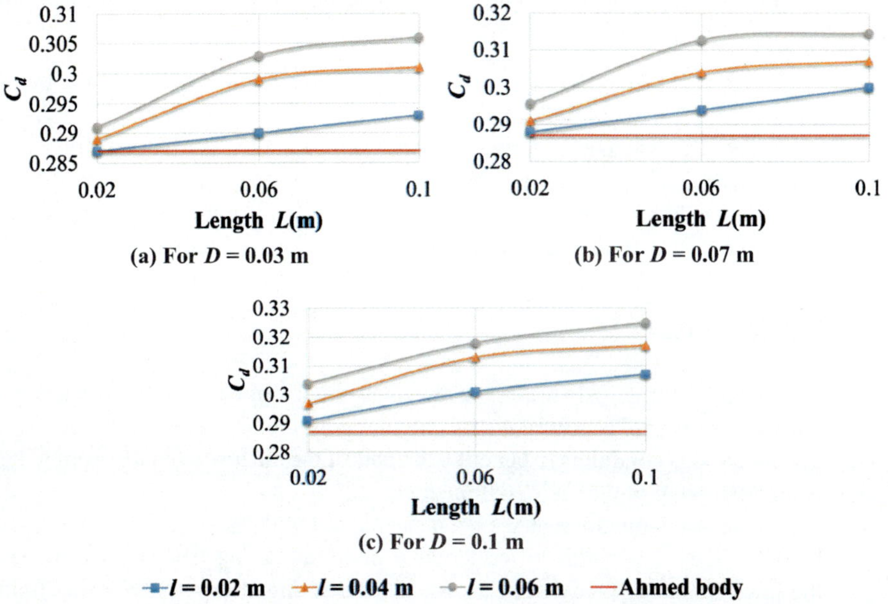

Fig. 7. Drag coefficient evolution with the distance D from the slant lower edge for various conduit dimensions

For this, this distance is considered in the next simulations. In these simulations, the effect of narrowing the exit of the conduit on drag coefficient is studied. Several configurations are studied based on the reduction ratios of the conduit dimensions R. The evolution of drag coefficient with the reduction ratio R is showed in the Fig. 8.

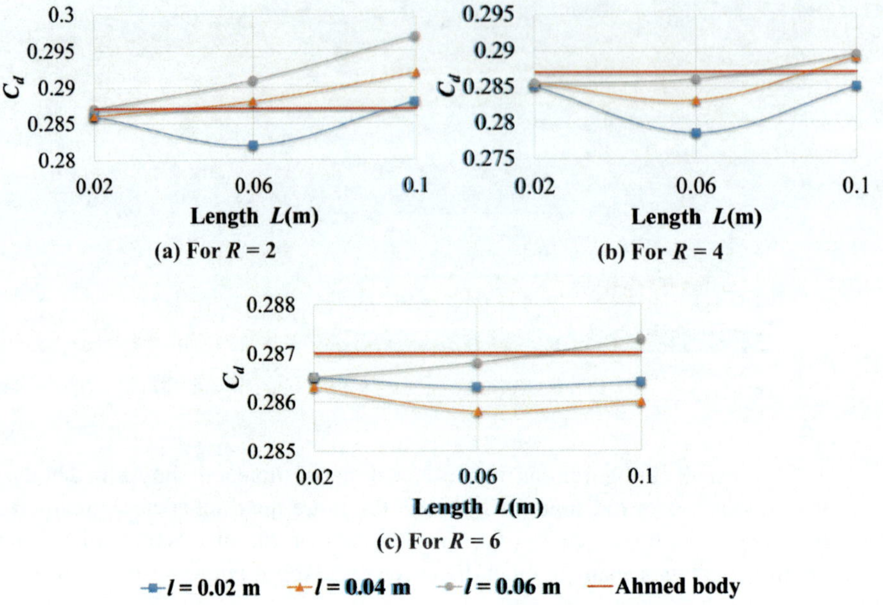

Fig. 8. Drag coefficient evolution with the reduction ratio R of the conduit dimensions

For $R = 2$, a maximum drag reduction up to 1.91% is obtained for the conduit of dimensions ($L = 0.06$ m, $l = 0.02$ m).

For $R = 6$, a maximum drag reduction up to 0.47% is obtained for the conduit of dimensions ($L = 0.06$ m, $l = 0.04$ m).

$R = 4$ represents the most favorable ratio especially for the conduit of dimensions ($L = 0.06$ m, $l = 0.02$ m). It allows a maximum drag reduction up to 3%.

5 Conclusion

In this study, numerical simulations have been performed to study impacts of changing the dimensions and position of a rectangular conduit perforated along the Ahmed body. The aim of this conduit is to benefit from part of the airflow from the front to be injecting behind the body in the recirculation zone.

Numerical simulations have been carried out with the CFD software ANSYS FLUENT. Various dimensions of rectangular conduits taking from various distances from the lower slant edge have been tested. For the configuration that allows minimum drag coefficient, the effect of reducing the conduit dimensions at the outlet has been studied. Various conduit configurations depending on the reduction ratio have been tested.

Results found in this study lead us to conclude that in certain configurations of the conduit, drag coefficient could be reduced. The favorable configuration allows a maximum drag reduction of 3%.

This work is a continuation of a two–dimensional previous study of this new drag reduction technique. In the perspective of understanding the physical mechanisms enabling the optimization of the conduit's position and dimensions. Future works will focalize on studying the conduit outlet inclination and its impacts on drag reduction.

References

Agriss, A., Agouzoul, M., Ettaouil, A.: Reducing drag on a flat plate subjected to incompressible laminar flow. MATEC Web Conf. **286**, 07006 (2019). https://doi.org/10.1051/matecconf/201928607006

Agriss, A., Agouzoul, M., Ettaouil, A., Mehdari, A.: Numerical study of new techniques drag reduction: application to aerodynamic devices. Int. J. Simul. Multidisci. Des. Optim. **12**, 16 (2021). https://doi.org/10.1051/smdo/2021015

Ahmed, S.R., Ramm, G., Faltin, G.: Some Salient features of the time-averaged ground vehicle wake. SAE Technical Paper 840300 (1984). https://doi.org/10.4271/840300

Brunn, A., Henning, L., Nitsche, W., King, R.: Application of slope–seeking to a generic car model for active drag control. In: 26th AIAA Applied Aerodynamics Conference, Honolulu, Hawaii: American Institute of Aeronautics and Astronautics (2008). https://doi.org/10.2514/6.2008-6734

Eulalie, Y.: Etude aérodynamique et contrôle de la traînée sur un corps de Ahmed culot droit. [Aerodynamic analysis and drag reduction around an Ahmed bluff body], Ph.D. dissertation, Université de Bordeaux, Bordeaux, France (2014)

Gilliéron, P., Kourta, A.: Aerodynamic drag reduction by vertical splitter plates. Exp. Fluids **48**, 1–16 (2010). https://doi.org/10.1007/s00348-009-0705-7

Guilmineau, E.: Computational study of flow around a simplified car body. J. Wind Eng. Ind. Aerodyn. **96**(6–7), 1207–1217 (2008). https://doi.org/10.1016/j.jweia.2007.06.041

Guilmineau, E., Deng, G.B., Leroyer, A., Queutey, P., Visonneau, M., Wackers, J.: Assessment of hybrid RANS–LES formulations for flow simulation around the Ahmed body. Comput. Fluids **176**, 302–319 (2018). https://doi.org/10.1016/j.compfluid.2017.01.005

Hirschel, E.H.: Numerical Flow Simulation III. Vol. 82. Notes on Numerical Fluid Mechanics and Multidisciplinary Design (NNFM), Springer Berlin Heidelberg, Berlin, Heidelberg (2003). https://doi.org/10.1007/978-3-540-45693-3

Kourta, A., Leclerc, C.: Characterization of synthetic jet actuation with application to Ahmed body wake. Sens. Actuator A Phys. **192**, 13–26 (2013). https://doi.org/10.1016/j.sna.2012.12.008

Mariette, K., Bideaux, É., Bribiesca-Argomedo, F., Ébérard, D., Sesmat, S., Haffner, Y., Borée, J., Castelain, T., Michard, M.: Wake symmetrization of a bluff Ahmed body based on sliding mode control. IFAC-PapersOnLine **53**(2), 7746–7751 (2020). https://doi.org/10.1016/j.ifacol.2020.12.1531

Menter, F.R.: Two–equation eddy–viscosity turbulence models for engineering applications. AIAA J. **32**(8), 1598–1605 (1994). https://doi.org/10.2514/3.12149

Oruç, V.: Strategies for the applications of flow control downstream of a bluff body. Flow Meas. Instrum. **53**, 204–214 (2017). https://doi.org/10.1016/j.flowmeasinst.2016.08.008

Pujals, G., Depardon, S., Cossu, C.: Drag reduction of a 3D bluff body using coheren streamwise streaks. Exp. Fluids **49**, 1085–1094 (2010). https://doi.org/10.1007/s00348-010-0857-5

Rossitto, G., Sicot, C., Ferrand, V., et al.: Influence of afterbody rounding on the pressure distribution over a fastback vehicle. Exp. Fluids **57**, 43 (2016). https://doi.org/10.1007/s00348-016-2120-1

Rouméas, M., Gilliéron, P., Kourta, A.: Drag reduction by flow separation control on a car after body. Int. J. Numer. Methods Fluids. **60**, 1222–1240 (2009). https://doi.org/10.1002/fld.1930

Tian, J., Zhang, Y., Zhu, H., Xiao, H.: Aerodynamic drag reduction and flow control of Ahmed body with flaps. Adv. Mech. Eng. **9**(7) (2017). https://doi.org/10.1177/1687814017711390

Urquhart, M., Varney, M., Sebben, S., Passmore, M.: Aerodynamic drag improvements on a square–back vehicle at yaw using a tapered cavity and asymmetric flaps. Int. J. Heat Fluid Flow **86**, 108–737 (2020). https://doi.org/10.1016/j.ijheatfluidflow.2020.108737

Check for updates

Effect of Load on Minimum Film Thickness in Non-Newtonian MEHD Parabolic Slider Bearing

M. Mouda[1(✉)], M. El Khlifi[2], and M. Nabhani[2]

[1] University of Cadi Ayyad, ENSA Safi, 46000 Safi, Morocco
m.mouda@uca.ac.ma

[2] Hassan II University of Casablanca, FST Mohammedia, 20650 Mohammedia, Morocco

Abstract. The minimum oil film thickness, at a given load, is calculated in this paper for Magneto-elastohydrodynamic (MEHD) parabolic slider bearings lubricated with non-Newtonian couple stress fluid. The modified MHD Reynolds equation of finite and infinite geometries is developed. It is discretized by finite difference technique and the resulting algebraic equations are iteratively solved using the Gauss–Seidel method. In particular, the value of minimum oil film thickness is decreasing with that of the load.

Keywords: Minimum Film Thickness · Magneto-elastohydrodynamics · Parabolic Slider Bearings · non-Newtonian lubricant

1 Introduction

MEHD pad thrust bearings are generally used to support axial loads in industrial applications. Many researchers have studied the MHD performance of slider bearings with classical Newtonian lubricant in the presence of applied magnetic field. (Snyder 1962) has already studied the characteristics of infinite inclined plane slider bearings. (Hughes 1963) analyzed the characteristics of step slider bearings using an electrically conducting liquid metal lubricant. (Lin et al. 2009) were interested in the study of the static and dynamic characteristics of 1D slider bearings lubricated with electrically conducting fluid. (Rao et al. 2019) analyzed a model of MHD partial slip laser textured slider bearings. In conclusion, the increase in the strength of applied magnetic field increases the load capacity of slider bearings compared to nonmagnetic case.

Applications of non-Newtonian lubricants have attracted much attention. Especially, the non-Newtonian couple stress fluid behaviour requires taking into account the particle size. As the effects of this latter are neglected by the classical continuum theory, a micro-continuum theory has been then developed by (Stokes 1966). Several researchers are interested in the study of the combined effect of MHD and couple stress, such as slider bearings by (Nabhani and El Khlifi 2014, 2015) and (Mouda et al. 2019), various shapes of finite plates by (Fathima et al. 2015), non-parallel squeeze film bearings by (Mouda et al. 2020) and journal bearing by (Kardoudi et al. 2022). Recently, (Anthony and

S. Aniss et al. (Eds.): CMM 2022, LNME, pp. 120–127, 2024.
https://doi.org/10.1007/978-3-031-46973-2_11

Elamparithi 2023) investigated in Reyleigh step slider bearing. These authors confirmed that the effects of non-Newtonian couple stress and MHD increase load capacity. All these studies used plane surfaces.

The minimum oil film thickness is an important operating parameters to surely protect slider bearings from intimate contact and the consequent danger of catastrophic failure. The effect of imposed load on this latter was investigated in this work for infinite and finite parabolic slider bearings. The 2D MEHD Reynolds equation is developed. The bearing deformation is calculated using the Winkler approach. Some results are validated by comparison against Newtonian previously published data.

2 Problem Position and Resolution

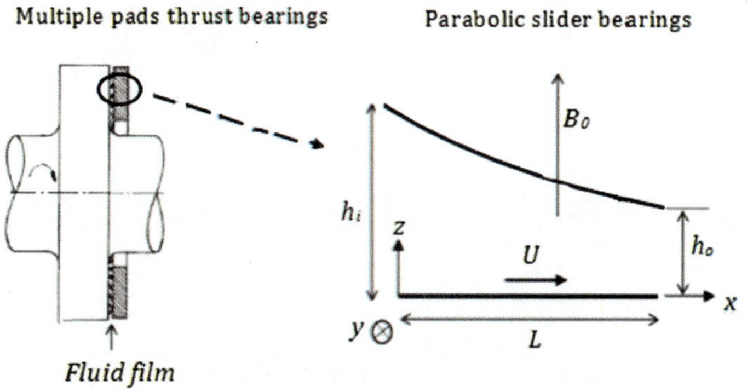

Fig. 1. Parabolic bearing configuration.

The studied configuration geometry (Fig. 1), of length L and width B is considered. In the film region, the lubricant is a conducting Stokes couple stress fluid. The plane lower surface is rigid and is moving at a constant velocity U while the upper surface is fixed and elastic. Perpendicular to the bearing lower surface a uniform transverse magnetic field B_0 is applied. The flow is considered laminar. The lubricant is taken to be an isothermal and incompressible electrically conducting fluid. Assume that the fluid film is thin, the body force is negligible except for the Lorentz force, and the induced magnetic field is very small compared to the applied field as described by (Zikanov and Thess 1998). So, the dimensionless modified Reynolds equation for finite slider bearings was given by (Mouda et al. 2019) as follows:

$$\frac{\partial}{\partial x^*}\left\{F^*\frac{\partial p^*}{\partial x^*}\right\}+\frac{1}{\beta^2}\frac{\partial}{\partial y^*}\left\{G^*\frac{\partial p^*}{\partial y^*}\right\}=6\frac{\partial h^*}{\partial x^*} \tag{1}$$

where

$$F^*=\frac{6}{M^2}\left(-2h^*+\frac{1}{l^*}\frac{\left(A_1^{*2}-B_1^{*2}\right)h^{*2}}{\frac{A_1^{*2}}{B_1^*}\tanh\left(\frac{B_1^*h^*}{2l^*}\right)-\frac{B_1^{*2}}{A_1^*}\tanh\left(\frac{A_1^*h^*}{2l^*}\right)}\right)$$

$$G^* = h^{*3} - 12\,l^{*2}\,h^* + 24\,l^{*3}\,\tanh\left(\frac{h^*}{2\,l^*}\right)$$

in which:

$$M = B_0 h_o \sqrt{\frac{\sigma}{\mu}}, l^* = \frac{l}{h_o}, l = \sqrt{\frac{\eta}{\mu}}, x^* = \frac{x}{L}, y^* = \frac{y}{B}, \beta = \frac{B}{L}$$

$$A_1^* = \frac{\sqrt{1+\sqrt{1-4M^2 l^{*2}}}}{\sqrt{2}}, B_1^* = \frac{\sqrt{1-\sqrt{1-4M^2 l^{*2}}}}{\sqrt{2}}$$

where x, y are the Cartesian coordinates, p^* is the dimensionless film pressure, l^* is the couple stress parameter, M is the Hartmann number, h_o is the minimal film thickness, σ is the electrical conductivity of the fluid, μ is the dynamic viscosity coefficient andη is the material constant of couple stress fluids.

The dimensionless film thickness h^*, taking into account the dimensionless elastic deformation of immobile surface calculated according to the (Winkler 1867) model, can be formulated as follows:

$$h^* = h_m^* + C_s p^* \tag{2}$$

in which:

$$h_m^* = \alpha + (\alpha - 1)\left(x^{*^2} - 2x^*\right), \alpha = \frac{h_i}{h_0} C_s = \frac{\mu d L U}{E h_o^3} \frac{(1+v)(1-2v)}{(1-v)}$$

where v, E, h_i, d, and α are the Poisson ratio, the Young's modulus, inlet film thickness, thin layer thickness and the profile parameter of a parabolic slider bearing respectively and C_S represents the dimensionless compliance coefficient.

In order to get the dimensionless load capacity W_{cal}^* calculated by integration of the dimensionless pressure through the fluid film for a given film thickness h^*, the finite difference technique is adopted to discretize Eq. (1). Beginning by an initial guess of $h_o = 2\mu m$, W_{cal}^* is compared with the imposed load W_{imp}^* on the slider bearing. If the difference between the imposed dimensionless load and the calculated load doesn't meet the accuracy requirement of numeric calculation, Eq. (3) is applied to define a new initial minimal film thickness h_o^* (=h_o normalized by a value of 2 μm). Figure 2 shows the computational flow chart, describing numerical procedure steps, developed in this study.

$$h_o^* = h_o^* + \omega\left(W_{cal}^* - W_{imp}^*\right) \tag{3}$$

Where ω is under-relaxation factor whose optimal value is 1.8 to ensure an accelerated solution convergence.

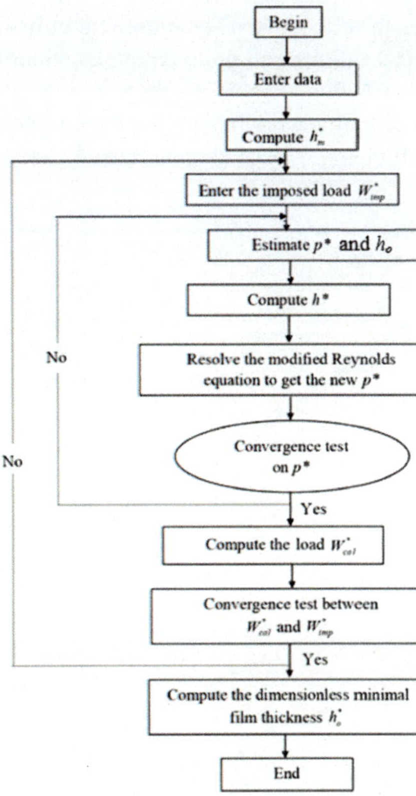

Fig. 2. Flow chart for numerical computing.

3 Results

With the aim to validate the present computer program, the load capacity is given in Table 1 to compare with (Naduvinamani et al. 2017) in the case of an imposed minimum film thickness of Newtonian MHD parabolic infinite $\left(\beta = \frac{B}{L} \to \infty\right)$ slider bearing. The results obtained in the current analysis show excellent agreement with the results obtained by the latter authors.

Table 1. Dimensionless load capacity of parabolic slider bearingsfor different given values of M with $l^* = 0$ and $Cs = 0$.

M	(Naduvinamani et al. 2017)	Present analysis
0	0.1722	0.1702
2.5	0.2087	0.2063
5	0.2885	0.2849

For the first time, the effect of load on minimum film thickness considering the elastic deformation for MHD infinite and finite parabolic slider bearings lubricated with non-Newtonian couple stress fluids in the presence of externally applied magnetic fields has been mainly analysed in this work by varying three dimensionless parameters. Namely, the compliance coefficient C_s, Hartmann number M and the couple stress parameter l^*.

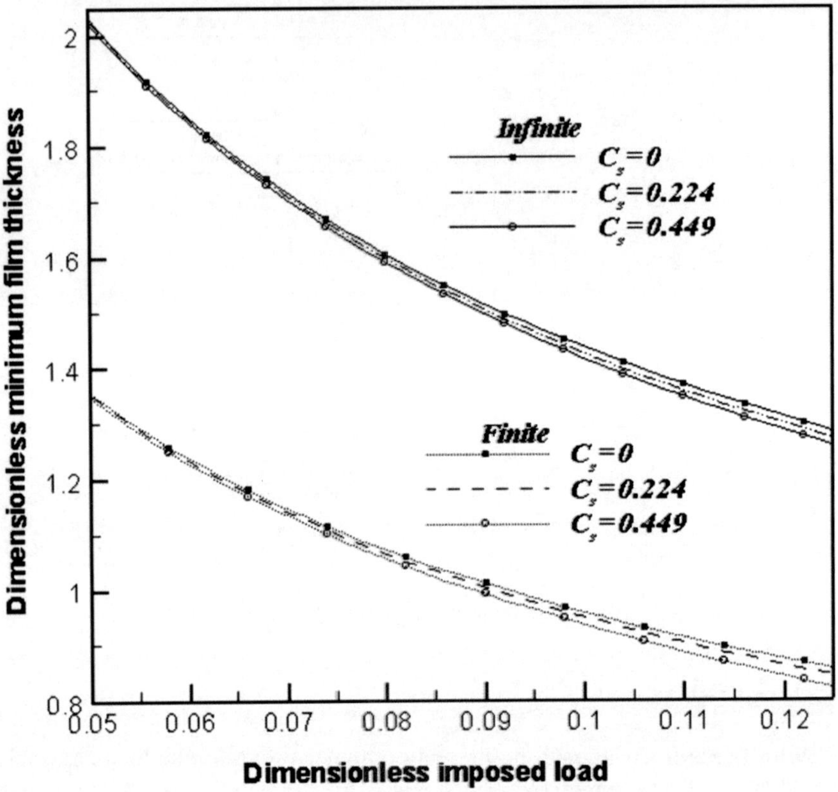

Fig. 3. Dimensionless minimum film thickness versus imposed load at $M = 2$ and $l* = 0.1$ for different compliance coefficients.

Figure 3 presents the variation of dimensionless minimum film thickness of finite and infinite parabolic slider bearings against the imposed load for different values of Cs using a non-Newtonian couple stress lubricant $l^* = 0.1$ and Hartmann number $M = 2$. It can be seen that increasing the imposed load on the MHD bearing reduces the dimensionless minimum film thickness for all the considered values of Cs. In addition, the results show that the dimensionless minimum film thickness decreases with increasing values of Cs compared to rigid case ($Cs = 0$) for any given value of the imposed load. Thus, considering surface deformation is advised to predict correctly slider bearings performance. In addition, the minimum film thickness of the infinite parabolic slider bearings is higher than that of the finite slider bearings. This is useful for bearings design in many industrial applications because higher minimum film thickness will protect lubricated surfaces from direct contact in addition to the fact that infinite parabolic slider bearings support higher load capacity.

The effects of imposed load on the minimum film thickness considering the couple stresses effect using a compliance coefficients $Cs = 0.224$ and Hartmann number $M = 2$ is depicted in Fig. 4. Using couple stresses lubricants increase the dimensionless minimum film thickness compared to Newtonian lubricant ($l^* = 0$) for any given value of the imposed load. Neglecting non-Newtonian effects could underestimate slider bearings performance characteristic. In addition, for each fixed imposed load value the minimum film thickness of the infinite parabolic slider bearing is higher than that of the finite slider bearing.

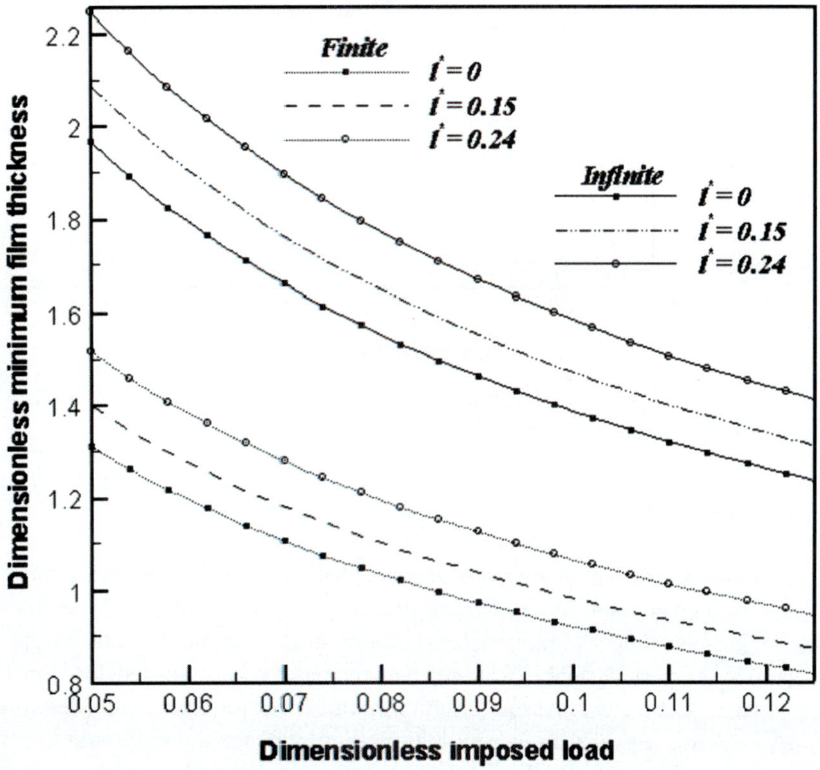

Fig. 4. Dimensionless minimum film thickness versus imposed load at $Cs = 0.224$ and $M = 2$ for different couple stress parameter.

Figure 5 presents the relationship between the load and the minimum film thickness considering the magnetic fields effects using a couple stresses lubricant ($l^* = 0.1$) with a compliance coefficient ($Cs = 0.224$). The transverse magnetic field increases the minimum film thickness of MEHD parabolic slider bearings compared to non conducting lubricant case. This might be useful for design these types of bearings, as it is of great importance in the construction of high frequency transducers.

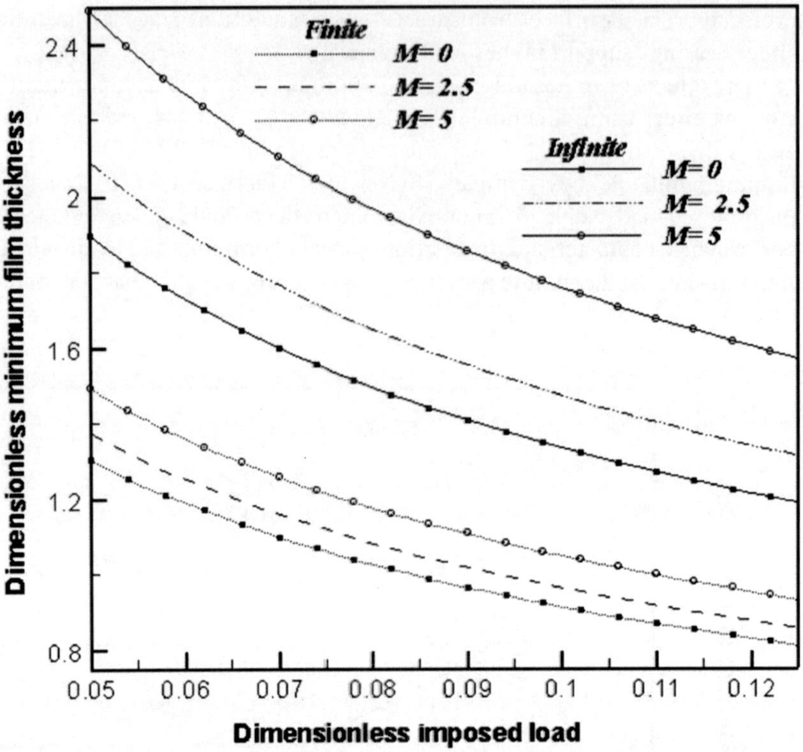

Fig. 5. Dimensionless minimum film thickness versus imposed load at $Cs = 0.224$ and $l^* = 0.1$ for different Hartmann number.

4 Conclusion

In this work, a new numerical model is established to simulate the minimum oil film thickness, at a given load, for MHD parabolic slider bearings lubricated with non-Newtonian couple stress fluid including bearing deformation using the Winkler model.

The numerical analysis shows that increasing the imposed load on the bearings reduces the dimensionless minimum film thickness for all the considered values of Cs, M, l^*. Also, considering elastic deformation influence the minimum film thickness of parabolic slider bearings. In addition, using couple stresses lubricants increase the dimensionless minimum film thickness com-pared to traditional lubricant for any given value of the imposed load. Under a certain load, the transverse magnetic field increases the minimum film thickness of MHD parabolic slider bearings compared to non-conducting lubricant case. These findings are desirable to provide more useful information for engineers in selecting and designing bearing systems.

References

Snyder, W.T.: The magnetohydrodynamic slider bearing. J. Basic Eng. **84**, 197–204 (1962)

Hughes, W.F.: The Magnetohydrodynamic finite step slider bearing. J. Basic Eng. **85**, 129–136 (1963)

Lin, J.R., Hung, C.R., Hsu, C.H., Lai, C.: Dynamic stiffness and damping characteristics of one-dimensional magneto-hydrodynamic inclined-plane slider bearings. Part J: J. Eng. Tribol. **223**, 211–219 (2009)

Rao T., Rani A.M.A., Mohamed N.M., Ya H.H., Awang M., Hashim F.M.: Analysis of magneto-hydrodynamic partial slip laser bump texture slider and journal bearing. Part J: J. Eng. Tribol. **0**, 1–18 (2019)

Stokes, V.K.: Couple Stresses in Fluids. Phys. Fluids **9**, 1709–1715 (1966)

Nabhani, M., El Khlifi, M.: Inertia effects on inclined slider bearings lubricated by a couple stress fluid. Tribol. Trans. **57**, 515–523 (2014)

Nabhani, M., El Khlifi, M.: Inertial MHD couple stress effects on infinitely wide slider bearings. Tribol. Trans. **58**, 374–383 (2015)

Mouda, M., Nabhani, M., El Khlifi, M.: Effect of non-Newtonian magneto-elastohydrodynamic on performance characteristics of slider-bearings. Indust. Lubric. Tribol. **71**, 1158–1165 (2019)

Fathima, S.T., Naduvinamani, N.B., Hanumagouda, B.N., Kumar, J.S.: Modified reynolds equation for different types of finite plates with the combined effect of MHD and couple stresses. Tribol. Trans. **58**, 660–667 (2015)

Mouda, M., Nabhani, M., El Khlifi, M.: Surface roughness effects on non-Newtonian MHD non-parallel squeeze film bearing. Indust. Lubric. Tribol. **73**, 45–51 (2020)

Kardoudi, M., Nabhani, M., El Khlifi. M.: Non-Newtonian couple stress lubrication of MHD elastic journal bearings. Industrial Lubrication and Tribology (2022)

Anthony, J., Elamparithi, S.: Effect of MHD and surface roughness on porous step-slider bearing lubricated with couple-stress fluid. Int. J. Heat Technol. **41**, 135–142 (2023)

Zikanov, O., Thess, A.: Direct numerical simulation of forced MHD turbulence at low magnetic Reynolds number. J. Fluid Mech. **358**, 299–333 (1998)

Winkler, E.: Die Lehre von Elastizitat und Festigkeit (on Elasticity and Fixity), Theil 1–2. H Dominicus, Prag (1867)

Naduvinamani, N.B., Siddangouda, A., Patil, S.: A comparative study of static and dynamic characteristics of parabolic and plane inclined slider bearings lubricated with MHD couple stress fluids. Tribol. Trans.. Trans. **60**, 1–11 (2017)

Linear Stability Analysis of Viscoelastic Fluids in a Plane Channel Flow

M. Lamine(✉), S. Aniss, and A. Hifdi

Laboratory of Mechanics, Faculty of Sciences Aïn-Chock,
Hassan II University of Casablanca, Casablanca, Morocco
mustapha.lamine@gmail.com

Abstract. In this paper, the linear hydrodynamic stability analysis of a viscoelastic fluid in a plane channel flow driven by a constant pressure gradient is investigated numerically. A nonlinear eigenvalue problem for the perturbed state is obtained from a generalized Orr-Sommerfeld equation. The Chebyshev spectral collocation method with expansions in Lagrange's polynomials are used for the numerical calculation. The complex wave speed, the critical Reynolds and wavenumber numbers are calculated for different parameters. We mainly examine the combined effects of relaxation time (via the Deborah number) and delay time (via the elasticity number) on the onset of instability. Results show that the relaxation time has a destabilizing effect and the retardation time has a stabilizing effect, with a displacement towards the long-wave region. Moreover, an important feature occurs when the Reynolds number is equal to the ratio of both characteristic times, the Jeffrey's fluid model behaves like the Newtonian model.

Keywords: Hydrodynamic instabilities · Viscoelastic fluids · Nonlinear eigenvalues problem · Spectral methods

1 Introduction

Since the experimental works of Osborne Reynolds in 1883, the stability of channel flows has been considered as one of the major topics in fluid mechanics research because of its great importance in fundamentals and applications. Several experimental, numerical and theoretical investigations have been devoted to analyzing the stability of these flows in various situations as summarized in the reviews [1–5].

In a pioneering study, Orszag [1] obtained numerically, using expansions in Chebyshev polynomials to approximate the Orr-Sommerfeld equation solutions, a critical Reynolds number, $Re_{cp} = 5772.22$ for the plane channel flows of Newtonian fluids. The stability analysis of non-Newtonian fluid flows are more difficult compared to the Newtonian one. This is related to the additional complexity introduced by the non-linear rheological behavior of these fluid models. An important non-Newtonian character of this fluid family is the viscoelastic behavior. This character combines both viscous and elastic properties and gives the

S. Aniss et al. (Eds.): CMM 2022, LNME, pp. 128–137, 2024.
https://doi.org/10.1007/978-3-031-46973-2_12

rheological behavior to many polymer solutions: oil, toothpaste, molten polymers, clay...

One of the most general viscoelastic fluid models is that developed by James Gardner Oldroyd, called the Oldroyd-B model. It is considered the best model for describing physiological fluids. This fluid model is an extension of the generalized Maxwell fluid and allows the effects of solvent viscosity to be taken into account. The linear model of the Oldroyd-B fluids is known as the Jeffrey fluid, since, it does not include nonlinear upper convective terms in stress tensor. It was used to modelled a wide range of fluids. One can cite for example, bitumens, blood, pastes, lubricants, concrete, pastes, the earth's crust, polymers and their solutions as well as aqueous solutions of polyethylene oxide at well-defined concentrations [3]. The stability of Jeffrey fluid flows is a subject that has been analyzed in several studies since the last century. In this context, Saasen and Hassager [6] studied the Rayleigh-Taylor instability with the Jeffrey fluid. They showed that the elastic propriety can either delay or advances the instability. They also proved that the Jeffrey fluid behaves like a Newtonian fluid in the short-wave regime. Later, Renardy and Tech [7] performed a nonlinear stability analysis of Jeffrey model flows. They showed that, only in the case of the linearized motion equations and for weakly elastic fluid, the flow is linearly stable with respect to infinitesimal disturbances. Later, Bergström [8] analyzed the transient growth of small disturbances in a Jeffrey fluid flow in pipes. He showed the dependence between the fluid rheological parameters, relaxation and retardation times, and the transient amplification of the disturbance. Thereafter, Barra and Afkhami [9] studied the linear stability of the Jeffrey fluid film in interaction with Van Der Walls waves. It was found that Jeffrey rheological model can be used to describe the linear stability of the interface in the long-wave limit. In the particular case of Jeffrey fluids corresponding to the linear Maxwell model, characterized only by relaxation time, Lamine et *al.* [10] studied the effect of relaxation time on the hydrodynamic stability of plane Poiseuille flow subject to a uniform cross-flow. It was shown that the destabilizing effect is due to the fluid elasticity. New type of bi-critical state has been observed by these authors and has been called as "mode competition". The latter corresponds to a state having two critical wave numbers and the same critical Deborah and cross-flow numbers.

In this paper, we investigate the linear hydrodynamic stability of a viscoelastic liquid in parallel shear flow driven by a constant pressure gradient. The Jeffrey fluid model taking into account the relaxation and retardation times is considered. This study aims to examine the combined effects of both characteristic times on the threshold of hydrodynamic instability and also to identify the nature of the governing modes. The corresponding stability problem is formulated into a nonlinear eigenvalue problem, and the stability characteristics of the flow are obtained numerically using the Chebyshev spectral collocation method. This paper is organized as follows. Section 2 is devoted to presenting the physical model and the studied configuration. This section also presents the steady base flow solution and the linear stability formulation, as well as the numerical method used to solve the stability problem. In Sect. 3, pertinent results are discussed quantitatively and in the Sect. 4, the concluding remarks are presented.

2 Problem Formulation

2.1 Physical Model and Governing Equations

Consider a viscoelastic incompressible fluid flow of density ρ in a plane channel formed by two parallel rigid plates of length L and separated by a fixed distance $2h$ with $(L >> h)$. The fluid is set in motion by a constant pressure gradient in the **x** direction.

The following quantities are used as scales to express the governing equations in their non-dimensional forms: channel half-height h for lengths, maximum streamwise velocity U for velocities, and U/h for time, ρU^2 for pressure and stresses. The flow is mathematically modeled by the continuity equation

$$\nabla \cdot \mathbf{V} = 0 \tag{1}$$

the Cauchy momentum equation

$$\frac{\partial \mathbf{V}}{\partial t} + (\mathbf{V} \cdot \nabla)\mathbf{V} = -\nabla P + \nabla \cdot \tau \tag{2}$$

where $\mathbf{V}(V_x, V_y)$, P and t are, respectively, the velocity vector, the pressure and time. Further, τ is the stress tensor of the Jeffrey fluid, whose constitutive relation is given as follows [12]

$$\tau + De\frac{\partial \tau}{\partial t} = \frac{1}{Re}\gamma + E\frac{\partial \gamma}{\partial t} \tag{3}$$

where $\gamma = \nabla\mathbf{V} + (\nabla\mathbf{V})^T$ is the rate strain, $De = \frac{\lambda_1 U}{h}$ is the Deborah number, $Re = \frac{Uh}{\nu}$ is the Reynolds number and $E = \frac{\nu\lambda_2}{h^2}$ is the elasticity number. λ_1, λ_2 and ν represent, respectively, the relaxation time, the retardation time and the kinematic viscosity. It should be noted that Jeffrey fluid recovers linear Maxwell fluid when $E = 0$ and Newtonian fluid when $De = E = 0$.

The boundary conditions at the walls $(y = \pm 1)$ are

$$V_x(x, y = \pm 1) = 0 \tag{4}$$

2.2 Base State and Linear Stability Analysis

For the steady flow of Jeffrey fluid, Eq. (3) reduces to that of the Newtonian fluid $\tau = Re^{-1}\gamma$. In the context of the unidirectional flows, using the assumption $(L >> h)$, the analytical resolution of Eq. (2) associated to the no-slip boundary conditions (4) results in the base pressure and velocity of the classical Poiseuille flow P_b and V_b [11].

In order to study the linear stability, the infinitesimal perturbations $(V^{'}, P^{'})$, in the normal modes form, are superimposed to the base flow (V_b, P_b) as follows

$$\begin{cases} V_x(x,y,t) = V_b(y) + \psi_x(y)\exp(i\alpha(x-ct)) \\ V_y(x,y,t) = \psi_y(y)\exp(i\alpha(x-ct)) \\ P(x,y,t) = P_b(x) + p(y)\exp(i\alpha(x-ct)) \end{cases} \quad (5)$$

where $\psi_x(y)$, $\psi_y(y)$ and $p(y)$ are, respectively, the complex amplitudes of the perturbations $u^{'}$, $v^{'}$ and $P^{'}$. Here α, is the wavenumber, c is the complex wave speed and $i^2 = -1$.

By substituting the quantities (5) into the governing Eqs. (1) and (2) and by subsequently neglecting all nonlinear terms, we get:

$$i\alpha\left[\alpha V_b - c)(\frac{\partial^2}{\partial y^2} - \alpha^2) - \alpha\frac{\partial^2 V_b}{\partial y^2}\right]\psi_y = \alpha^2 c De\left[(c - V_b)(\frac{\partial^2}{\partial y^2} - \alpha^2) + \frac{\partial^2 V_b}{\partial y^2}\right]\psi_y + \left[(Re^{-1} - i\alpha c E)(\frac{\partial^2}{\partial y^2} - \alpha^2)^2\right]\psi_y \quad (6)$$

The boundary conditions on ψ_y at $y = \pm 1$ are

$$\psi_y = \frac{\partial \psi_y}{\partial y} = 0 \quad (7)$$

Equation (6) is the Orr-Sommerfeld equation with two additional terms reflecting the intrinsic properties, λ_1 and λ_2, of the Jeffrey fluids model. λ_1 is involved in the Deborah number De while λ_2 is involved in the elasticity number E. Equations (6) and (7) constitute a problem with nonlinear eigenvalues c as follows:

$$\mathbf{A}\psi_y + c\mathbf{B}\psi_y + c^2\mathbf{C}\psi_y = 0 \quad (8)$$

where $\mathbf{A}$, $\mathbf{B}$ and $\mathbf{C}$ are the operators characterized by N, α, Re, De and E. The stability of the disturbed flow depends on the sign of the imaginary components c_i of the eigenvalue. The marginal stability condition is defined by $c_i = 0$.

To solve the eigenvalues problem, we use a numerical approach based on the Chebyshev spectral collocation method, in which the function ψ_y is expanded as a finite sum of Lagrange polynomials discretized at N-collocation points of

Gauss-Labatto [14]. The numerical code has been checked through comparison with the results of Orszag [1] for the Newtonian case and also with those obtained by Lamine et *al.* [10] for the Maxwellian case.

3 Results and Discussion

In this section, we focus our attention on elucidating the combined effects of relaxation and retardation times on the eigenspectrum structure and the diagram marginal stability ($c_i = 0$). All the numerical results presented in this paper converge at $N = 120$ collocation points.

3.1 Eigenspectrum

In this subsection, we focus our interest on the effects of the relaxation and retardation times on the disturbance modes of the base flow. For this purpose, we have to examine the behavior of the eigenspectrum versus the Deborah and elasticity numbers at the critical values $\alpha = \alpha_{cp} = 1.02056$, $Re = Re_{cp} = 5772.22$ related to the onset of instability of classical Poiseuille flow [1].

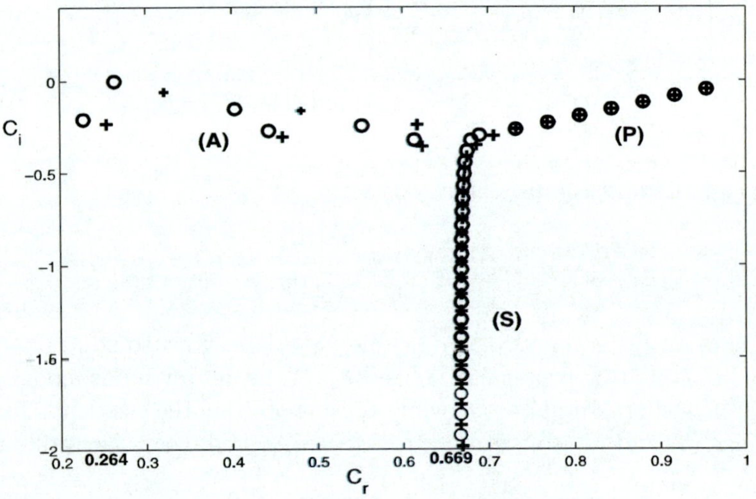

Fig. 1. Orr-Sommerfeld eigenspectrum for $De = 0$, $E = 0$ (Newtonian fluid) and for $Re = 5772.22$, $\alpha = 1.02056$. (o): even modes, (+): odd modes.

The eigenspectrum of the Orr-Sommerfeld operator for plane Poiseuille flow of a Newtonian fluid ($De = 0$ and $E = 0$ [see Fig. 1]) is divided into three distinct branches, A, P and S [15]. Branch A represents modes concentrated near the fixed walls (wall modes) and corresponds to a low phase velocity. Branch P corresponds to the modes in channel center where the phase velocity, c_r, is close to the maximum velocity. The S branch presents the average modes and is characterized by a constant phase velocity, $c_r \approx 0.66845$, close to the mean velocity.

In order to examine the effect of different intrinsic fluid properties on the eigenspectrum structure of Eq. (6), we track in Figs. 2 (a and b) the evolution of the 44 least stable eigenmodes as a function of De and E, respectively. The results in Fig. 2(a), where $E = 0$ corresponds to the Maxwellian case, the initial value, denoted ($\circ$), describes the classical Orr-Sommerfeld spectrum. It turns out that, the relaxation time accelerates the phase velocity associated to mean modes: the S modes shift from $c_r = 0.66845$ at $De = 0$ to $c_r = 1.3041$ at $De = 0.5$. That leads to enhance the stability of mean modes, then to dampen it when De tends towards 0.22. The P-modes stabilize monotonically with a slight decrease in phase velocities c_r as De increased. Such behavior has been produced by cross-flow [16] and magnetic field [5]. For the A modes, we see the least stable mode of plane Poiseuille flow of Newtonian fluid destabilizes with an increase in the phase velocity by 0.023% when De increases from 0 to 0.5.

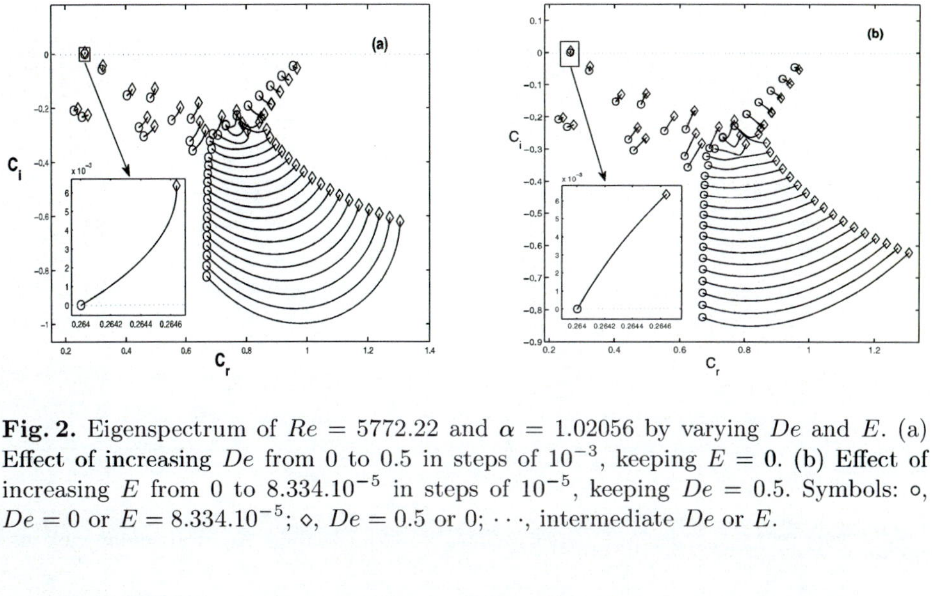

Fig. 2. Eigenspectrum of $Re = 5772.22$ and $\alpha = 1.02056$ by varying De and E. (a) Effect of increasing De from 0 to 0.5 in steps of 10^{-3}, keeping $E = 0$. (b) Effect of increasing E from 0 to $8.334.10^{-5}$ in steps of 10^{-5}, keeping $De = 0.5$. Symbols: $\circ$, $De = 0$ or $E = 8.334.10^{-5}$; $\diamond$, $De = 0.5$ or 0; $\cdots$, intermediate De or E.

Figure 2(b) shows the qualitative effect of increasing E at $De = 0.5$. The initial value, denoted ($\diamond$), represents the Maxwellian spectrum where $De = 0.5$. The results show that the elasticity number acts to recover the classical Orr-Sommerfeld spectrum form: by increasing E. The A, P, and S modes decelerate and the unstable A mode associated with both walls stabilizes monotonically. In fact, this mode becomes marginal ($c_i = 0$) at a specific value of E. This result reveals an important behavior that occurs during the competition between relaxation and retardation times to govern flow stability. It concerns the existence of values of De and E for which the Jeffrey fluid model behaves like the Newtonian model. This feature appears when the Reynolds number is equal to the ratio of relaxation time and retardation time, $Re = De/E$.

Referring to Figs. 3(*a-d*), where the eigenspectrums are plotted in both cases, the Newtonian case is represented by (o) and the particular case of Jeffrey's fluid is identified by (x). It seems that the eigenspectrum of both fluid models is perfectly the same. Moreover, this particular behavior of the Jeffry fluid is independent of the wavenumber values.

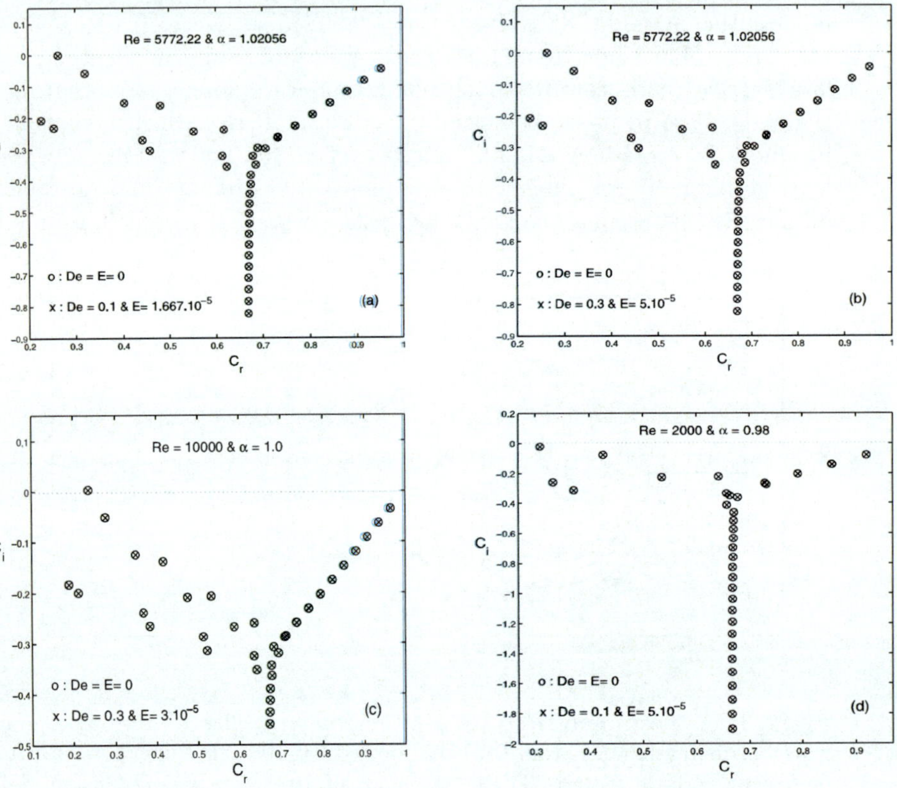

Fig. 3. Eigenspectrum structure for: Newtonian fluid (o). Particular case of Jeffrey fluid (x).

3.2 Marginal Stability

The combined effect of relaxation and retardation times on the marginal stability is depicted in Figs. 4 and 5.

Figure 4 displays the neutral curves at $De = 0.2$ for different values of elasticity number, E. The curves indicate that the critical Reynolds number, Re_c, increase and the unstable perturbations regions get smaller as the elasticity number increases. In addition, the retardation time decreases the critical wavenumber α_c. It means that the unstable disturbances get dumped with retardation time and the mechanism of stability is governed by the long wavelengths regime.

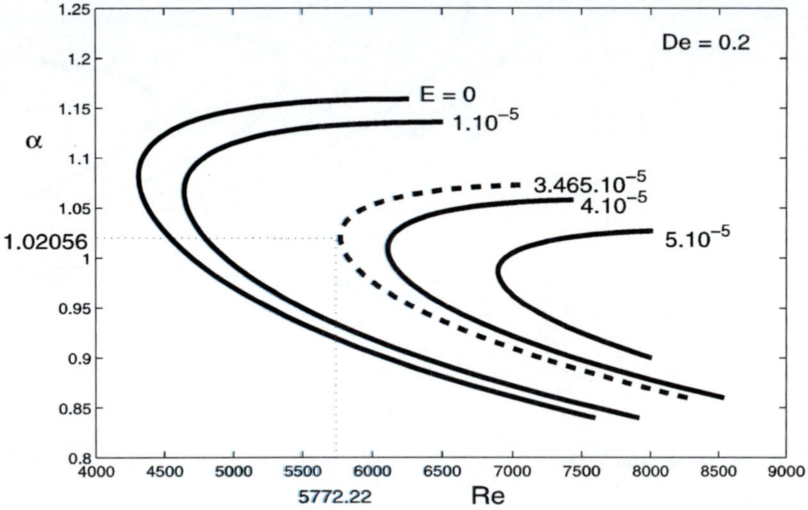

Fig. 4. Typical neutral curves for various elasticity number E at $De = 0.2$

Setting $E = 3.465.10^{-5}$, the effect of relaxation time on the marginal stability diagram ($c_i = 0$) in the (Re, α)-plane is explored in Fig. 5. It can be observed that, as the value of the Deborah number De increases, the critical Reynolds number, Re_c, decreases and the critical wavenumber, α_c, increases. Indeed, the decrease in the critical Reynolds number leads to a widening of the unstable the disturbance zones, which extend towards the short wavelengths regime. It means that, the relaxation time has a destabilizing effect on the base flow.

As it was shown in Fig. 4 and 5, there is a specific value of the Reynolds number, $Re = (De/E)$, for which the Jeffrey fluid model behaves as the Newtonian one in the plane Poiseuille flow. The dashed curves in Figs. 4 and 5 clarify more this behavior. As one can see, for $De/E = 5772.22 = Re_{cp}$, the Jeffrey fluid model and the Newtonian fluid have an identical marginal stability diagram. In addition, the corresponding critical wavenumbers, α_c, obtained in this particular case is also identical to that of the classical plane Poiesuille flow, $\alpha_c = \alpha_{cp} = 1.02056$ [1], which proves the observed particular behavior at $Re = De/E$.

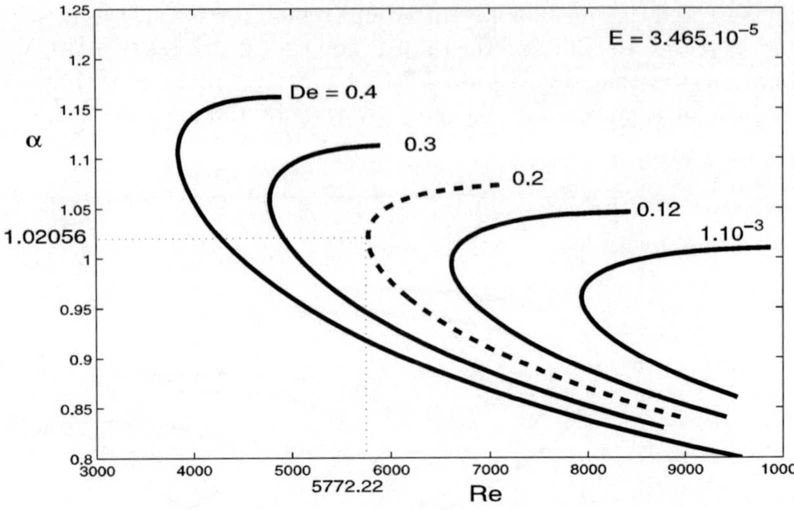

Fig. 5. Typical neutral curves for various Deborah number De at $E = 3.465.10^{-5}$.

4 Conclusion

We have studied a linear stability of plane Poiseuille flow in a Jeffrey fluid. The linear problem has been reduced to a generalized Orr-Sommerfeld equation. The nonlinear eigenvalue problem is solved numerically using the Chebyshev spectral collocation method. We have focused our analysis on the effect of the elasticity number E and that of the Deborah number De on the onset of the flow instability. The numerical results have been discussed using two approaches, where both of them show the stabilizing effect of the relaxation time and the destabilizing effect of the retardation time. These effects are governed by the A mode associated with the walls. Furthermore, an important behavior is obtained when the Reynolds number equals ratio of Deborah and Elasticity numbers, in this situation the Jeffrey fluid behaves as the Newtonian one. This behavior is independent of the disturbance wavenumbers.

References

1. Orszag S.A.: Accurate solution of the Orr-Sommerfeld stability equation. J. Fluid Mech. **50**, 689–703 (1971)
2. Gao, P., Lu, X.-Y.: Instability of channel flow with oscillatory wall suction/blowing. Phys. Fluids **18**, 034102 (2006)
3. Baranovskii, E.S.: An inhomogeneous boundary value problem for the stationary motion equations of Jeffreys viscoelastic medium. J. Appl. Ind. Math. **7**(1), 22–28 (2013)
4. Lamine, M., Hifdi, A.: Stability analysis of plane Couette-Poiseuille flow with porous walls under transverse magnetic field. J. Therm. Eng. **2**(4), 807–813 (2016). https://doi.org/10.18186/jte.27837

5. Lamine, M., Hifdi, A.: Dominant modes in hydromagnetic stability of channel flow with porous walls. MATEC Web Conf. **286**, 07009 (2019). https://doi.org/10.1051/matecconf/201928607009
6. Saasen, A., Hassager, O.: Gravity waves and Rayleigh-Taylor instability on a Jeffrey-fluid. Rheol. Acta **30**, 301–306 (1991)
7. Renardy, M., Tech, V.: Nonlinear stability of flows of Jeffreys fluids at low Weissenberg numbers. Research Report No. 94-NA-022 (1994)
8. Bergström, L.B.: Transient growth of small disturbances in a Jeffrey fluid flowing through a pipe. Fluid Dyn. Res. **32**, 29–44 (2003)
9. Barra, V., Afkhami, S.: Linear stability analysis of thin viscoelastic liquid of Jeffreys type with van der Waals interaction. In: FACM, New Jersey Institute of Technology, Newark (2014). https://doi.org/10.13140/2.1.1915.8409
10. Lamine, M., Riahi, M., Hifdi, A.: Hydrodynamic stability of plane Poiseuille flow in Maxwell fluid with cross-flow. Eur. Phys. J. Plus **135**(2), 189 (2020). https://doi.org/10.1140/epjp/s13360-020-00165-6
11. Drazin, P.G., Reid, W.H.: Hydrodynamics stability. Cambridge University Press, Cambridge, p. 154 (1981)
12. Fontaine, A.: Étude de l'hydrodynamique de fluides rhéologiquement complexes dans un mélangeur Maxblend par vélocimétrie laser. Mémoire de maîtrise, École Polytechnique de Montréal (2012)
13. Canuto, C., Hussaini, M.Y., Quarteroni, A., Zang, T.A.: Spectral methods in fluid dynamics. New York, Springer-Verlag (1988). https://doi.org/10.1007/978-3-642-84108-8
14. Weideman, J.A., Reddy, S.C.: A MATLAB differentiation matrix suite. ACM Trans. Math. Softw. **26**, 465–519 (2000)
15. Mack, L.M.: A numerical study of the temporal eigenvalue spectrum of the Blasius boundary layer. J. Fluid Mech. **73**, 497 (1976)
16. Guha, A., Frigaard, I.A.: On the stability of plane Couette-Poiseuille flow with uniform cross-flow. J. Fluid Mech. **656**, 417 (2010)

Faraday Instability in Hele-Shaw Cell with Two Commensurate Frequencies Forcing

A. Rachik(✉), M. Assoul, and S. Aniss

Faculty of Sciences Aïn-Chock, Laboratory of Mechanics,
University Hassan II of Casablanca, Casablanca, Morocco
{azeddine.rachik-etu,said.aniss}@etu.univh2c.ma

Abstract. We investigate the Faraday instability with an air-liquid interface in a Hele-Shaw cell subjected to a periodic vertical oscillation with two commensurate frequencies, Ω_1 and Ω_2. The linear stability formulation allows to reduce the governing equations to a damped Mathieu equation describing the evolution of the interface amplitude. Thereafter, spectral methods are used to establish the marginal stability curves in terms of reduced critical forcing amplitude as a function of wavenumber, which allows us to examine the effects of frequency ratios, the amplitudes of the excitations, and the effect of phase shift of the two superimposed accelerations. The instability diagrams are presented in the form of tongues which are resonances that can be harmonic or sub-harmonic with the existence of several bi-critical points. We show that the phase shift has a stabilizing effect. Also, for each frequency ratio, $\omega = \frac{\Omega_2}{\Omega_1}$, the instability has maximum at a finite frequency Ω_1 while for the low and high fresquencies stabilizing effect is observed. The decrease of the superimposed acceleration amplitudes has also a stabilizing effect.

Keywords: Parametric instability · Mathieu equation · Hele-Shaw cell

1 Introduction

When a horizontal fluid layer with a free surface is subject to a vertical periodic oscillation, Faraday [1] observed that a free surface instability occurs in the form of a standing wave appearing in the centre of the fluid layer with a frequency equal to half that of the oscillation. Thereafter, with experiments similar to those of Faraday, Matthiessen [2] noticed that the wave can also have the same frequency as the oscillation. Later, Rayleigh [3] carried out a series of experiments on the Faraday problem and gave qualitative explanations, notably that the wave at the surface is of parametric type. More later, Benjamin and Ursell [5] performed a linear stability study and reduced the governing equations to a system of Mathieu equations describing the elevation of the free surface from equilibrium. Kumar and Tuckerman [4] carried out a complete linear stability analysis in the configuration of two superimposed layers of viscous fluid of different heights confined between two rigid walls of infinite extension in the horizontal directions. They showed that the effect of viscosity is notable on wavelength

S. Aniss et al. (Eds.): CMM 2022, LNME, pp. 138–147, 2024.
https://doi.org/10.1007/978-3-031-46973-2_13

selection. Recently, Li et al. [6] have been interested in Faraday waves in a Hele-Shaw cell through experimental, numerical and theoretical studies. Inspired by the Kelvin-Helmholtz-Darcy theory, they produced stable standing waves at half the frequency of the forced external vibration. Additionally, the case of a system consisting of three fluid layers with two interfaces was studied by Li et al. [7]. In their work, a system of two coupled Faraday waves was experimentally observed at the two interfaces of the three superimposed fluid layers (air, pure ethanol, and silicon oil) in a closed Hele-Shaw cell in the presence of periodic vertical vibration.

The Faraday instability with oscillation at two frequencies has been examined previously by several authors. For instance, Edwards and Fauve [11] studied experimentally the Faraday instability in a low-depth system with a viscous fluid and with two simultaneous forcing frequencies. They observed that the frequency ratio and amplitudes, as well as the phases of the two sinusoidal components, have a significant influence on the pattern that occurred just above the instability threshold. Beyer and Friedrich [10] numerically studied the linear stability of the Faraday instability for the case of two excitation frequencies and proposed a comparison of their results with those of Edwards and Fauve [11]. In parallel, Besson et al. [12] had extended the stability analysis of Kumar and Tuckerman to the case of two-frequency forcing. They took into account the effects of viscosity and finite depth and compared their results to the experiment to within 2 percent for stability thresholds. Recently, Batson et al. [13] studied theoretically and experimentally the excitation of Faraday waves in small containers using two commensurate frequencies.

The motivation of this work is that the spraying of a liquid through injectors uses the Faraday instability. The size of the droplets is related to the wavelength of the instability. Thus, in light of the above work, attention is focused on the Faraday instability in a Hele-Shaw cell. The oscillations have two commensurate frequencies. We will examine the effect of the frequency and amplitude ratios as well as the effect of the phase shift between the two superimposed accelerations on the behaviour of parametric resonances.

1.1 Mathematical Formulation

Consider a liquid layer of height h confined in a Hele-Shaw cell of thickness e and length L (see Fig. 1). The Hele-Shaw cell is subjected to periodic oscillation of the form $Z(t) = \alpha_1 \cos(\Omega_1 t) + \alpha_2 \cos(\Omega_2 t + \chi)$, where Ω_1 and Ω_2 are two commensurate frequencies and α_1 and α_2 are the amplitudes.

1.2 Governing Equations

The Navier-Stokes equation and the continuity equation are written in a relative frame linked to the Hele-Shaw cell:

$$\frac{\partial \mathbf{U}}{\partial t} + (\mathbf{U} \cdot \nabla)\mathbf{U} = -\frac{1}{\rho}\nabla p - \tilde{g}\,\mathbf{k} + \nu \nabla^2 \mathbf{U} \tag{1}$$

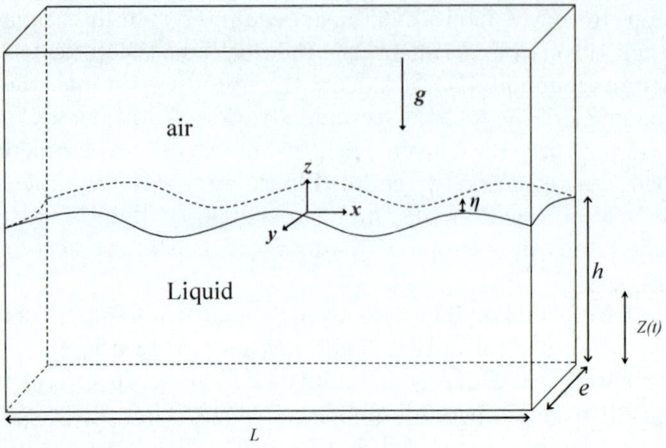

Fig. 1. Hele-Shaw cell filled with a Newtonian liquid and subjected to periodic vertical oscillation.

$$\nabla \cdot \mathbf{U} = 0 \tag{2}$$

where $\mathbf{U}(x, y, z, t)$ is the velocity field, ρ is the density of the fluid, p is the pressure, ν is the kinematic viscosity and $\mathbf{k}$ is the unit vector in the vertical upward direction. We denote by $\tilde{g} = g\,[1 - F_1 \cos(\Omega_1 t) - F_2 \cos(\Omega_2 t + \chi)]$ the acceleration that the liquid feels in the reference frame of the Hele-Shaw cell. We designate by $F_i = \alpha_i \Omega_i^2/g$ $(i = 1, 2)$ the reduced amplitudes of the two superimposed acceleration and by χ the phase shift between these two superimposed acceleration. In the Hele-Shaw formulation , the velocity profile is considered to be parabolic in the thickness interval e, $\mathbf{U} = \frac{3}{2}(1 - (2\frac{y}{e})^2)\mathbf{u}(x, z, t)$ [9], and the bidimensional velocity field is denoted by $\mathbf{u}(x, z, t) = (u, w)$. Thereafter, the Navier-Stokes equation is integrated over the interval $[-\frac{e}{2}, \frac{e}{2}]$. The most important part of the viscous dissipation occurs in the direction normal to the walls (in the y direction), and for the studied wavelengths, the attenuation time is much smaller than the viscous diffusion time, $\nu\nabla_2\mathbf{u} << \gamma\mathbf{u}$. Thus, we obtain:

$$\frac{\partial \mathbf{u}}{\partial t} + \frac{6}{5}(\mathbf{u} \cdot \nabla_2)\mathbf{u} = -\frac{1}{\rho}\nabla p + \tilde{g}\,\mathbf{k} - \gamma \mathbf{u}, \tag{3}$$

$$\nabla_2 \cdot \mathbf{u} = 0, \tag{4}$$

where $\nabla_2 = \frac{\partial^2}{\partial x^2} + \frac{\partial^2}{\partial z^2}$, $\gamma = \frac{12\nu}{e^2}$ is the external Rayleigh viscosity.

1.3 Boundary Conditions

We denote by $\eta(x, t)$, the deformation of the interface with respect to its equilibrium position, $z = 0$. At the bottom of the cell, the impermeability condition is:

$$w(x, z, t) = 0 \quad \text{at} \quad z = -h, \tag{5}$$

At the free interface, $z = \eta$, the kinematic condition is given by:

$$\frac{\partial \eta}{\partial t} + \frac{\partial \eta}{\partial x} u - v = 0 \quad \text{at} \quad z = \eta. \tag{6}$$

Ignoring the viscous terms, the continuity of the normal stresses at the interface is written as follows:

$$p - p_\infty = \sigma \nabla \cdot \mathbf{n} \quad \text{at} \quad z = \eta, \tag{7}$$

where $p_\infty = p_{atm}$ is the pressure of air, σ is the surface tension and $\mathbf{n} = \frac{1}{\sqrt{1+(\frac{\partial \eta}{\partial x})^2}}(1, -\frac{\partial \eta}{\partial x})$ represents the normal vector for the interface.

1.4 Linear Problem Formulation

The vorticity is considered damped due to the dissipation term $\gamma \mathbf{u}$. For very long time scales, the vorticity can be neglected and the potential flow can be assumed. Thus, considering irrotational flow and using the potential function, $\mathbf{u} = \nabla \phi$. Equation (3) is integrated and rewritten as a dynamic Bernoulli equation. Only solutions of small amplitudes close to rest, $\phi = 0$ and $\eta = 0$, are considered and for which the non-linear terms are negligible. Consequently, the boundary conditions are not satisfied at $z = \eta$ but at $z = 0$. Thus, the following linearised system is obtained:

$$\frac{\partial^2 \phi}{\partial x^2} + \frac{\partial^2 \phi}{\partial z^2} = 0 \quad \text{for} \quad -h < z < \eta, \tag{8}$$

$$\frac{\partial \phi}{\partial z} = 0 \quad \text{at} \quad z = -h, \tag{9}$$

$$\frac{\partial \eta}{\partial t} - \frac{\partial \phi}{\partial z} = 0 \quad \text{at} \quad z = 0, \tag{10}$$

$$\frac{\partial \phi}{\partial t} - \frac{\sigma}{\rho} \frac{\partial^2 \eta}{\partial x^2} + \tilde{g} \eta + \gamma \phi = 0 \quad \text{at} \quad z = 0. \tag{11}$$

1.5 Normal Mode Analysis

The solutions of the system of Eqs. (8)–(11) are considered periodic in the horizontal direction, then we use the following modes:

$$\phi(x, z, t) = K(t) \varphi(z) \sin(kx), \tag{12}$$

$$\eta(x, t) = \zeta(t) \cos(kx). \tag{13}$$

Using Eqs. (8), (9) and (10), the following expression of $\phi(x, z, t)$ is obtained:

$$\phi(x, z, t) = \frac{\partial \zeta}{\partial t} \frac{\cosh(k(h+z))}{k \sinh(kh)} \sin(kx) \tag{14}$$

Thereafter, Eqs. (13), (14) and (11) are reduced to the damped Mathieu equation:

$$\ddot{\zeta} + \gamma\dot{\zeta} + gk\tanh(kh)\big(1 + \frac{\sigma}{\rho g}k^2 - F_1\cos(\Omega_1 t) - F_2\cos(\Omega_2 t + \chi)\big)\zeta = 0, \quad (15)$$

Using the variable change $\tau = \Omega_1 t$, Eq. (15) is written as follows:

$$\ddot{\zeta} + 2\mu\dot{\zeta} + \left[\delta + \epsilon\left(cos(\tau) + r\,\omega^2\cos(\omega\tau + \chi)\right)\right]\zeta = 0, \quad (16)$$

where $\delta = gk\tanh(kh)\left[1 + \sigma k^2/(\rho g)\right]/\Omega_1^2$, $\mu = \gamma/(2\Omega_1)$, $\epsilon = -gk\tanh(kh)F_1/\Omega_1^2$. We denote by $r = \alpha_2/\alpha_1$ the ratio of amplitudes and by $\omega = \Omega_2/\Omega_1$ the ratio of frequencies. Note that $\frac{F_1}{F_2} = \frac{\alpha_1\Omega_1^2}{\alpha_2\Omega_1^2}$.

2 Results and Discussion

Using numerical spectral methods [14], Eq. (16) is solved and marginal stability curves, in the (F_1, k) plane, are obtained. These marginal stability curves correspond to parametric resonances which are in the form of tongues, as shown in Fig. 2 for an interface water-air. The tongues correspond alternately to subharmonic (SH) and harmonic (H) periodic temporal solutions. Here, the threshold of convection occurs at a called bi-critical point where the minimum of the curves, $F_{1,c} = 0.953$, corresponds to two different wavenumbers $k_{c1} = 1.16\,\mathrm{cm}^{-1}$ and $k_{c2} = 2.67\ cm^{-1}$.

Table 1. Critical values of the dimensionless amplitude of forcing F_c versus the phases, χ for $e = 1$ mm, $h = 1$ cm $\omega = 3$ $\Omega = 10/2\pi = 4$ Hz and for different values of the depth h

χ	F_c
0	3.7422
30	3.6947
60	3.6361
90	3.7422
120	3.6947
180	3.6361

Figure 3 illustrates the marginal stability curves for different values of the phase shift. The increase in the phase creates new zones of instability called 'islands'. The Islands are also shown in the stability charts of Besson et al. [12]. Note that when increasing the phase shift, the wavenumber remains unchanged, However, there is a slight stabilization effect that manifests itself as a slight increase and then decrease in the magnitude of the critical forcing $F_{1,c}$, as can be seen in Table 1. This stabilizing effect is shown in detail in Fig. 4 where the

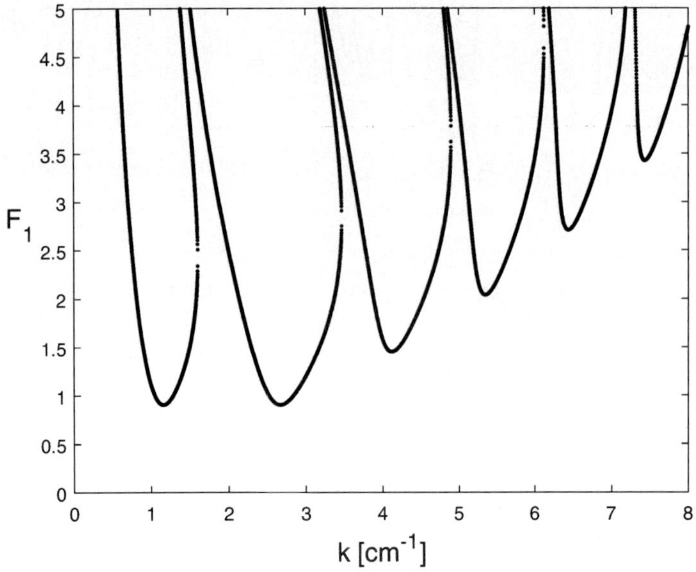

Fig. 2. Water-air interface. Marginal stability curves in the plane amplitude forcing versus wavenumber, (F_1, k), with $\Omega_1/2\pi = 9.7Hz$, $\omega = 2$, $r = 0.1$, $h = 1$cm and $e = 1$mm.

variations of the critical forcing amplitude as a function of the phase shift are presented for different excitation frequencies Ω_1. Indeed, it should be noted that the destabilization is maximum at $\chi = 0$, after that, the amplitude of the critical forcing increases with the phase shift until a phase shift beyond which it starts to decrease. This phase shift where there is a break of the curves corresponds to bi-critical points which depend here on the excitation frequency $\Omega_1/2\pi$. Let us note here also that the decrease of the dimensional frequency Ω_1 has a stabilizing effect.

Figure 5 shows the evolution of the critical forcing amplitude, $F_{1,c}$, as a function of the dimensional frequency, Ω_1, for different values of the frequency ratio, ω, and when amplitudes of the superimposed accelerations are equal, $\alpha_1 \Omega_1^2 = \alpha_2 \Omega_2^2$. As can be seen, modulation with two commensurate frequencies has two effects, one stabilising and the other destabilising. For the frequency ratio $\omega = 1/2$, the destabilization is maximum at the frequency $\Omega/2\pi = 23.0$ Hz with $F_{1,c} = 0.5230$ ($k_c = 3.15\text{cm}^{-1}$), however the system is less unstable in the limit of low and high frequency. The same behaviour takes place for $\omega = 2$, which has also a single minimum at a lower $\Omega/2\pi = 11.5$ Hz with $F_{1,c} = 0.5231$ ($k_c = 3.15\text{cm}^{-1}$). For the frequency ratio ω =3, there are two minimums, at $\Omega_1/2\pi = 27.1$ and $\Omega_1/2\pi = 9.6$, where instability is potential with $F_{1,c} = 0.589$ ($k_c = 3.7\text{cm}^{-1}$) and $F_{1,c} = 0.598$ ($k_c = 3.14\text{cm}^{-1}$). It should be noted here beyond the frequency corresponding to the minimum of these curves, increasing the value of the frequency leads to re-stabilization but accompanied with

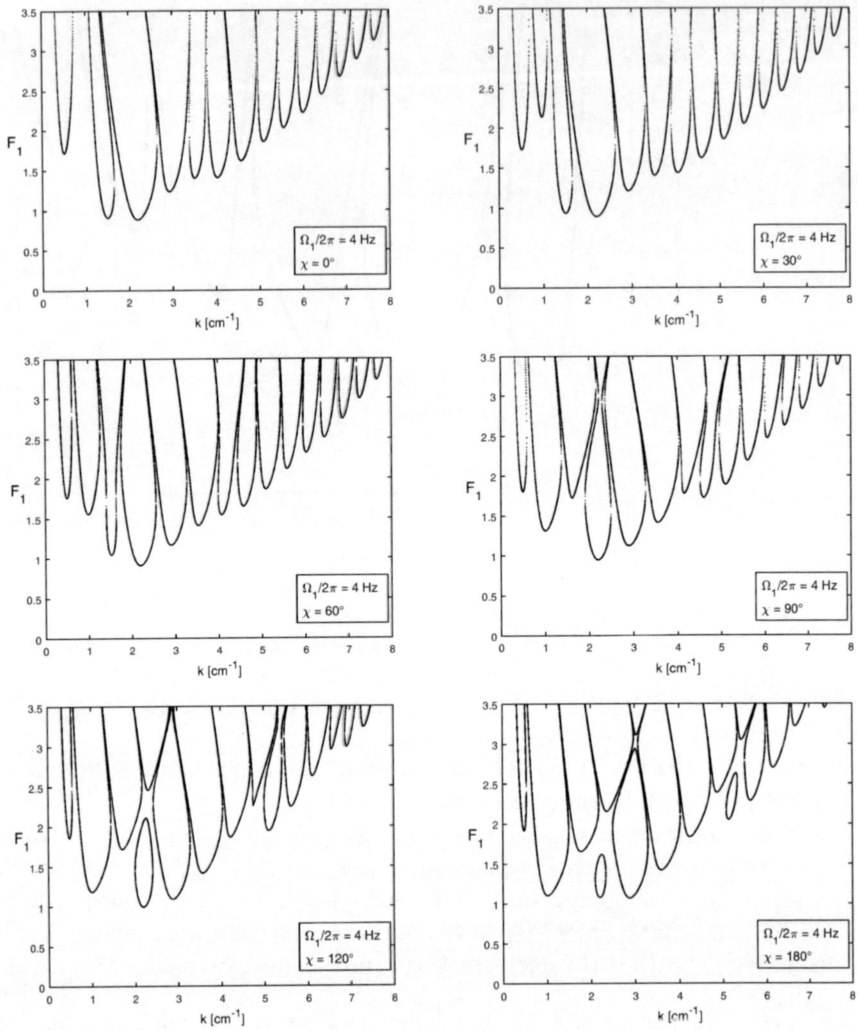

Fig. 3. Water-air interface. Marginal stability curves in the plane amplitude forcing versus the wavenumber, (F_1, k), for different phase shift, χ, and for dimensional frequency $\Omega_1/2\pi = 4$ Hz, frequency ratio $\omega = 3$, height $h = 1$ cm and thickness $e = 1$ mm.

a feature where there is a break in the slope of the curve $F_{1,c}(\Omega_1/2\pi)$ corresponding to bi-critical points. Also for high and low frequency, $\Omega_1/2\pi$, when the frequency ratio ω is greater than 2, this one has no effect on the instability threshold, indeed the curves become merged. We illustrate in Fig. 6, for different amplitude ratios of the superimposed acceleration amplitudes, r, the evolution of the critical forcing as a function of the excitation frequency $\Omega_1/2\pi$. We note that the decrease in the amplitude ratio causes the curves to shift upward, indicating a stabilizing effect. For each amplitude ratio, there is also a frequency

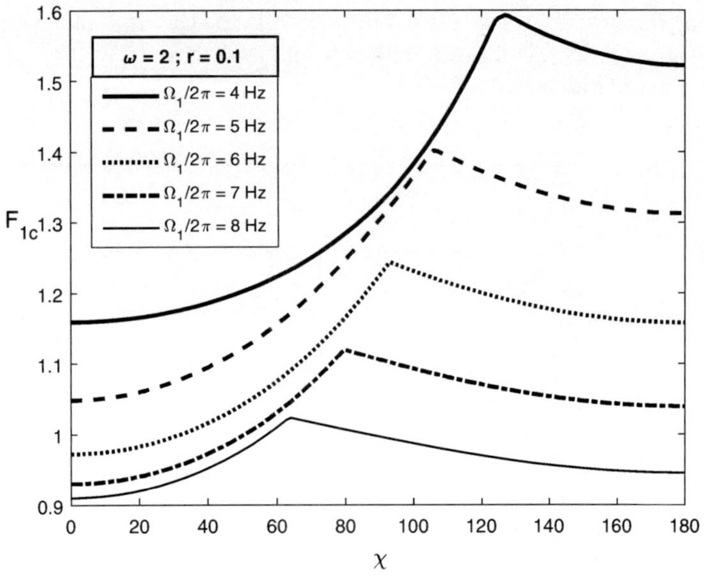

Fig. 4. Water-air interface. Critical forcing amplitude, $F_{1,c}$ versus the phase shift χ in degrees, for different values of the dimensional frequency, $\Omega_1/2\pi$, and for $r = 0.1$, $h = 1$ cm and $e = 1$ mm.

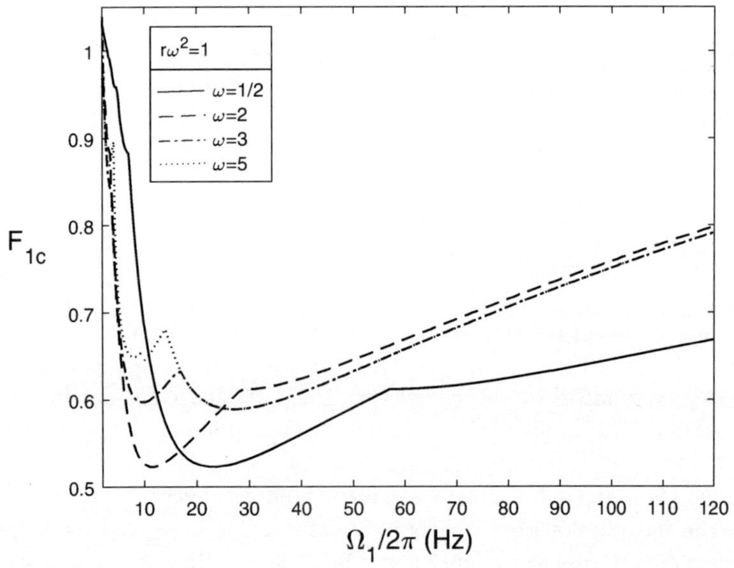

Fig. 5. Water-air interface with equal superimposed acceleration amplitude, $\alpha_1 \Omega_1^2 = \alpha_2 \Omega_2^2$. Critical forcing amplitude, $F_{1,c}$, versus excitation frequency , $\Omega/2\pi$ **for different** frequency ratio, ω, and for $h = 1$ cm and $e = 1$ mm.

where the destabilization is maximal, while for the high and low frequencies, the instability is less. A bi-critical point exist for $r = 0.1$ where there is a break of the slope of the curve.

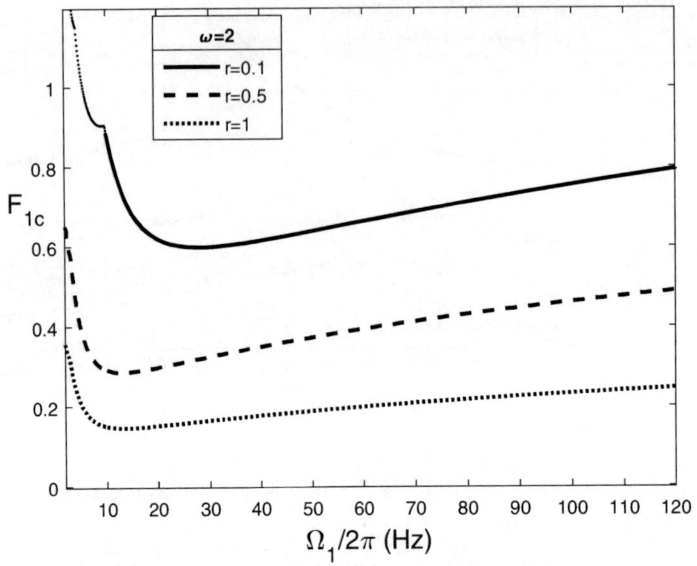

Fig. 6. Water-air interface. Critical forcing amplitude, $F_{1,c}$ versus the dimensional frequency, $\Omega_1/2\pi$, for different values of the superimposed acceleration amplitudes, r, and for $\omega = 2$, $h = 1$ cm and $e = 1$ mm.

3 Conclusion

We investigated the Faraday instability of a liquid layer confined in a Hele-Shaw cell subjected to vertical oscillating motion having two commensurate frequencies. A reduction of the conservation equations leads to the damped Mathieu equation. The numerical solutions of this equation are determined using spectral methods. First, the numerical results focused on the phase shift where the study showed that it has a stabilizing effect. Second, the results also focused on the marginal stability curves and the critical reduced forcing amplitude as a function of dimensional frequency for different frequency and amplitude ratios. It was shown that the destabilization is maximum near a finite frequency, however stabilization occurs in the limit of low and high frequency. Also, increasing the amplitude ratio has a destabilizing effect while increasing the frequency ratio has a stabilizing effect.

References

1. Faraday, M.: Phil. Trans. R. Soc. Lond. **121**, 299–340 (1831)
2. Matthiessen, L.: Ann. Phys. (Leipzig) **134**, 107 (1868); Matthiessen, L.: Ann. Phys. (Leipzig) **141**, 375 (1870)
3. Rayleigh, L.: Philos. Mag. **15**(5), 235 (1883); Rayleigh, L.: Philos. Mag. **16**(5), 50 (1883)
4. Kumar, K., Tuckerman, L.S.: Parametric instability of the interface between two fluids. J. Fluid Mech. **279**, 49–68 (1994)
5. Benjamin, T. B., Ursell: The Stability of the Plane Free Surface of a Liquid in Vertical Periodic Motion (22 September 1954). The Royal Society (1958)
6. Li, J., Li, X., Chen, K., Xie, B., Lia, S.: Faraday waves in a Hele-Shaw cell. Phys. Fluids **30**(4), 042106 (2018)
7. Li, X., Li, X., Liao, S.: Observation of two coupled Faraday waves in a vertically vibrating Hele-Shaw cell with one of them oscillating horizontally. Phys. Fluids **30**(1), 012108 (2018)
8. Zounes, R.S., Rand, R.: Transition curves for the quasiperiodic Mathieu equation. SIAM J. Appl. Math. **58**, 1094–1115 (1998)
9. Gondret, P., Rabaud, M.: Shear instability of two-fluid parallel flow in a Hele-Shaw cell. Phys. Fluids **9**(11), 3267–3274 (1997)
10. Beyer, J., Friedrich, R.: Faraday instability: linear analysis for viscous fluids. Phys. Rev. E **51**, 1162 (1995)
11. Edwards, W.S. Fauve, S.: Patterns and Quasi-patterns in the Faraday Experiment. Cambridge University Press (1994)
12. Besson, T., Edwards, W.S., Tuckerman, L.S.: Two-frequency parametric excitation of surface waves. Phys. Rev. E **54**, 507–514 (1996)
13. Baston, T., Zoueshtiagh, F., Narayanan, R.: Two-frequency excitation of single-mode Faraday waves. J. Fluid Mech. **764**, 538–571 (2015)
14. Weideman, J.A.C., Reddy, S.C.: A Matlab differentiation matrix suite (1998). http://osu.orst.edu/~weidemaj/differ.html

Heat Transfer, Mass Transfer, Renewable Energy and Environment

Numerical Study of Latent Heat Discharge of a Phase Change Material Shell-and-Tube Thermal Energy Storage System

K. Oudaoui(✉) and M. Faraji

Renewable Energy and Dynamic System Laboratory – Renewable Energy and Laser Plasma Group, Physics Department, FSAC, Hassan II University of Casablanca, Casablanca, Morocco
kenzaoudaoui@gmail.com

Abstract. Solar energy is currently the most abundant and cleanest source of renewable energy, making it a good substitute for increasing efficiency. Solar water heaters can significantly reduce the cost of electricity consumption and help reduce greenhouse gas emissions. The solar heat exchanger, a phase-change heat storage system, is designed in this work. A numerical analysis was carried out to examine the impact of various cooling temperatures on the solidification thermal cycle of the phase change material. Phase Change Material (PCM) can be used to effectively regulate temperature. Depending on the PCM and its storage capacity, the temperature of the produced water is controlled. Through the tubes, a fluid used for heat transfer from the solar collector transfers heat from the sun's rays to the PCM. During the discharge phase, the heat that has been stored in the PCM is then transferred to the water, heating it up. In order to quantitatively evaluate the thermal performance of the latent heat storage unit, a numerical simulation using the finite volume method was developed. To determine the best design, several simulations were carried out for the different types of PCM (Paraffin, n-eicosane and Rubitherm Paraffin) and also to improve their discharge time.

Keywords: Phase Change Material · Discharge time · Solidification · Solar heat exchanger

1 Introduction

Thermal energy storage is a useful and vital approach to save and improve the utilization of available energy. As a result, using thermal energy storage devices to address the mismatch between energy supply and demand is a viable option [1]. Phase change materials (PCMs), which are utilized in latent heat storage units, are potential prospects for solar energy, electronic heating and cooling and other applications due to their high energy storage and isothermal phase transition [2]. It should be noted that due to the phase change material's low thermal conductivity, the rate of heat transfer in a heat exchanger of heat storage unit is limited during charging and discharging processes. The energy produced by the PCM's solidification must be transmitted from the solid-liquid interface to the heat transfer medium via the growing solid layer around the

S. Aniss et al. (Eds.): CMM 2022, LNME, pp. 151–159, 2024.
https://doi.org/10.1007/978-3-031-46973-2_14

transfer surface. The thermal conductivity of the solid PCM at the discharge, which is frequently the dimensioning procedure in many applications, limits the overall transfer coefficient in this manner. The interaction between the heat transfer fluid and the storage medium is limited by the low thermal conductivity of most PCMs. As a result, developing strategies to improve thermal performance is necessary for designing and efficient and cost-effective storage material. Several review studies have identified the design of the heat exchanger, as well as phase change material qualities and operating conditions, as elements that influence the performance of TES systems. Furthermore, shell and tube heat exchangers are the most researched geometric configuration, with the phase change material occupying either the shell or tube side. One of the most effective ways to improve heat exchanger performance is to use multiple PCMs or more. Despite the HTF temperature drop, this application can extract and preserve more inlet energy and facilitate successful melting of the PCM near the tube exit. In a solar power plant, Li et al. [3] investigated the effect of incoming water temperature on a system running with three different PCM melting temperature, in which PCMs are restricted in a tube with walls. Agarwal and Sarviya [4] looked at how the temperature distribution, heat charging and discharging fractions and melting were affected by the HTF inlet temperature and fluid mass flow rate. Wang et al. [5] investigated another heat exchanger configuration, the zigzag configuration of multiple PCMs. Six sets of various PCMs with varied melting temperatures were used in the experiment. Agyenim et al. [6] also experimented with erythritol as phase transition material in an energy storage system. Finless systems and two-systems with circular and longitudinal fins were tested. The system with longitudinal fins produced the best results, with greater thermal reactivity during the charging phase and lower cooling during the discharging process, according to the findings. Abduljalil et al. [7] examined the solidification of a PCM in a triplex exchanger tube numerically as well. The main purpose of this paper is to provide a mathematical model that can be used to study the effects of water inlet temperature on the solidification of PCM_1, PCM_2 and PCM_3 and to investigate the optimum discharge times of the multiple PCMs on the thermal performance of the exchanger.

2 Physical and Mathematical Model

The studied system is a heat exchanger with a central tube filled with multiple Phase Change Materials (PCMs) and supported by an internal tube, as shown in Fig. 1. Inside the inner tube, water flows through. When a PCM undergoes solidification, it releases latent heat, maintaining a constant temperature during this phase transition. However, as solidification progresses, the temperature difference between the PCM and its surroundings may decrease. During the flow process, the water interacts with the PCMs. The heated water exists the system from the right end, extracting the thermal energy previously stored in the PCMs. The heat transfer fluid's initial temperature upon entry is intentionally set below the melting point of the first PCM ($T_i < T_{melt}$), initiating the solidification process of the PCMs within the system.

In order to simplify the mathematical model, the following assumptions were made:

- The thermal resistance of the inner tube is negligible;
- Viscous dissipation and axial conduction in the fluid are negligible;

- The effect of natural convection during melting process is considered by using an efficient thermal conductivity of the PCM liquid phase;
- The thermophysical properties of the water and the PCM will be independent of temperature.

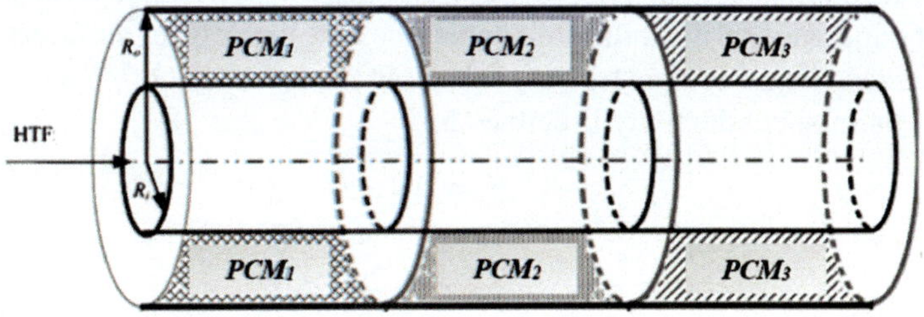

Fig. 1. Schematic representation of the studied system with multiple PCM

The appropriate Water and PCMs equations are expressed as follows:
For the Water:

$$\frac{\partial \theta_f}{\partial t} = -\frac{\dot{m_f}}{\rho_f \pi R_i^2} \frac{\partial \theta_f}{\partial x} - \frac{2U}{(\rho C_p)_f R_i} (\theta_f - \theta^*) \tag{1}$$

where: $\theta_f = T_f - T_{m_3}$ and $\theta^* = T(x, r = R_i, t) - T_{m_3}$
For PCMs, the equation can be written as follows:

$$(\rho C_p)_p \frac{\partial h}{\partial t} = \frac{\partial}{\partial x}\left(k_p \frac{\partial h}{\partial x}\right) + \frac{1}{r}\frac{\partial h}{\partial r}\left(k_p \frac{\partial h}{\partial r}\right) - \rho_p \Delta H \frac{\partial f}{\partial t} \tag{2}$$

The liquid fraction is given as follows:

$$\begin{cases} f = 0 if \theta < \theta_p \\ 0 < f < 1 if \theta = \theta_p \\ f = 1 if \theta > \theta_P \end{cases} \tag{3}$$

In order to solve the Eqs. (1) and (2), Voller et al. [8] proposed a numerical model using the enthalpy formulation used in the previous equations, also the initial and boundary conditions are required:

The initial and boundary conditions are:

$$\theta_f(x, r, t = 0) = T_i - T_{m3} \tag{4}$$

With:

$$\begin{cases} i = 1 0 \le x \le L_1 \\ i = 2 L_1 \le x \le L_2 \\ i = 3 L_2 \le x \le L_3 \end{cases} \tag{5}$$

$$\left.\frac{\partial\theta}{\partial t}\right|_{x=0,r,t} = \left.\frac{\partial\theta}{\partial t}\right|_{x=L,r,t} = 0 \quad (6)$$

$$\left.\frac{\partial\theta}{\partial r}\right|_{x,r=R_0,t} = 0 \quad (7)$$

The thermal conductivity of the phase change material (PCM) at the interfaces refers to the measure of its ability to conduct heat at the contact surfaces. It quantifies how effectively heat can transfer between the PCM and neighboring materials or structures. This thermal conductivity is defined by:

$$k_r = fk_e + (1-f)k_s \quad (8)$$

$$\text{With}: \frac{k_e}{k_l} = CR_a \quad (9)$$

The effective thermal conductivity, denoted as k_e accounts for heat transfers through natural convection during the solidification phase of the PCM.

Table 1. Thermophysical properties [9]

Propriety	Paraffin 53	Rubitherm Paraffin	$C_{20}H_{42}$
ρ [kg/m^3]	830	930	800
k_{liquid} [W/m°K]	0.19	0.21	0.16
k_{solid} [W/m°K]	0.28	0,2	0.212
C_p [J/kg.°C]	2385	812	2050
Q_{Lat} [kJ/Kg]	164	214	237.4
T_{fusion} [°C]	53	60	36.49

3 Results and Discussion

The solar heat exchanger is responsible for absorbing solar energy and storing it in phase change materials (PCMs). The heat exchanger is designed to integrate three different PCMs, with different thermo-physical properties, as described in Table 1 [9]. We conducted numerical simulations to investigates the behavior of the heat exchanger system. By analyzing the numerical data, we gained valuable insights into the thermal performance and heat storage capabilities of the system with the integrated PCMs. The numerical simulation results illustrate how the temperature of the water is affected as it interacts with the PCM/Water heat exchanger. We interpreted these findings to understand how the variation in thermos-physical properties among the multiple PCMs influences the overall system's thermal behavior and energy storage capacity. In the following section, we present a comprehensive analysis and interpretation of the numerical simulation results, highlighting the significance of each PCM's contribution to the heat transfer and storage processes.

3.1 Time Wise Variation of HTF's Temperature

The incorporation of multiple PCMs in the solar heat exchanger enhances the efficiency of energy storage and release. Each PCM is selected based on its specific temperature range, allowing for optimized heat transfer and increased storage capacity within the system. When water enters the heat exchanger at an initial temperature of 25 °C, it starts extracting heat from the fully charged PCMs. As a result, the PCMs begin to solidify and release the stored thermal energy in the form of sensible heat. During this solidification process, the PCMs transfer their heat to the water and after approximately 56 min (as it shown in Fig. 2), all multiple PCMs have solidified and transferred their heat to the water. At this point, the heat exchange between the water and PCMs continues for an additional 80 min. Throughout this duration, the PCMs gradually release their stored energy, sustaining the heat transfer to the water until the desired outlet temperature is achieved. The entire process of heat transfer, solidification, and energy release is cyclic. As one PCM solidifies completely, the water continues to extract heat from the remaining PCMs until they also solidify. This sequential discharge allows for a continuous and controlled supply of thermal energy to the water. As a result, the heat exchanger sustains its efficient operation, ensuring a consistent heat output for the desired duration. By effectively utilizing the distinct melting temperatures of the PCMs and their efficient heat exchange with the water, the solar heat exchanger maximizes its energy storage capacity and provides a reliable and sustainable source of thermal energy. This continuous cycle of heat transfer and discharge persists until all three PCMs have been fully discharged, and the water reaches the desired outlet temperature.

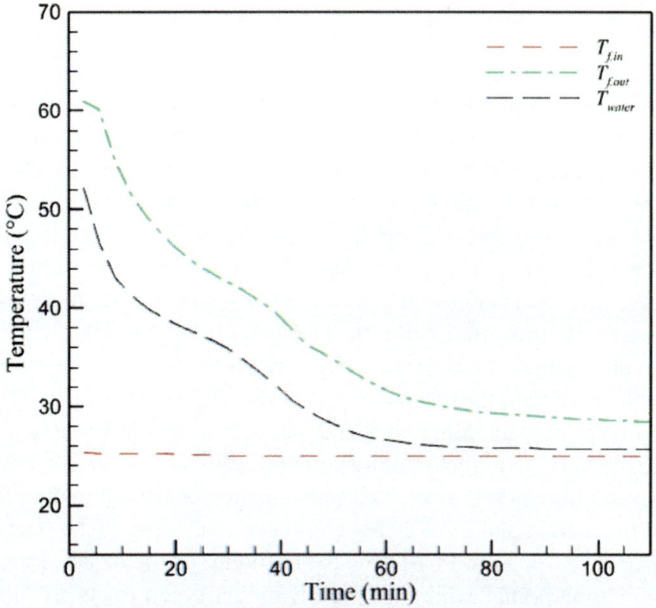

Fig. 2. Time wise variation of the HTF's temperature

3.2 Time Wise Variation of HTF Charge Rate, Q

During the discharge phase, the water continues to circulate within the heat exchanger, while the phase change materials (PCMs) are now in a liquid state. As the water flows through the exchanger, it extracts heat from the PCMs, causing them to solidify. The heat exchange increases gradually, starting from 15 kJ and reaching 25 kJ over a duration of 60 min. This process represents the transfer of thermal energy from the PCMs to the water. As the PCMs solidify, they release the stored thermal energy, resulting in an increase in heat transfer between the water and the PCMs. This heat transfer rate rises from 0 kJ to 100 kJ over a period of 128 min. The amount of energy transferred is substantial during this phase, allowing the water to absorb a significant amount of thermal energy from the solidifying PCMs. The rate of heat transfer from the water during the discharge phase is influenced by several factors. The water's temperature plays a vital role in determining the heat transfer rate. A higher water temperature results in a faster rate of heat extraction from the PCMs. Additionally, the temperature difference between the PCMs and the water also contributes to the high heat transfer rate during discharge. A larger temperature difference facilitates a more efficient transfer of thermal energy from the PCMs to the water. Overall, the discharge phase involves the release of the stored thermal energy as the PCMs solidify. The water's continuous circulation and its ability to extract heat from the solidifying PCMs lead to a gradual increase in heat transfer. The rate of heat transfer is influenced by the water temperature and the temperature difference between the PCMs and the water, both of which impact the efficiency of energy transfer during this phase (Fig. 3).

3.3 Time Wise Variation of the PCM's Liquid Fraction, *f*

It is observed that the first PCM solidifies after 29 min, followed by the second PCM solidifying after 120 min. In contrast, the third PCM remains in the solidification phase for 38 min, and it takes approximately 120 min after the second PCM to solidify completely. This behavior can be attributed to the fact that the phase change process requires latent heat to initiate, and PCMs with higher melting points require more latent heat to start the phase change. The optimal value of the liquid fraction is slightly higher for PCMs with higher melting points. Specifically, the liquid fraction at the beginning of the discharge process is approximately 1 for the second PCM. As a result, the discharge of the third PCM starts after 120 min. The latent heat stored in the two PCMs (PCM_1 and PCM_3) during their solidification phase slows down the heat extraction process for the PCM_2. In summary, the discharge process of PCMs in the heat exchanger is influenced by their respective melting points and the requirement of latent heat to initiate the phase change. PCMs with higher melting points require more time to accumulate enough latent heat for solidification to start. This phenomenon is evident in the delay of the solidification of the second and third PCMs compared to the first PCM. The optimal value of the liquid fraction for each PCM affects the timing of their discharge, with PCMs having higher melting points exhibiting slightly higher liquid fractions. Ultimately, the storage of latent heat during the solidification phase of PCM_1 and PCM_3 impacts the rate of heat extraction and subsequent solidification of the third PCM (Fig. 4).

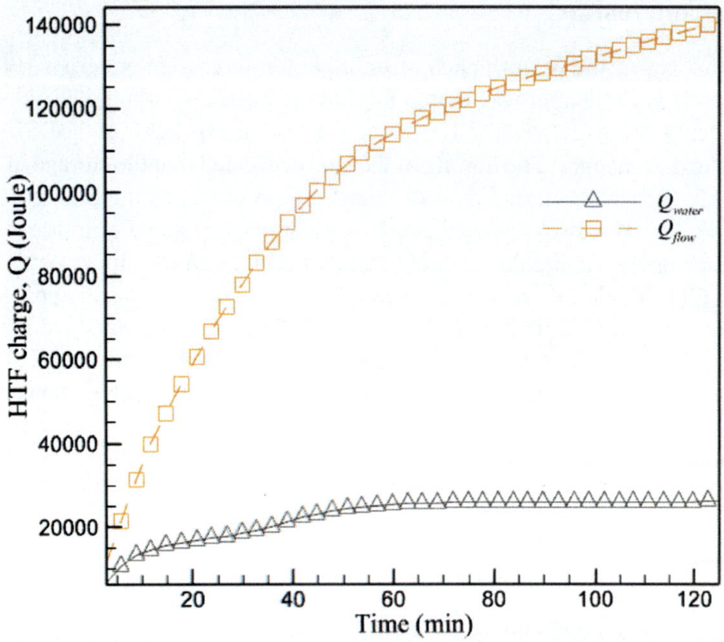

Fig. 3. Time wise variation of HTF rate charge, Q

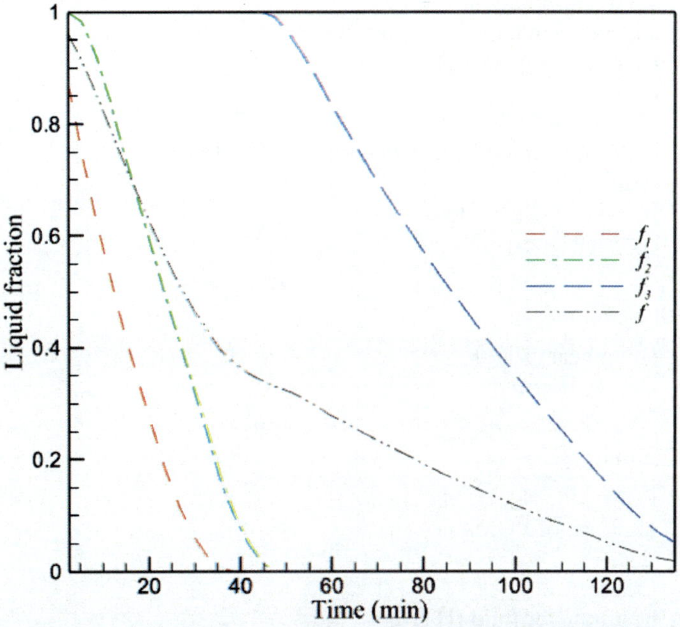

Fig. 4. Time wise variation of the multiple PCM liquid fraction

4 Conclusion

In this paper, the performance of multiple phase change materials (PCMs) in a PCM-water heat exchanger was thoroughly investigated and numerically simulated. The aim objective was to enhance and optimize the discharge process of the PCMs integrated with the heat exchanger. The finding of the study revealed that the storage of latent heat in the PCM_1 and PCM_3 during their discharge phase impacted the heat extraction process of the PCM_2. The discharge of the latter took about 120 min. For future research, the focus will be on improving the discharge time of each PCM to achieve better efficiency in the overall heat release process. This could involve exploring novel approaches to facilitate more uniform solidification of each PCM. The ultimate objective is to create a more robust and effective system that maximizes the utilization of multiple PCMs, leading to increased energy storage capacity and improved thermal energy management.

Acknowledgments. I would like to thanks the National Scientific Research and Technology Center (CNRST) for providing me a national Doctoral Research Scholarship.

Nomenclature

f	liquid fraction
k	thermal conductivity (w/m·°C)
L	length of the tube (m)
t	time (min)
R_a	Rayleigh number
x	axial coordinate, m
C_p	specific heat capacity of water (J/kg·°C)
m_f	mass flow rate (kg/s)
Q	heat (J)
T	temperature (°C)
U	local convection heat transfer coefficient (w/m^2·°C)
R	radius, m
r	radial coordinate, m
C	constant in Eq. (9)

Greek Symbols

ΔH	latent heat (J/Kg)
ρ	density (kg/m^3)

Subscripts

1,2,3	PCM_1, PCM_2, PCM_3
i	interface of PCMs
in	Inlet
o	Outlet
f	heat transfer fluid (HTF)
l, s	liquid phase and solid phase
out	outlet
p	phase change material

References

1. Wang, Z., Yang, Z., Fang, R., Yan, Y., Ran, J., Zhang, L.: A state-of-the-art review on action mechanism of photothermal catalytic reduction of CO2 in full solar spectrum. Chem. Eng. J. (2022). https://doi.org/10.1016/j.cej.2021.132322
2. Faraji, H., Benkaddour, A., Oudaoui, K., El Alami, M., Faraji, M.: Emerging applications of phase change materials: a concise review of recent advances. Heat Transfer. (2020). https://doi.org/10.1002/HTJ.21938
3. Li, M.-J., Li, M.-J., Xue, X.-D., Li, D.: Optimization and design criterion of the shell-and-tube thermal energy storage with cascaded PCMs under the constraint of outlet threshold temperature. Renewab. Energy (2021). https://doi.org/10.1016/j.renene.2021.09.086
4. Ashish Agarwal, R.M.S.: An experimental investigation of shell and tube latent heat storage for solar dryer using paraffin wax as heat storage material. Eng. Sci. Technol. Int. J. (2016). https://doi.org/10.1016/j.jestch.2015.09.014
5. Wang, L., et al.: Sandwich-zigzag structure enhanced erosion resistance of TiN coatings. Mater. Lett. (2022). https://doi.org/10.1016/j.matlet.2021.131496
6. Agyenim, F., Knight, I., Rhodes, M.: Design and experimental testing of the performance of an outdoor LiBr/H2O solar thermal absorption cooling system with a cold store. Sol. Energy (2010). https://doi.org/10.1016/j.solener.2010.01.013
7. Al-Abidi, A.A., Mat, S., Sopian, K., Sulaiman, M.Y., Mohammad, A.Th.: Numerical study of PCM solidification in a triplex tube heat exchanger with internal and external fins. Int. J. Heat Mass Transf. (2013). https://doi.org/10.1016/j.ijheatmasstransfer.2013.02.030
8. Voller, V., Prakash, C.: A fixed grid numerical modelling methodology for convection-diffusion mushy region phase-change problems. Int. J. Heat Mass Transf. **30**(8), 1709–1719 (1987). https://doi.org/10.1016/0017-9310(87)90317-6
9. Kenisarin, M., Makhamov, K.: Solar energy storage using phase change materials. Renew. Sustain. Energy Rev. (2007). https://doi.org/10.1016/j.rser.2006.05.005

Numerical Evaluation of the Petroleum Coke and Biomass Co-combustion in Rotary Cement Kilns

H. Fetian, Z. Ngadi(✉), and M. L. Lahlaouti

Faculty of Sciences Tetouan, University Abdelmalek Essaadi, 93002 Mhannech II, Morocco
zakia.ngadi@gmail.com

Abstract. The partial substitution of petroleum coke with alternative solid fuels, such as biomass, is advantageous for mitigating greenhouse gases and energy consumption in cement production. This study evaluates the effect of using two alternative biomass-type fuels, olive pomace powder and crushed almond shells, on the cement rotary kiln's combustion performance by comparing the two particles' co-combustion behavior to the case of combustion of petroleum coke alone using the Ansys Fluent Computational Fluid Dynamics (CFD) software. According to the results, the maximal temperatures achieved by the co-combustion of olive pomace and almond shells are lower than the temperature of coke combustion alone. However, on the other hand, it has noticed a difference in combustion behavior in terms of the char particle fractions.

Keywords: Co-combustion · CFD modeling · Biomass · Rotary kiln

1 Introduction

Co-combustion is the combustion of fossil fuels with alternative fuels from different resources. In fact, the co-combustion represents a practical solution for high-consumption energy, such as in the cement industry. In addition, co-combustion reduces the average global carbon emission of cement kilns [1, 2], so it is interesting to consider alternative, less expensive, and CO_2-neutral alternative fuels. However, if the co-combustion is not well controlled, that can generate undesirable pollution emissions and impact the combustion efficacy. Therefore, the co-combustion process causes an essential impact on cement material quality and a significant change in combustion behavior. We emphasize that the combustion behavior of fossil fuels is well-known, contrary to alternative fuel combustion. This difference is due to the alternative fuel composition, geometry, and source.

Alternative fuels used in cement succeed from several sources. Essentially, we can find refused-derived fuel, used tires, and biomass [3]. Low heating value (LHV), Carbon and volatiles are the main essential characteristics influencing combustion. The alternative fuels from biomass sources have significant LHV, rich in organic but contain more volatile than fossil fuels. Research has shown that replacing 20% of thermal energy in

S. Aniss et al. (Eds.): CMM 2022, LNME, pp. 160–169, 2024.
https://doi.org/10.1007/978-3-031-46973-2_15

the cement kiln with agricultural biomass can be achieved without significant investment [4]. Actually, many investigations have focused on biomass co-combustion in cement kilns.

Several experiments have highlighted the importance of co-firing biomass fuel with coal. Actually, understanding biomass co-combustion, including its thermal, chemical, and fluid mechanical complexities, remains restricted. Recent efforts over the last decade have been devoted to comprehending key features such as ignition processes, emissions analysis, furnace performance, and the influence of biomass co-combustion on combustion properties, species emission, and unburned Carbon in ash [5–7]. This research advances knowledge on biomass co-combustion processes and improves their use in energy production and environmental control.

Therefore, the present study evaluates the effect of using two alternative bio-mass-type fuels, olive pomace (OP) and almond shell (AS), on the kiln's combustion performance by comparing the two particles' co-combustion behavior case of coke's combustion alone. The substitution rate used in this study is about 20% according to the substitution ratio calculated based on Petroleum coke (PC) thermal units. In a series of published works, authors study the performance of turbulence models, including the k-ε standard, Realizable, and Re-Normalization Group, combined with hybrid finite rate/eddy dissipation and basic eddy dissipation models. Based on various parameters such as mesh accuracy, convergence time, temperature distribution, and concentrations of essential chemical elements, we identify that the most suitable simulation model is the combination of the standard k-ε model with the hybrid finite rate/eddy dissipation model, which provides improved co-combustion predictions and lowers computational requirements for 2-D modeling [8, 9]. In this paper, the focus is to evaluate the heat input and the results obtained, including temperature and species concentrations, to provide valuable conclusions allowing the optimization of the stable operation of the furnace in the case of biomass and petroleum coke co-combustion.

2 Characterization of Alternative and Fossil Combustibles

Vegetal-based fuels are carbon-neutral because the Carbon released during their combustion is extracted from the atmosphere by the plant during its growth phase [10]. Since the growth of this biomass and its use as fuel takes place over a short time, the whole cycle would have no net impact on carbon emissions into the atmosphere.

This study selected two types of biomass, olive pomace (OP) and almond shells (AS), and each has its particularities. These biomasses were chosen as alternative fuel primarily for their high energy value (the LHV), availability, composition (as depicted in Table 1), and suitability for pulverization. Furthermore, Table 1 summarizes the chemical and physical properties of the three fuels: olive Pomace (OP), almond shells (AS) as biomass fuels, and petroleum coke as fossil fuel reference.

Table 1. Alternative and fossil fuels properties as received basis

Fuel properties	PC	OP	AS
	Ultimate Analysis (wt. %)		
Carbon	88.6	51.6	49.68
Hydrogen	3.74	6	8.18
Oxygen	1.4	33.7	37.38
Sulfur	3.98	0.35	1.04
Nitrogen	1.62	1.89	3.75
	Proximate Analysis (wt. %)		
Volatile	10.6	57	66.93
Fixed Carbone	87.02	16.4	20.01
Ashes	0.58	4.7	3 359
Moisture	1.58	21.9	9.7
	Physical properties		
Low Heat value LHV (kJ/kg)	34805	15349	29860
Dry particle density DPD (kg/m3)	900	657	832.04
Mean particle diameter MPD (µm)	15	200	20

3 CFD Model for Co-combustion in Rotary Cement Kiln

Co-combustion begins with injecting combustible particles from the kiln burner into the furnace, convoyed by pressurized air. The flow regime within the rotary kiln is characterized as gas-solid turbulent flow involving chemical combustion reactions. The Eulerian-Lagrangian approach is frequently used and recommended when simulating dilute gas-solid fluxes in pulverized fuel (PF) combustion applications. The following equations govern the Eulerian approach of continuous motion (the gas phase).

$$\nabla \cdot (\rho \vec{u}) = S_m \tag{1}$$

The source S_m in Eq. (1) is the mass added to the continuous phase due to the combustion of particles. The velocity component transport equation is

$$\nabla(\rho \vec{u}\vec{u}) = -\nabla P + \nabla \cdot (\overline{\overline{\tau}}) + \rho \vec{g} + \vec{F} \tag{2}$$

Here, P is the static pressure and $\overline{\overline{\tau}}$ is the stress tensor. The third term on the right-hand side (RHS) indicates the gravitational body force, while the last term is the drag force source term.

$$\nabla \cdot (\rho \vec{u}_i Y_i) = -\nabla \cdot \vec{J}_i + R_i + S_i \tag{3}$$

Equation (3) is the conservation equation for the mass fraction of the ith species, where R_i is the chemical reaction production rate of species i (to be computed using the (FR/ED) turbulence-chemistry interaction models as explained in Sect. 3.2) and S_i denotes the source term contributed by the dispersed phase more details are given in [10]. In addition, $\overrightarrow{J_i}$ represents the turbulent diffusion flow of species i [11].

3.1 Standard k-ε Model

The two-equation models are often used to predict the characteristics of turbulent flow. The standard k-ε two equations model has proven to be a robust approach to predicting flow velocities in turbulence mediums over the years [12–14]. In this model, turbulent viscosity is modeled in terms of turbulent kinetic energy k and its dissipation rate ε. Thus the transport equations k and ε are presented as follows [15, 16].

$$\rho(\overline{u}.\nabla)k = \nabla\left[\left(\mu + \frac{\mu_t}{\sigma_k}\right)\nabla k\right] + P_k - \rho\varepsilon \tag{4}$$

$$\rho(\overline{u}.\nabla)\varepsilon = \nabla\left[\left(\mu + \frac{\mu_t}{\sigma_\varepsilon}\right)\nabla\varepsilon\right] + C_{\varepsilon 1}\frac{\varepsilon}{k}P_K - C_{\varepsilon 2}\rho\frac{\varepsilon^2}{k} \tag{5}$$

In Eqs. (4 and 6), the μ_t is the turbulent viscosity, P_k is the rate of turbulent kinetic energy production and constant $C_{\varepsilon 1} = 1.44$, $C_{\varepsilon 2} = 1.92$, $C_\mu = 0.09$, $\sigma_k = 1.0$ and, $\sigma_\varepsilon = 1.3$ as available in the literature for similar cases [17].

3.2 Eddy Dissipation Model FR/ED

Pulverized fuel combustion occurs in a highly turbulent environment when the fluctuations affect the source term R_i in Eq. (3). To this aim, the interaction between turbulent flow and reaction chemistry must be accurately modeled. For this study, the FR/ED model is used. The FR/ED model establishes a relationship between the reaction rate and the dissipation rate of the vortex structures present in the products and reactants [11, 18]. So, in this model, three reaction rates are calculated, and we retain the smallest one. Arrhenius's equation leads to calculating the first reaction rate using the following:

$$R'_{i,r} = A_r T \exp\left(-E/RT_g\right) \tag{6}$$

In Eq. (6), Ar and E are the reaction rate parameters, and R and T are the ideal gas constant and gas temperature, respectively. The next equation calculates the second and third reaction rates representing the dissipation rate of the vortex structures of reactants and products.

$$R_{i,r} = \nu_{i,r'}M_{w,i}A\rho\frac{\varepsilon}{k}\min\left(\min_R \frac{Y_R}{\nu_{R,r'}M_{w,R}}, B\frac{\sum_P Y_P}{\sum_j^N \nu''_{j,r}M_{w,j}}\right) \tag{7}$$

R and P represent the reactants and the index of the products, respectively, in Eq. (7). M is the molecular mass, ν is the stoichiometric coefficient, and Y is the mass fraction. A and B are the mixing rate. In the case of co-combustion, A, and B take a value of 0.4 and 5, respectively.

4 Operatory Entry Conditions

Table 2 represents the entry conditions for the case of co-combustion and combustion. The geometry of the rotary kiln used in this study is assimilated to a real kiln 46 m long and 3.8 m in diameter. Due to the intricacy of the co-combustion phenomena, we apply geometric simplifications to reduce computation time and expense; specifically, only 40m of the kiln is considered. The authors' previous work shows more details on the geometry, calculation strategy, and boundary condition [8].

The substitution rates and the physical characteristics and entry conditions for olive pomace, almond shells, and petroleum coke are similar to the real case data.

Table 2. Operating condition data

	Co-combustion			Combustion
	OP	AS	PC	PC
Flow rate (Kg/s)	0.5	0.5	0.7	1.2
T (K)	375			
Dmin (m)	1.10^{-4}	5.10^{-6}	5.10^{-6}	
Dmax	9.10^{-4}	3.10^{-4}	1.10^{-4}	
Dmoy	2.10^{-4}	$1{,}5.10^{-5}$	$1{,}5.10^{-5}$	
ρ (Kg/m^3)	675	832	900	
Cp_s	Polynomial function of the temperature [19]			
Cg	1.4	1.2	1.2	

5 Results and Discussion

The substitution rate used in this study is approximately 20%, based on coke thermal units, using two different types of biomass co-combusted with pulverized petroleum coke in a cement production kiln. The aim is to assess the thermal input and the results, including temperature and species concentrations, to evaluate the energy contribution of the use of biomass. Figure 1 presents the temperature prediction for the three cases.

The highest temperature obtained in the case of substitution by the almond shell is 200 K higher than the substitution by olive pomace case and is lower than in the combustion of PC alone case. Indeed, the industrial test revealed increased combustion air needs when biomass was substituted. The high requirement air flow, as well as the high humidity rate of olive pomace 21.9% compared to that of almond shells 9.7%, increase the exhaust airflow, resulting in a reduction of the temperature of the products of combustion of the olive pomace compared to the almond shells as a substitute fuel case. On the other hand, the high ash and moisture content in olive pomace also contributed to the temperature reduction. The ash and evaporated water must reach the equilibrium temperature. However, note that the profile in a real cement kiln is quite different due to the interaction of solid raw materials with airflow and non-uniform heat loss through the kiln wall. Therefore, it is expected that a decrease in temperature after the maximum point is reached.

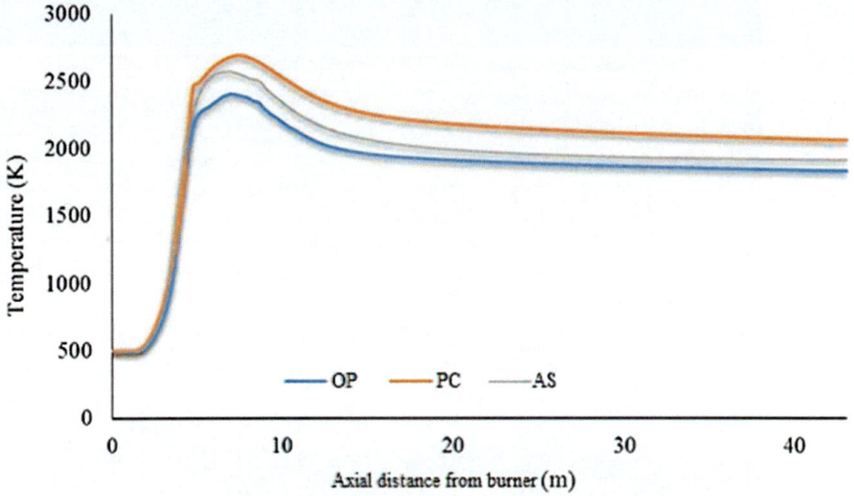

Fig. 1. Prediction of axial temperature profile for the three fuels.

Figure 2 shows the mass fraction distribution of PC, OP, and AS volatile substances from top to bottom. Having a high volatile matter content compared to PC (see Table 2), olive pomace and almond shells show intense devolatilization near the burner in the case of almond shells and extend further in the case of olive pomace compared to petroleum coke particles. The difference in the devolatilization kinetics of the fuels can also explain the different concentration profiles of the volatile substances obtained.

In fact, according to Table 2, the size of olive pomace particles is quite large compared to coke and almond shells (the largest olive pomace particle is nine times larger than the largest coke particle, and the average diameter is 13 times greater than the average diameter of coke) and these larger particles gradually pass out of the combustion zone due to the effect of gravity. The smaller particles spread through the entire furnace cross-section and devolatilize in symmetry as they are much less affected by gravity due to their low mass. This effect is similar for coke and almond shells, but a much more significant impact can be observed for olive pomace, as the diameter of these particles extends over a wide range. Consequently, the devolatilization of the larger particles continues in the inner part of the furnace, far from the burner, producing more volatile substances.

On the other hand, the almond shells tend to devolatilize quickly as soon as they leave the burner (Fig. 2). The composition of almond shells in the volatile matter is 66.93% higher than that of coke, and their small diameter compared to olive pomace Table 2 explains their devolatilization behavior. In addition, their low moisture content and small diameters facilitate that AS particles reach the devolatilization temperature faster than the olive pomace particles.

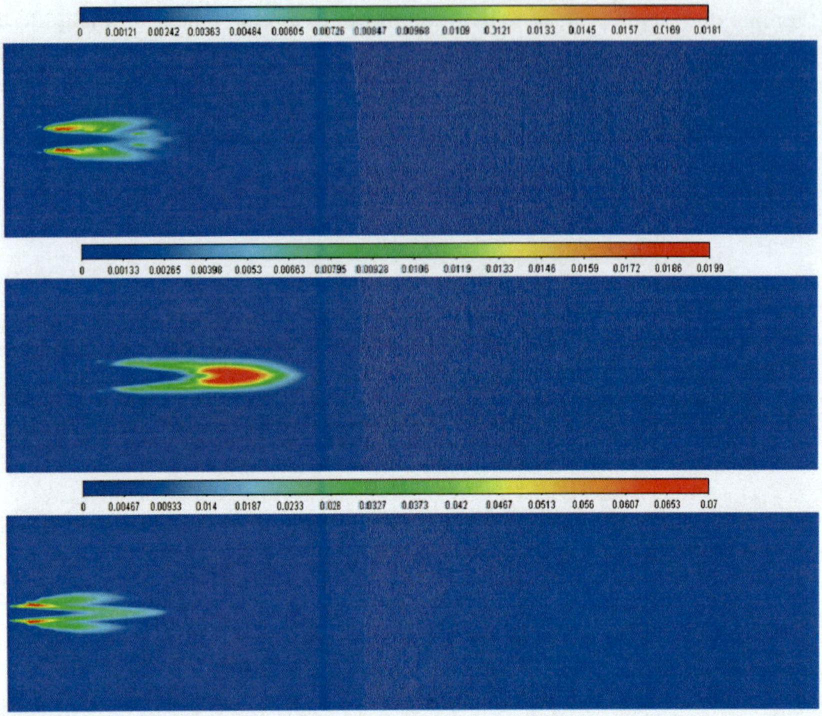

Fig. 2. Distribution of volatile substances mass fraction from top to bottom: PC, OP, and AS

Char combustion depends primarily on particle temperature, initial fuel particle diameter, and oxygen availability. The fixed carbon content of the fuel is also an important parameter. Figure 3(a, b, and c) shows the char mass fraction of five random particles from the olive pomace, almond shells, and petroleum coke. The randomness of the five particles is only due to a different particle trajectory history, i.e., the effect of stochastic particle tracking.

In all three cases, the char mass fraction typically increases when the volatiles are released, then begins to decrease, and eventually reaches zero when all char is entirely burnt. Moisture content and small diameters facilitate that AS particles get the devolatilization temperature faster than the olive pomace particles.

However, in the case of coke and almond shells, the mass fraction of the char remains constant over a significant time during combustion (Figs. 3b and c), showing that the combustion process of the coke and almond shells is relatively sluggish. A similar observation was made for olive pomace particles. Furthermore, the char particles' trajectories for coke and almond shells terminate at 12 m and 14 m, respectively, as opposed to the char particles' trajectories for olive pomace, which finishes at 9 m; this is because coke and almond shells have a higher fixed carbon content than olive pomace.

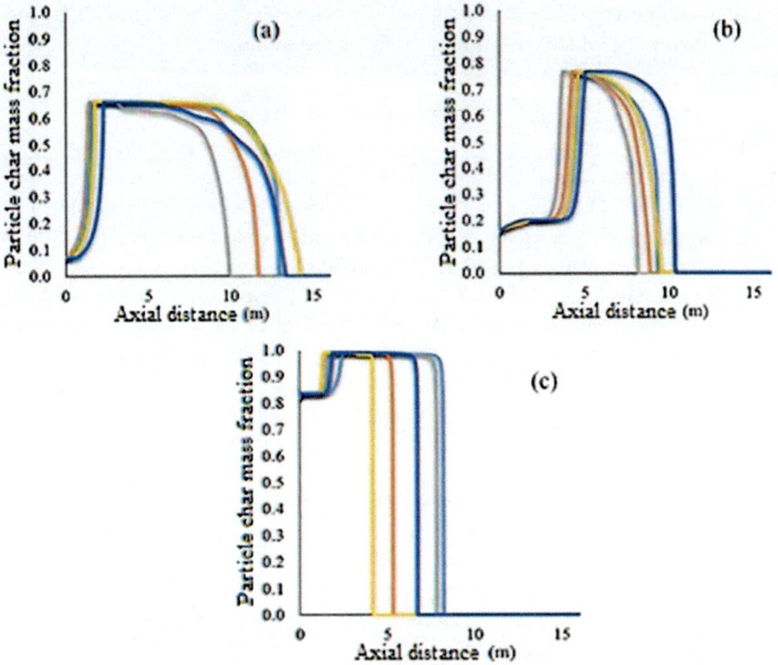

Fig. 3. Particle char mass fraction variation of random particle streams of (a) OP case, (b) AS case, and (c) PC case.

6 Conclusion

We recall that this study shows the impact of using two alternative fuels, olive pomace and almond shells, in co-combustion with petroleum coke on the combustion performance in cement kilns. We use the standard k-ε model coupled with the FR/ED model. We compared the co-combustion and combustion of Petrolium coke alone results. The temperature of the co-combustion products is lower than the combustion temperature of the coke-alone case. The lower temperature is mainly due to the excess air demand, ash, and higher moisture content of olive pomace and almond shells compared to coke. Biomass particles are generally larger and differ in their physical and chemical composition from fossil fuel particles, which explains the difference in volatile production and char combustion behavior.

We adopted the real case data for the kiln geometry and entry conditions to do explicit calculations. As mentioned above, the adopted CFD model is validated with real case data. The low temperature in co-combustion scenarios can be explained by biomass requiring more preparation before being placed into the kiln. As a result, we emphasize that biomass co-combustion is still favorable. A detailed setup, on the other hand, must correspond to a good balance between the biomass mass fractions and the burner's performance.

Acknowledgments. The authors gratefully acknowledge the helpful comments and suggestions of the reviewers, which have improved the presentation.

References

1. Huntzinger, D.N., Eatmon, T.D.: A life-cycle assessment of Portland cement manufacturing: comparing the traditional process with alternative technologies. J. Clean. Prod. **17**, 668–675 (2009). https://doi.org/10.1016/J.JCLEPRO.2008.04.007
2. Aranda Usón, A., López-Sabirón, A.M., Ferreira, G., Llera Sastresa, E.: Uses of alternative fuels and raw materials in the cement industry as sustainable waste management options. Renew. Sustain. Energy Rev. **23**, 242–260 (2013). https://doi.org/10.1016/J.RSER.2013.02.024
3. M, P.M.: Alternative Fuels in Cement Manufacturing. In: Alternative Fuel. InTech (2011)
4. Demirbaş, A.: Sustainable cofiring of biomass with coal. Energy Convers. Manage. **44**, 1465–1479 (2003)
5. Li, J., et al.: Study on the feasibility and co-combustion mechanism of mixed injection of biomass hydrochar and anthracite in blast furnace. Fuel **304**, 121465 (2021). https://doi.org/10.1016/j.fuel.2021.121465
6. Werther, J.: Potentials of biomass co-combustion in coal-fired boilers. In: Proceedings of the 20th International Conference on Fluidized Bed Combustion, pp. 27–42 (2009). https://doi.org/10.1007/978-3-642-02682-9_3
7. Yang, X., et al.: Experimental and numerical investigation of the combustion characteristics and NO emission behaviour during the co-combustion of biomass and coal. Fuel (2021). https://doi.org/10.1016/j.fuel.2020.119383
8. Ngadi, Z., Lahlaouti, M.L.: Coal and biomass co-combustion: CFD prediction of velocity field for multi-channel burner in cement rotary kiln. Proceedings **63**, 18 (2020). https://doi.org/10.3390/proceedings2020063018
9. Ngadi, Z., Lahlaouti, M.L.: CFD modeling of petcoke co-combustion in a real cement kiln: the effect of the turbulence-chemistry interaction model applied with K-ϵ variations. Int. Rev. Appl. Sci. Eng. (2021). https://doi.org/10.1556/1848.2021.00326
10. Intergovernmental Panel on Climate Change. IPCC Guidelines for National Greenhouse Gas Inventories Third Authors/Experts Meeting: Industrial Processes and Product Use (2006)
11. Spalding, D.B.: Mixing and chemical reaction in steady confined turbulent flames. Symp. Int. Combust. **13**, 649–657 (1971). https://doi.org/10.1016/S0082-0784(71)80067-X
12. Guessab, A., Aris, A., Bounif, A.: Simulation of turbulent piloted methane non-premixed flame based on combination of finite-rate/eddy-dissipation model. Mechanika **19**, 657–664 (2013). https://doi.org/10.5755/j01.mech.19.6.6000
13. Tabet, F., Fichet, V., Plion, P.: A comprehensive CFD based model for domestic biomass heating systems. J. Energy Inst. **89**, 199–214 (2016). https://doi.org/10.1016/J.JOEI.2015.02.003
14. Chaney, J., Liu, H., Li, J.: An overview of CFD modelling of small-scale fixed-bed biomass pellet boilers with preliminary results from a simplified approach. Energy Convers. Manage. **63**, 149–156 (2012). https://doi.org/10.1016/J.ENCONMAN.2012.01.036
15. Launder, B.E., Spalding, D.B.: The numerical computation of turbulent flows. Comput. Methods Appl. Mech. Eng. **3**, 269–289 (1974). https://doi.org/10.1016/0045-7825(74)90029-2
16. Jones, W., Launder, B.: The prediction of laminarization with a two-equation model of turbulence. Int. J. Heat Mass Transf. **15**, 301–314 (1972). https://doi.org/10.1016/0017-9310(72)90076-2

17. Launder, B.E., Spalding, D.B.: Lectures in Mathematical Models of Turbulence. Academic Press, London, New York (1972)
18. Magnussen, B.F.: The Eddy dissipation concept a bridge between science and technology (2005)
19. Peters, A.A.F., Weber, R.: Mathematical modeling of a 2.25 MWt swirling natural gas flame. Part 1: Eddy break-up concept for turbulent combustion; probability density function approach for nitric oxide formation. Combust. Sci. Technol. **110–111**, 67–101 (1995). https://doi.org/10.1080/00102209508951917

Impact of the Improvement of the Performance of Thermoelectric Modules on Its Efficiency

S. Jennah[1(✉)], N. Belouaggadia[2], R. Lbibb[3], and M. Ezzine[4]

[1] Laboratory of Engineering and Materials (LIMAT), Faculty of Sciences Ben M'Sick, Hassan II University of Casablanca, B.P 7955, Sidi Othmane, Casablanca, Morocco
jennahsara@gmail.com

[2] Laboratory of Signals, Distributed Systems and Artificial Intelligence, ENSET, Hassan II University, Mohammedia, Morocco
n.belouaggadia@gmail.com

[3] Laboratory of Nanostructure and Advanced Materials, Mechanics and Thermofluid Faculty of Sciences and Technics, Hassan II University, Mohammedia, Morocco

[4] Laboratory of Physico-chemistry of Applied Materials (LPCMA), Faculty of Sciences Ben M'Sick, Hassan II University of Casablanca, B.P 7955, Sidi Othmane, Casablanca, Morocco

Abstract. Improvement of the efficiency of thermoelectric modules requires the improvement of their performance. Thus, an optimisation method will be studied in this paper, as well as the analysis of the management of the operating conditions. By reversing the flow of hot and cold heat transfer fluids, we can reduce the temperature rise at the cold junctions and improve the performance. Water is the heat transfer fluid we have chosen for this study. The hot heat transfer fluid will flow in a channel to heat the hot junctions, and the cold heat transfer fluid will flow in a channel to cool the cold junctions. In the first case, the flows of the two heat transfer fluids are in the same direction. Subsequently, a second study will be carried out by opposing the flow direction of the heat transfer fluids. The results obtained show that by opposing the flow direction of the cold heat transfer fluid, we obtain a greater generated power than that obtained in the case of a flow in the same direction, which will allow us to have a better efficiency.

Keywords: Thermoelectric · Heat transfer fluid · Hot welds · Cold welds · Counter-current

Nomenclature

L	Tube width, $(\boldsymbol{cm})$
L_2	Tube height, $(\boldsymbol{cm})$
A	Thermopile surface, $(\boldsymbol{mm^2})$
$L_t h$	Thermopile thickness, $(\boldsymbol{mm})$

S. Aniss et al. (Eds.): CMM 2022, LNME, pp. 170–180, 2024.
https://doi.org/10.1007/978-3-031-46973-2_16

ρ_N	N-electrical junction's resistivity, $(\boldsymbol{\Omega m})$
ρ_P	P-electrical junction's resistivity, $(\boldsymbol{\Omega m})$
λ_N	Thermal conductivities of the N junctions, $(\boldsymbol{W/m.K})$
λ_P	Thermal conductivities of the P junctions, $(\boldsymbol{W/m.K})$
S_N	Seebeck coefficients for N, $(\boldsymbol{\mu V/K})$
S_P	Seebeck coefficients for P, $(\boldsymbol{\mu V/K})$
n	Number of thermocouples
R	Electrical resistivity, $(\boldsymbol{\Omega})$
η	Efficiency
P_u	Useful power, $(\boldsymbol{W})$
Q_h	Thermocouple's hot end's heat flux, $(\boldsymbol{W})$
Q_c	Thermocouple's hot end's heat flux, $(\boldsymbol{W})$
I	Electric current, $(\boldsymbol{A})$
T_h	Hot weld temperatures, $(\boldsymbol{K})$
T_c	Cold weld temperatures, $(\boldsymbol{K})$
c_f	Capacity of the heat transfer fluid, $(\boldsymbol{J/kg.K})$
λ_f	Thermal conductivity of the heat transfer fluid, $(\boldsymbol{W/m.K})$
μ_f	Dynamic viscosity of the heat transfer fluid, $(\boldsymbol{kg/m.s})$
v	Velocity of the heat transfer fluid
Re_f	Reynolds number
Pr_f	Prandtl number
Nu_f	Nusselt number

1 Introduction

Food and heating are vital needs of the human being. For a very long time, humanity has sought to diversify its thermal energy production capacities, and then to diversify the forms of energy. Today, human beings are mainly looking for ways to manage energy and its uses in a sustainable development perspective. It is within this framework that thermoelectricity is currently taking off, absolutely admitting to optimizing energy efficiency. Today, the demand for power generation in extreme environments by thermo-electric generators is relatively low, with about a hundred generators for space and about ten thousand for terrestrial applications. Initially, the main source of thermal energy was biomass with fire. This has evolved throughout history with, for example, the use of fossil fuels. The sources then diversified with the discoveries and technical progress to be multiple today. This multitude of sources allows thermal energy to be abundant and relatively well distributed on the Earth. Nevertheless, the production of heat is not always desired. A large part of it is unemployed and rejected directly into the atmosphere. This waste energy can be recovered to enhance the energy efficiency of systems. For example, the heat released by a process can be recovered and directed to areas that need to be heated. Direct heat recovery for heating is not always appropriate, mainly because the amount of energy and the temperature level are not sufficient. Another solution is to convert this waste heat into electricity using thermal/electrical conversion devices. Among other conversion systems, thermoelectric generators (TEGs) are beginning to be developed on a large scale. Their low efficiency does not allow them to be competitive for large thermal powers. However, they have significant advantages at low power levels, such as excellent robustness and very good integration into compact systems. The properties of thermoelectric materials [1] are also strongly correlated with the heat exchanger performance. Zhao et al. [2] optimised an exhaust thermoelectric generator by inserting a perforated plate, increasing the output power by 73.4%. The voltage homogeneity coefficient reached a maximum of 0.967 with the plate in the middle of the third row of modules. Increasing the exhaust temperature improved the net power output by up to 79.5% at $T_f = 550\,\mathrm{K}$. The optimum mass flow rate of 45 g/s generated maximum net power output, with an increase of up to 90.5% at m = 30 g/s. These results guide the optimisation of thermoelectric generators for improved energy efficiency. In their experimental study, Xing et al. [3] investigated the impact of finned vortex generators on gas-liquid thermoelectric generators. The heat exchanger with non-uniform finned vortex generators increased the power output by 189.1% on average compared to a smooth exchanger, with a 177.4% improvement in net power output. A matched load resistor achieved maximum power, and an increase in hot air temperature of 30 K increased energy output by 28% and net output by 30%. This cascade approach improves the efficiency of thermoelectric generators. The authors, Gomaa et al. [4], studied waste heat recovery in a cement plant using thermoelectric generators (TEGs). Active water cooling was found to be the most efficient, improving TEG output by 4.4% compared to forced air cooling. The shell of the rotary kiln was identified as a potential source of electrical energy, with a waste heat of 10 to 12.5 $\mathrm{kW/m^2}$ at a temperature of 375 °C or more. The thermoelectric generator used has a high power of 17.6 W and a conversion efficiency of 5.85%. Future research will focus on optimising the TEG with different cooling technology parameters. In the case of large temperature variations, the use of different types of thermoelectric materials is relevant. On the other hand, for higher temperatures at 300 °C, the material Bi_2Te_3 begins to degrade. Espinosa et al. [5] examined a straightforward thermoelectric architecture of Bismuth Tellurium materials along a fixed heat exchanger and Mg_2Si/Zn_4Sb_3 (for high temperature). Therefore, optimizing the contact thermal resistance [6–8] or the impact of heat exchangers [9,10] is frequently necessary.

In this study, we drew inspiration from previous research which has mainly focused on adding features to improve system performance. Instead, we have chosen a different approach by exploring the operating parameters of our system. Our work follows on from our previous paper [11], and our aim is to improve the performance of the system by using a thermopile heated at the hot junction by a hot fluid and cooled at the cold junction by a low-temperature fluid. We have therefore chosen to play with the flows of the heat transfer fluids, by making them flow in the same direction and then exploring the potential advantages of opposing the direction of flow of the heat transfer fluids. The aim of this innovative approach is to increase the overall efficiency of the system and achieve better energy performance.

2 Configurations: Co-current and Counter-Current

Thermoelectric generators, are a means of direct conversion between thermal and electrical energy by using a heat flow through a temperature gradient to feed an electrical load through the external circuit, by associating thermoelectric modules with adapted heat exchangers, which to produce electricity from a hot source and a cold source. In our case, the heating and cooling of the junctions will be done by a heat transfer fluid (water), for heating we will make it flow at a temperature of 353 K and for cooling at a temperature of 300 K. The interface between the thermal cell and the heat transfer fluid is considered a thin layer with little thermal resistance and electrical insulation. In the first scenario, shown in Fig. 1(a), we'll assume that the two heat transfer fluids' flow directions are identical; this is known as the co-current scenario. And in the second case Fig. 1(b), we will consider that the flow of the two heat transfer fluids is of opposite direction and it is the counter-current case.

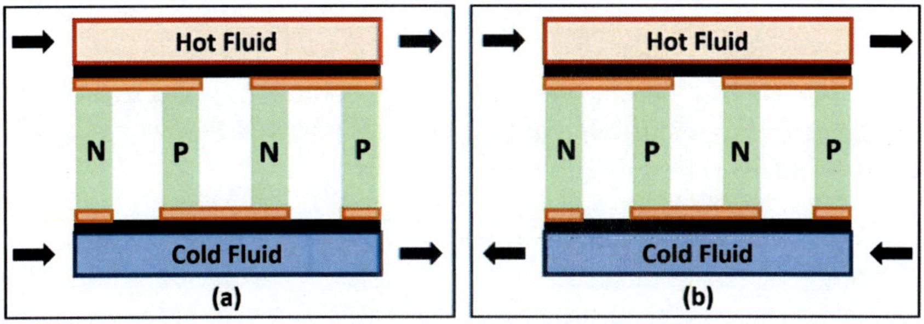

Fig. 1. (a) Case where the fluid flows are in the same direction; (b) Case where the fluid flows are opposite.

3 Mathematical Model

The thermoelectric generator (TEG) heat-to-electricity efficiency is calculated by [11]:

$$\eta = \frac{Pu}{Q_h} \tag{1}$$

where P_u is the useful power and Q_h is the hot end flux.

The thermocouple's hot end's heat flux is given by [11]:

$$Q_h = n[I(S_P - S_N)T_h + \lambda(T_h - T_c) - \frac{RI^2}{2}] \tag{2}$$

The thermocouple's cold end's heat flux is given by [11]:

$$Q_c = n[I(S_P - S_N)T_c + \lambda(T_h - T_c) + \frac{RI^2}{2}] \quad (3)$$

The hot and cold weld temperatures are thus T_h and T_c, the thermal conductivity is λ, the electric current is I, the electrical resistivity of the thermoelectric generator is R and the number of thermocouples is n. The thermoelectric material's Seebeck coefficients for P and N are S_P and S_N.

The useful power produced is given by [11]:

$$Pu = Q_h - Q_c \quad (4)$$

$$Pu = n[I(S_P - S_N)(T_h - T_c) - RI^2] \quad (5)$$

When a thermoelectric generator supplies a load, the following current is delivered [11]:

$$I = \frac{n(S_P - S_N)(T_h - T_c)}{R_E + nR} \quad (6)$$

The current becomes equal to the generator internal resistance $R_E = nR$ when the maximum value of the electrical power is reached, and the load resistance equals [11]:

$$I = \frac{(S_P - S_N)(T_h - T_c)}{2R} \quad (7)$$

With:

$$P_{max} = \frac{n(S_P - S_N)^2(T_h - T_c)^2}{4R} \quad (8)$$

The thermal conductivity λ is given by the following relation [11]:

$$\lambda = \frac{A(\lambda_P + \lambda_N)}{2L_{th}} \quad (9)$$

λ_P and λ_N are the thermal conductivities of the P and N junctions, and A and L_{th} are the thermopile's surface and thickness. According to the geometry and electrical characteristics, the electrical resistivity R may be characterized [11]:

$$R = \frac{2L_{th}(\rho_P + \rho_N)}{A} \quad (10)$$

where ρ_P is the P-electrical junction's resistivity and ρ_N is the N-electrical junction's resistivity. Convection is used to transmit heat from the fluid to the thermopile, and the convection coefficient h_f is related to the Nusselt number Nu_f, which is given by [11]:

$$h_f = \frac{Nu_f \lambda_f (L + L_2)}{2LL_2} \quad (11)$$

The rectangular fluid tube's width and height are L and L_2, and the heat transfer fluid's thermal conductivity is λ_f. There is a given correlation between the Prandtl Pr_f, Reynolds Re_f, and Nusselt Nu_f numbers [12]:

$$Nu_f = 4.4 + \frac{0.00398(\frac{2\rho v L L_2}{\mu_f(L+L_2)} * \frac{c_f \mu_f}{\lambda_f}(\frac{L_2}{L}))^{1.66}}{1 + 0.0114(\frac{2\rho v L L_2}{\mu_f(L+L_2)} * \frac{c_f \mu_f}{\lambda_f}(\frac{L_2}{L}))^{1.12}} \quad (12)$$

Such that, c_f, μ_f, ρ and v are the capacity, dynamic viscosity, density and velocity of the heat transfer fluid respectively (Table 1).

Table 1. Properties of the materials used in the TEG and the heat transfer fluid [13–18].

Parameters	values	units
L	1	cm
L_2	1	cm
c_f	4185	$J/kg.K$
λ_f	0.636	$W/m.K$
μ_f	0.0005	$kg/m.s$
A	2	mm^2
L_{th}	5	mm
ρ_N	E−5	Ωm
ρ_P	0.8E−5	Ωm
λ_N	1.4	$W/m.K$
λ_P	1.4	$W/m.K$
S_N	−200	$\mu V/K$
S_P	180	$\mu V/K$

4 Results and Discussion

First, we assume that the flow directions of the hot and cold fluids are the same. Then, as an improvement, we have the hot and cold fluids flow in opposite directions in the second stage. We will first investigate the impact of the thermocouples number and the flow rate of the heat transfer fluids. We will set the cold and hot heat transfer fluids flow rates of the cold and hot heat transfer fluids to a flow rate of $\dot{m} = 0.2$ kg/s in order to select the number of thermocouples that make up the thermopile (Fig. 2).

The Fig. 2 shows the useful power as a function of the number of thermocouples. The curve of the evolution of the useful power shows a maximum at

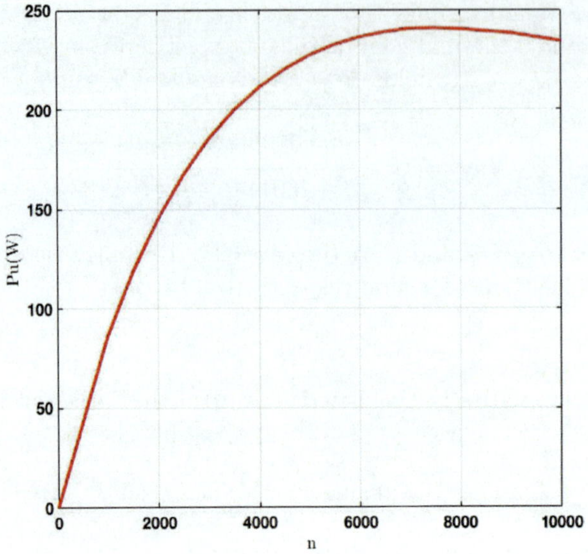

Fig. 2. Variation of the useful power according to the number of thermocouples.

$n = 7500$. For $n = 5000$ the thermopile develops a power of 227 W, while the power gain between $n = 5000$ and $n = 7500$ is only 13.5 W. For a good ratio cost on the generated power we choose a thermopile made up of 5000 thermocouples which corresponds a surface equal to $1\,m^2$.

We computed the useable power for various flow rates to select the heat transfer fluid flow rates that would be most effective. We will set the cold heat carrier flow rate $\dot{m}_c = 0.2\,kg/s$ in order to select the ideal flow rate for the hot fluid and the flow rate of the hot heat carrier to $\dot{m}_h = 0.2\,kg/s$ in order to select the ideal flow rate for the cold fluid. The power outputs as a function of the flow rates of the hot and cold heat transfer fluids are shown in the figures below.

In the Fig. 3 we see that as the flow rates of the hot heat transfer fluid increase, the slope of variation of the useable power drops. We search for the ideal flow rate to lower these costs because the flow of the heat transfer fluid is related to energy costs. So for this, the ideal flow rate of the hot heat transfer fluid is $\dot{m}_h = 0.1\,kg/s$. For the Fig. 4 the useful power decreases in significantly for high flow rates of the cold heat transfer fluid. Therefore, to reduce the energy expenditure as much as possible, a flow rate of $\dot{m}_c = 0.2\,kg/s$ is chosen for the cold heat transfer fluid.

Figure 5 demonstrates how the hot and cold thermocouple welds and the heat transfer fluids temperatures change in the co-current case with hot heat transfer fluid flow rates $\dot{m}_h = 0.1\,kg/s$ and $\dot{m}_h = 0.2\,kg/s$ from an inlet temperature of $T_h = 353\,K$ and cold heat transfer fluid flow rates $\dot{m}_c = 0.1\,kg/s$ and $\dot{m}_c = 0.2\,kg/s$ from a temperature of $T_c = 300\,K$. The improvement in the temperature

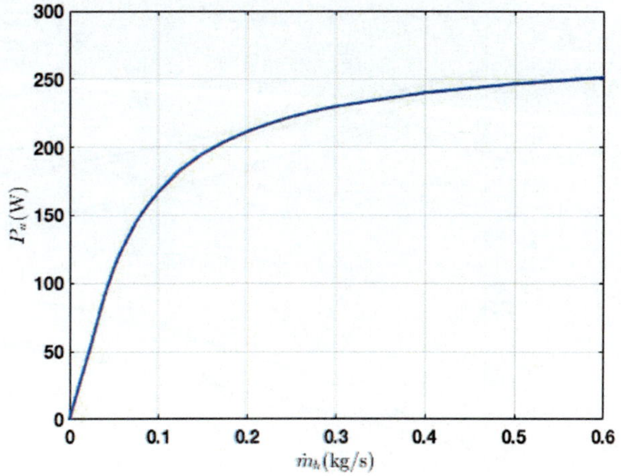

Fig. 3. Variation of the useful power according to the flow rate of the hot heat transfer fluid.

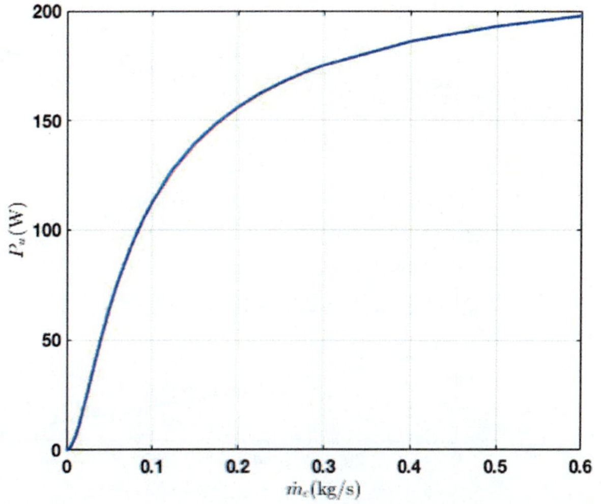

Fig. 4. Variation of the useful power according to the flow of the cold heat transfer fluid.

gap between the hot and cold solder is demonstrated by an increase in the useful power, which is 156 W, and the thermopile efficiency, which is equal to $\eta = 1.4\%$, as shown by the evolution of the heat transfer fluid and hot and cold solder temperatures relative to $\dot{m}_h = 0.1\,\text{kg/s}$ and $\dot{m}_c = 0.2\,\text{kg/s}$.

Figure 6 represents the evolution of the hot and cold thermocouple welds temperatures and the hot and cold heat transfer fluids for the counter-current case at a hot heat transfer fluid flow rate of $\dot{m}_h = 0.1\,\text{kg/s}$ and a cold heat transfer fluid flow rate of $\dot{m}_c = 0.2\,\text{kg/s}$, respectively, with an inlet temperature

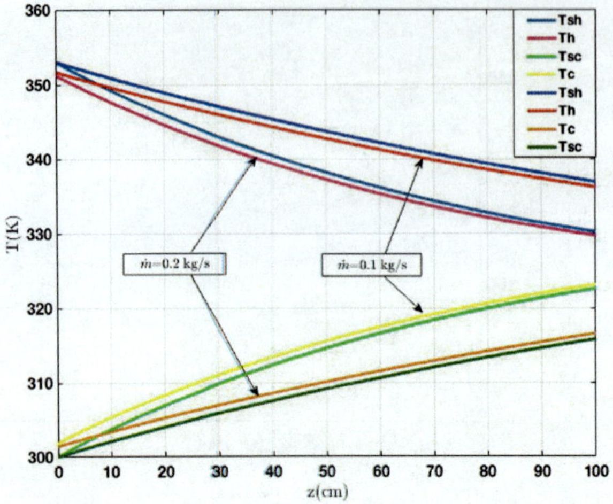

Fig. 5. Temperature profile of junctions and heat transfer fluids for the co-current case as a function of z.

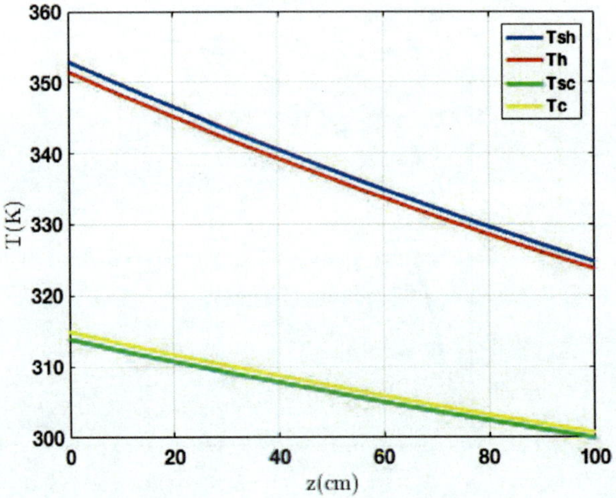

Fig. 6. Temperature profile of junctions and heat transfer fluids for the countercurrent case as a function of z.

of $T_h = 353\,\mathrm{K}$ and $T_c = 300\,\mathrm{K}$. The thermopile's performance is then discovered to have improved, with a power increase of 11% ($P_{max} = 156$) W and an efficiency of 1.46%. Compared to the co-current flow, the hot heat transfer fluid's exit temperature marginally dropped (325) K. Therefore, we could enhance the performance of our TEG using the countercurrent flow.

5 Conclusion

In demanding high-value installations, it is essential to have a solution that is both efficient and reliable. In this regard, thermoelectric generators unquestionably remain the most performant, ensuring excellent reliability. The results obtained in this study highlight several important points that reinforce this assertion. Firstly, it has been conclusively demonstrated that the use of a thermopile composed of 5000 thermocouples offers an advantageous cost-to-power ratio. This optimal number of thermocouples enables achieving balanced performance in terms of efficiency and cost, making it a preferred choice for real-world applications. Furthermore, the optimal flow rates of the heat transfer fluids have been carefully determined, with a flow rate of $\dot{m}_h = 0.1\,\text{kg/s}$ for the hot fluid and $\dot{m}_c = 0.2\,\text{kg/s}$ for the cold fluid. These optimized flow rates maximize heat transfer and maintain suitable thermal conditions for power generation. Lastly, a significant finding of this study is that adopting the counterflow configuration, where heat flows are opposed, results in a notable improvement in power generation. This configuration has led to an 11% increase in power output, with an achieved efficiency of 1.46%. These results underscore the crucial importance of considering flow configuration to optimize the performance of thermoelectric generators.

References

1. Crane, D.T.: An introduction to system-level, steady-state and transient modeling and optimization of high-power-density thermoelectric generator devices made of segmented thermoelectric elements. J. Electron. Mater. **40**(5), 561–569 (2011). https://doi.org/10.1007/s11664-010-1451-6
2. Zhao, Y., Lu, M., Li, Y., Wang, Y., Ge, M.: Numerical investigation of an exhaust thermoelectric generator with a perforated plate. Energy **263**, 125776–125776 (2023). https://doi.org/10.1016/j.energy.2022.125776
3. Lu, X., Yu, X., Qu, Z., Wang, Q., Ma, T.: An introduction to system-level, steady-state and transient modeling and optimization of high-power-density thermoelectric generator devices made of segmented thermoelectric elements. Energy Convers. Manage. **150**, 403–414 (2017). https://doi.org/10.1016/j.enconman.2017.08.030
4. Gomaa, M.R., Murtadha, T.K., Abu-jrai, A., Rezk, H., Altarawneh, M.A., Marashli, A.: Experimental investigation on waste heat recovery from a cement factory to enhance thermoelectric generation. Sustainability **14**(16), 10146 (2022). https://doi.org/10.3390/su141610146
5. Espinosa, N., Lazard, M., Aixala, L., Scherrer, H.: Modeling a thermoelectric generator applied to diesel automotive heat recovery. J. Elecron. Mater. **39**(9), 1446–1455 (2010). https://doi.org/10.1007/s11664-010-1305-2
6. Min, G., Rowe, D.: Improved model for calculating the coefficient of performance of a Peltier module. Energy Convers. Manage. **41**(2), 1982–1992 (2000). https://doi.org/10.1016/S0196-8904(99)00102-8
7. Lee, K.H., Kim, O.J.: Analysis on the cooling performance of the thermoelectric micro-cooler. Int. J. Heat Mass Transfer **50**(9–10), 1982–1992 (2007). https://doi.org/10.1016/j.ijheatmasstransfer.2006.09.037

8. Omer, S., Infield, D.: Design optimization of thermoelectric devices for solar power generation. Solar Energy Mater. Solar Cells **53**(1–2), 67–82 (1998). https://doi.org/10.1016/S0927-0248(98)00008-7
9. Chein, R., Chen, Y.: Performances of thermoelectric cooler integrated with microchannel heat sinks. Int. J. Refrig. **28**(6), 828–839 (2005). https://doi.org/10.1016/j.ijrefrig.2005.02.001
10. Duffie, J.A., Beckman, W.A., Blair, N.: Solar Engineering of Thermal Processes, Photovoltaics and Wind. Wiley, New York (2020)
11. Jennah, S., Belouaggadia, N., Lbibb, R., Ezzine, M.: Performance improvement of a TEG by a heat transfer fluid. In: International Conference on Innovative Research in Applied Science, pp. 1–4 (2022). https://doi.org/10.1109/IRASET52964.2022.9738018
12. Duffie, J.A., Beckman, W.A.: Solar Engineering of Thermal Processes, pp. 250–330. Wiley, New York (1991)
13. Rosengarten, G., Mutzenich, S., Kalantar-Zadeh, K.: Integrated microthermoelectric cooler for microfluidic channels. Exp. Therm. Fluid Sci. **30**(8), 821–828 (2006). https://doi.org/10.1016/j.expthermflusci.2006.03.011
14. Kikas, N.: Laminar flow distribution in solar systems. Solar Energy **54**(4), 209–217 (1995). https://doi.org/10.1016/0038-092X(94)00125-W
15. Tondeur, D., Fan, Y., Luo, L.: Flow distribution and pressure drop in 2D meshed channel circuits. Chem. Eng. Sci. **66**(1), 15–26 (2011). https://doi.org/10.1016/j.ces.2010.09.024
16. Shah, R.K., London, A.L.: Laminar Flow Forced Convection in Ducts, vol. 1. Academic Press, New York (1978)
17. Yang, W., et al.: An experimental investigation on the performance of TEGs with a compact heat exchanger design towards low-grade thermal energy recovery. Appl. Therm. Eng. **194**, 117119–117119 (2021). https://doi.org/10.1016/j.applthermaleng.2021.117119
18. Ritzer, T.M., Lau, P.G.: Economic optimization of heat sink design. In: AIP Conference Proceedings, vol. 316, no. 1, pp. 177–180 (1994). https://doi.org/10.1063/1.46788

Analytical and Numerical Accurate Solutions of the Steady Convection-Diffusion in Potential Flow Around a Circular Cylinder Heated with Uniform Heat Flux

M. Darif, E. Essaghir, A. Oubarra, and J. Lahjomri(✉)

Laboratory of Mechanics, Faculty of Science Ain Chock, Hassan II University, 20100 Casablanca, Morocco
lahjomri@hotmail.com

Abstract. The fundamental problem of the steady convection-diffusion in potential flow around a heated circular cylinder with uniform heat flux is solved both analytically and numerically. The analytical solution is derived in closed form by using the method of Green's function and presented as an integral. This solution is useful to validate the numerical codes of several others more complex problems considering Neumann boundary condition. It is thereby used to check the present numerical solution with potential flow. The asymptotic solution, at a great distance in the thermal wake region behind the cylinder is also obtained. It divulges the existence of singularities which lead to a deterioration of the accuracy of the numerical solution. An accurate numerical solution is then obtained by solving efficiently theses singularities. Its validity and its accuracy are supported by a comparison with the results derived from the exact analytical solution with relative error less than 0.1%, notably for the thermal characteristics. A useful correlation formula giving the overall Nusselt number as function of Péclet number is also proposed.

Keywords: potential flow · Green's function · convection-diffusion · Neumann boundary condition · singular behavior · asymptotic solution · thermal wake

1 Introduction

The problem of convection diffusion in potential flows with Dirichlet wall boundary condition (DBC), has been extensively investigated in the theory of solidification, melting, and freezing, Kornev and Mukhamadullina [1], Cummings et al. [2]. The theory with a potential flow and (DBC) shows that the problem can be reduced to the solution of the Fredholm's integral equation of the first kind expressed in terms of the usual Péclet number $\mathrm{Pe} = U_\infty R/\alpha$, the streamline coordinate $X = \phi/2$ (ϕ is the velocity potential function) and the modified Bessel function K_0:

$$\int_{-1}^{1} exp[\mathrm{Pe}(X-\xi)]K_0[\mathrm{Pe}|X-\xi|]q_w(\xi)d\xi = \pi \quad -1 < X < 1 \tag{1}$$

S. Aniss et al. (Eds.): CMM 2022, LNME, pp. 181–190, 2024.
https://doi.org/10.1007/978-3-031-46973-2_17

The problem consists of determining the distribution of the unknown flux density q_w satisfying the integral Eq. (1). We note that there is no general theory to solve (1) explicitly. The asymptotic solutions of this equation have been found by Choi et al. [3] for low and high Pe. They obtained a formula for the distribution of q_w by overlapping the corresponding expansions solutions. By applying conformal maps technique, this distribution has been used to construct the exact solutions to the Navier-Stokes equations, Bazant and Moffat [4]. Recently Boulais and Gervais [5] have solved the problem of convection-diffusion in two-dimensional flows by applying the same technique for high and low Pe to obtain an effective solution in the whole domain of the flow. They showed that this solution can be used to model diffusion in confined geometries and microfluidic devices. Using three-dimensional finite element enthalpy approach, Erchiqui and Annasabi [6] solved numerically the nonlinear heat conduction problem in transient state involving phase changes. Their numerical results were successfully validated with those obtained analytically by Serkar et al. [7] in the case of steady state condition.

To our knowledge, no studies have been done considering a prescribed wall Neumann boundary condition (NBC) on the cylinder surface. We note that, for this case, in contrast to (DBC), as shown here, an analytical solution in closed form can be explicitly obtained for all Pe by using Green's function method. This method has been used earlier for heat conduction in anisotropic media by Chang et al. [8]. The objective of this study is to establish an analytical and numerical solutions for the steady convection-diffusion problem in potential flow around a circular cylinder heated with uniform known wall heat flux. Furthermore, the asymptotic analytical solution, at a great distance in the thermal wake region behind the cylinder is also obtained. This latter indicates that there are singularities at infinity which cause a great difficulty for the numerical computation when one approaches these singularities, notably for the accuracy and convergence of the numerical solution. The main goal of this work is to handle and solve these singular behaviors with the aim of obtaining an accurate numerical solution for this fundamental problem.

2 Mathematical Formulation and Analysis

2.1 Problem Description

We consider a steady two-dimensional potential flow of inviscid fluid around a long circular cylinder of radius R and diameter D with uniform velocity U_∞ and uniform temperature T_∞ at infinity upstream. Figure 1 shows the coordinates system used in the present study. We use cartesian coordinates $\left(x^{'}, y^{'}\right)$ or polar coordinates $(r^{'}, \theta)$ with origin at the center of the cylinder. The physical properties of the fluid are assumed independent of temperature; therefore, the natural convection is considered negligible. The cylinder is heated with a uniform flux density, $q_w = -k\partial T'/\partial r'\big|_{r'=R}$ on its surface (k is the thermal conductivity of the fluid). The problem consists of finding the temperature field, $T\prime(x^{'}, y^{'})$, advected in potential flow $\overrightarrow{U} = \overrightarrow{\nabla}\varphi$ and to establish the evolution law of overall averaged Nusselt number, with Péclet number, $Nu_m(\text{Pe})$. Since the flow is symmetrical with respect to the $x^{'}$-axis, thus we have only to consider the upper half of physical plane $(x^{'}, y^{'})$.

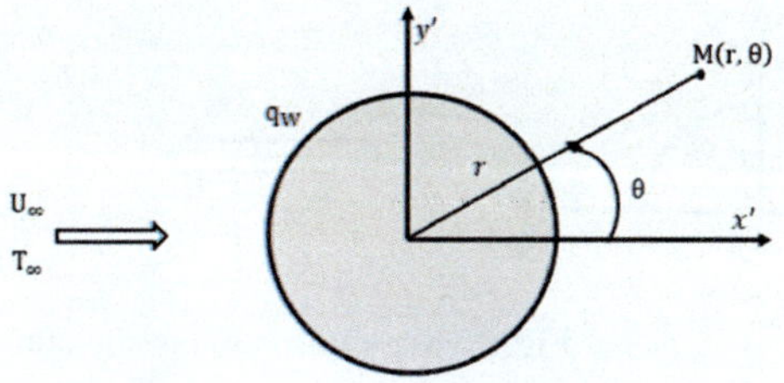

Fig. 1. Problem configuration and the system of coordinates used.

2.2 Energy Equation and Boundary Conditions

The energy equation in polar coordinates (r, θ) can be written in dimensionless form as:

$$u_r(r,\theta)\frac{\partial\Theta}{\partial r} + \frac{1}{r}u_\theta(r,\theta)\frac{\partial\Theta}{\partial\theta} = \frac{1}{\text{Pe}}\left(\frac{\partial^2\Theta}{\partial r^2} + \frac{1}{r}\frac{\partial\Theta}{\partial r} + \frac{1}{r^2}\frac{\partial^2\Theta}{\partial\theta^2}\right) \text{ for } r > 1\ 0 < \theta < \pi \tag{2}$$

where $u_r(r,\theta) = (1 - 1/r^2)\cos\theta$ and $u_\theta(r,\theta) = -(1 + 1/r^2)\sin\theta$ are the radial and orthoradial components of the dimensionless velocity $\vec{u}$. The dimensionless parameters and variables are defined as follows:

$$r = \frac{r'}{R}, x = \frac{x'}{R}, y = \frac{y'}{R}, \vec{u} = \frac{\vec{U}}{U_\infty} = \vec{\nabla}\phi, \Theta = \frac{T' - T_\infty}{\frac{q_w R}{k}}, \phi = \frac{\varphi}{U_\infty R}, \Psi = \frac{\psi}{RU_\infty} \tag{3}$$

ϕ and Ψ denote the velocity potential function and stream function. The boundary conditions associated to the problem written in nondimensional form are given by:

$$\begin{cases} \Theta \to 0 & \text{for } r \to \infty,\ 0 \le \theta \le \pi \\ \pi\frac{\partial\Theta}{\partial r} = -1 & \text{for } r = 1,\ 0 \le \theta \le \pi \\ \frac{\partial\Theta}{\partial\theta} = 0 & \text{for } r > 1,\ \theta = 0, \theta = \pi \end{cases} \tag{4}$$

The components of the velocity in (2) may be removed by introducing the streamline coordinates $X = \phi/2$ and $Y = \Psi/2$ taken as independent variables. With this transformation, the upper half flow region $(1 \le r \le \infty,\ 0 \le \theta \le \pi)$ is transformed into the upper (X, Y) plane and the circular cylinder of radius unity into the slit $Y = 0,\ |X| \le 1$ (Fig. 2), with dimensionless length 2, that is simply the change in X from the leading to the trailing edges.

In terms of the new coordinates (X, Y), the energy Eq. (2) becomes the Oseen equation:

$$\frac{\partial^2\Theta}{\partial X^2} + \frac{\partial^2\Theta}{\partial Y^2} - 2\text{Pe}\frac{\partial\Theta}{\partial X} = 0 \quad \begin{cases} -\infty < X < +\infty \\ Y > 0 \end{cases} \tag{5}$$

The boundary conditions in the (X, Y) plane are then given by:

$$\begin{cases} \Theta \to 0 & X^2 + Y^2 \to \infty \\ \frac{\partial\Theta}{\partial Y} = -\frac{1}{\sqrt{1-X^2}} & Y = 0^+,\ |X| < 1 \\ \frac{\partial\Theta}{\partial Y} = 0 & Y = 0^+,\ |X| \ge 1 \end{cases} \tag{6}$$

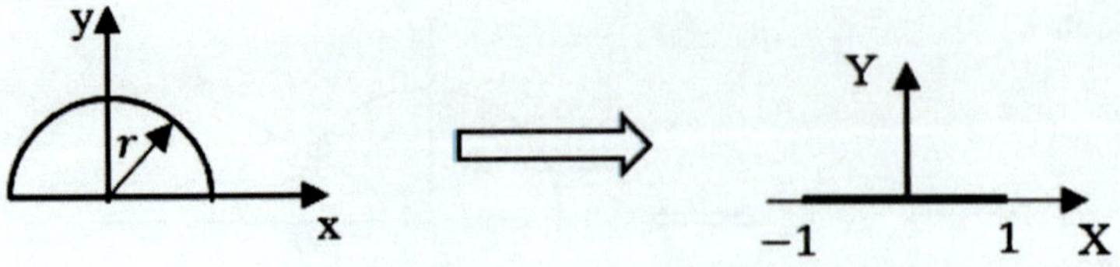

Fig. 2. Transformation of the domain

2.3 Analytical Solution

The first order derivative in Eq. (5) can be removed, by proceeding to the following change:

$$\Theta(X, Y) = e^{PeX} \mathrm{Q}(X, Y) \tag{7}$$

Therefore, Eq. (5) becomes the two-dimensional Helmholtz's Eq. (8) in the semi-infinite plane (Y > 0), where the boundary conditions fulfilled by the function Q are given in (9):

$$\frac{\partial^2 \mathrm{Q}}{\partial X^2} + \frac{\partial^2 \mathrm{Q}}{\partial Y^2} - \mathrm{Pe}^2\mathrm{Q} = 0 Y > 0, -\infty < X < +\infty \tag{8}$$

$$\begin{cases} \mathrm{Q} \to 0 & X^2 + Y^2 \to +\infty \\ \frac{\partial \mathrm{Q}}{\partial Y} = \frac{-e^{-PeX}}{\sqrt{1-X^2}} & Y = 0^+, |X| < 1 \\ \frac{\partial \mathrm{Q}}{\partial Y} = 0 & Y = 0^+, |X| \geq 1 \end{cases} \tag{9}$$

Now, we propose to formulate the solution of the boundary value problem (BVP) (8) and (9) by using the method of Green's function. The Green function associated to Helmholtz's equation in the semi-infinite plane (Y > 0) is available and given by Polyanin [9]:

$$\mathrm{G} = \frac{1}{2\pi}\{K_0\left[\mathrm{Pe}\sqrt{(X-\xi)^2 + (Y-\eta)^2}\right] + K_0\left[\mathrm{Pe}\sqrt{(X-\xi)^2 + (Y+\eta)^2}\right]\} \tag{10}$$

To solve the (BVP) (8) and (9), we need the solution representation in term of the Green function (10). Applying Green's formula by considering a large semi-circle (C) of surface area (S) in the upper half plane (Fig. 3) and taking the limit as the radius $r = \sqrt{X^2 + Y^2}$ tending to infinity and reversing the roles of the variables (X, Y) and (ξ, η), we obtain the function Q as a product of convolution of the Green function with the gradient of Q at $\eta = 0^+$ [9]:

$$Q(X, Y) = -\int_{-\infty}^{+\infty} G\big(X, Y; \xi, 0^+\big)\frac{\partial Q}{\partial \eta}|_{\eta=0^+} d\xi \tag{11}$$

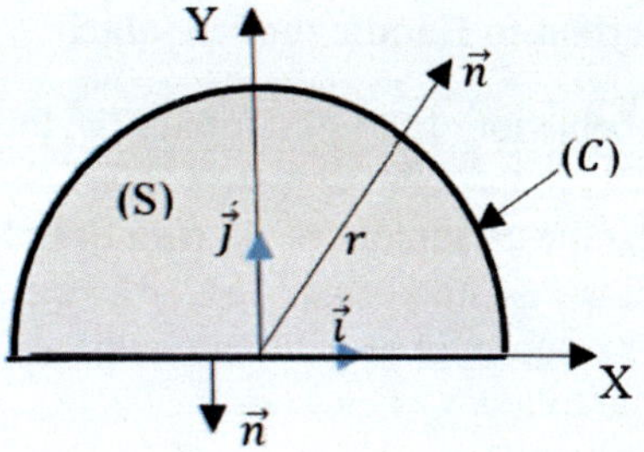

Fig. 3. Domain for the application of the Green formula (14).

Now, by using (7), (11), boundary conditions (9) and the expression (10) of G, we can obtain the solution of the problem in closed form, giving the temperature field Θ in the real space of the flow expressed in term of the cylindrical coordinates:

$$\Theta(r,\theta) = \int_{-1}^{+1} \frac{e^{\mathrm{Pe}\,[(r+1/r)\cos\theta/2-\xi])}}{\pi\sqrt{1-\xi^2}} K_0\left[\mathrm{Pe}\sqrt{[(r+1/r)\cos\theta/2-\xi]^2+[(r-1/r)\sin\theta/2]^2}\right] d\xi \tag{12}$$

3 Numerical Method

In this section we find a numerical solution of the (BVP) (2), (4) to validate it with the exact analytical solution (12). It should be noted that obtaining accurate numerical solution of this (BVP) by numerical method is not an easy task. Indeed, the numerical analysis divulges some serious difficulties. The difficulty is mostly due to the boundary condition at infinity and comes from the singular behavior of Θ at great distance behind the cylinder in the wake region. These singularities lead to the deterioration of the accuracy of numerical solution especially when Pe becomes large, $\theta = 0$ and as r approaches infinity. Therefore, before to proceed to the resolution of the problem by numerical method, we need to solve these singularities. We first begin the analysis by applying the algebraic transformation $\xi = 1/r$ used by Ingham [10] in another different problem to transform the semi-infinite domain $r \in [1, \infty[$ into a bounded domain $\xi \in [0, 1]$, and for the variable θ, we introduce a new variable η defined as: $\eta = \theta/\pi$. With this mapping, the (BVP) (2), (4) becomes:

$$\xi^3\frac{\partial^2\Theta}{\partial\xi^2} + \left[\xi^2 + \mathrm{Pe}\,\xi u_r(\xi,\eta)\right]\frac{\partial\Theta}{\partial\xi} + \frac{\xi}{\pi^2}\frac{\partial^2\Theta}{\partial\eta^2} - \frac{\mathrm{Pe}}{\pi}u_\theta(\xi,\eta)\frac{\partial\Theta}{\partial\eta} = 0 \tag{13}$$

$$\begin{cases} \Theta \to 0 \text{ for } \xi = 0, \quad 0 \le \eta \le 1 \\ \frac{\partial\Theta}{\partial\xi} = 1 \text{ for } \xi = 1, \quad 0 \le \eta \le 1 \\ \frac{\partial\Theta}{\partial\eta} = 0 \text{ for } 0 < \xi < 1, \quad \eta = 0, \quad \eta = 1 \end{cases} \tag{14}$$

3.1 Analytical Method to Handle the Singularity

Let us examine the behavior of the solution for the thermal field at great distance from the cylinder, especially in the thermal wake region behind the obstacle. We note that it is not easy to determine the asymptotical behavior of Θ from the exact analytical solution (12). As alternative, we use the original energy Eq. (2) to determine the leading term for the asymptotical solution of Θ valid in the wake region of large r. To do this, it is appropriate to use the following new coordinates (X_w, Y_w) defined as:

$$X_w = \sqrt{\mathrm{Pe}}\left(\sqrt{r} + \frac{1}{\sqrt{r}}\right)cos\frac{\theta}{2},\ Y_w = \sqrt{\mathrm{Pe}}\left(\sqrt{r} - \frac{1}{\sqrt{r}}\right)sin\frac{\theta}{2} \quad (15)$$

with $X_w > 0$ and $Y_w > 0$, $(0 < \theta < \pi)$ due to the symmetry. In terms of these new variables, the energy Eq. (2) can be transformed as:

$$\frac{\partial^2\Theta}{\partial {X_w}^2} + \frac{\partial^2\Theta}{\partial {Y_w}^2} + 2Y_w\frac{\partial\Theta}{\partial Y_w} - 2X_w\frac{\partial\Theta}{\partial X_w} = 0 \quad (16)$$

It can be shown that the fundamental solution of (16) which satisfies the boundary condition at infinity, i.e., $(r \to \infty)$, ${X_w}^2 + {Y_w}^2 \to \infty$ and the symmetry BC for $\theta = 0$, i.e., $\partial\Theta/\partial Y_w|_{Y_w=0} = 0$, (but not on the cylinder surface) can be obtained in the following form:

$$\Theta \approx K_1\frac{X_w}{{X_w}^2 + {Y_w}^2}\exp\left(-{Y_w}^2\right) \quad (17)$$

The constant K_1 in (17) can be found by applying the global conservation of energy in the whole field of the flow. Thus, as function of (ξ, η) coordinates, the leading term of the asymptotic solution in the thermal wake of the cylinder becomes:

$$\Theta \approx \frac{\sqrt{\pi}}{\sqrt{\mathrm{Pe}}}\frac{\sqrt{\xi}(\xi + 1)\cos(\frac{\pi\eta}{2})exp\left[\mathrm{Pe}(2 - \xi - \frac{1}{\xi})sin^2\frac{\pi\eta}{2}\right]}{4\xi\, cos^2\frac{\pi\eta}{2} + (\xi - 1)^2} \quad (18)$$

As $\xi \to 0$ the solution (18) simplifies to:

$$\Theta \approx \frac{\sqrt{\pi}}{\sqrt{\mathrm{Pe}}}\cos\left(\frac{\pi\eta}{2}\right)\sqrt{\xi}\,exp\left[-\frac{\mathrm{Pe}}{\xi}sin^2\frac{\pi\eta}{2}\right] \quad (19)$$

It is clear from (19) that for a fixed value of η, with $\eta \neq 0$, there is an essential type of singularity, i.e., $\Theta = O\left\{\sqrt{\xi}exp\left[-\frac{a\mathrm{Pe}}{\xi}\right]\right\}$ as $\xi \to 0$ in the thermal wake, where a is non-zero positive constant. While the singularity is of square root type $\Theta = O(\sqrt{\xi})$ when $\eta = 0$. This latter behavior is very difficult to treat in numerical method due to a diverging derivative at $\xi = 0$, $(r \to \infty)$. To remedy to these behaviours, we use the same transformation employed by Choi et al. [3] for a similar problem with (DBC):

$$\mathrm{T}(\xi, \eta) = \sqrt{\xi}exp\left[-\mathrm{Pe}(2 - \xi - \frac{1}{\xi})sin^2\frac{\pi\eta}{2}\right]\Theta(\xi, \eta) \quad (20)$$

Therefore, instead to solve the energy equation for Θ, we solve numerically that governing the new function T to deduce afterward the solution Θ. With this change, it is easy to verify that T becomes regular function of ξ, $T = O(\xi)$ as $\xi \to 0$ and satisfies the same first boundary condition of Θ given in (14). The interest to introduce this new function is illustrated in Figs. 4(a) and (b). These figures show how the singular behaviors for Θ (vertical slope at $\xi = 0$) is removed for the function T for which its slope is clearly reduced. These figures also show the comparison of the results found from numerical, analytical Eq. (12) and asymptotic (19) solutions in the thermal wake behind the cylinder. The agreement between the two first calculations is excellent for the two values of Péclet number. The relative error is less than 0.1%. The result found from asymptotic solution (19) agrees also well with those of analytical and numerical solutions in the region of its validity for small ξ; since (19) is not valid in the region near the cylinder.

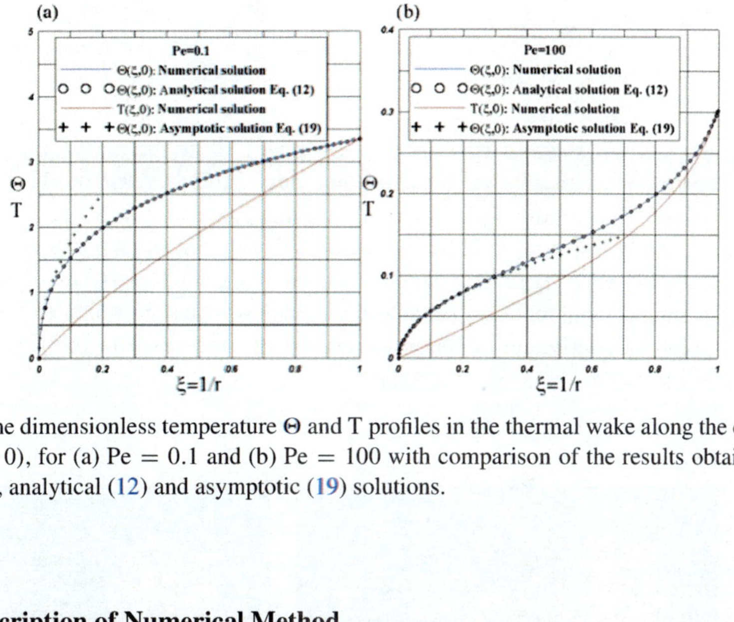

Fig. 4. The dimensionless temperature Θ and T profiles in the thermal wake along the centerline axis ($\eta = 0$), for (a) Pe = 0.1 and (b) Pe = 100 with comparison of the results obtained from numerical, analytical (12) and asymptotic (19) solutions.

3.2 Description of Numerical Method

The partial differential equation governing the new function T after using the transformation (20) (See Ref. [3]), is solved numerically by using a centered scheme of second order to discretize all derivatives via a finite difference method. The method uses the non-uniform variables step in the two directions of ξ and η with Gauss-Lobatto distributions at the nodal points. Therefore, the system of equations generated by the discretization is solved by Newton-Raphson method and using *LU* decomposition to calculate the Jacobian of the matrix of the system. All the results obtained by this numerical method are performed with a nodes $N_\xi = 160$ and $N_\eta = 160$ leading to a very accurate numerical solution. The test of convergence of the solution used, is defined when the Euclidian norm δ of the following unknown vector $\overrightarrow{T} = \left[T_1, T_2, .., T_{(N_\xi+1)(N_\eta+1)}\right]^t$ becomes less than the tolerance ε chosen to be equal to 10^{-7}, where $T_k = \mathrm{T}(i, j)$ and $k = j\big(N_\xi + 1\big) + i + 1$, $(i = 0..N_\xi, j = 0..N_\eta)$.

4 Results and Discussion

Figure 5(a) and (b) indicate the features of the isotherms in the flow for small and large values of Péclet number highlighting the competition between the effects of conduction and convection. For low Péclet number, Pe = 0.1, the effect of conduction dominates that of convection, the isotherms in Fig. 5(a) show that the thermal perturbation propagates away from the cylinder in the form of concentric circles. For high Péclet number, i.e., Pe = 100, in Fig. 5(b), the convection effect dominates that of conduction. Therefore, a thermal boundary layer develops from upstream stagnation point, giving rise to a thermal wake behind the cylinder characterized by a thin thermal layer of width $\mathrm{O(Pe^{-1/2})}$. Within this layer the higher temperatures are confined in which the hot flashes coming from the rear stagnation point of the cylinder are convected. Figure 6, shows the effect of Péclet number both on the local Nusselt number $\mathrm{N_u}(\theta) = q_w D/k(T_w - T_\infty) = 2/\Theta_\mathrm{w}(\theta)$, where $\Theta_\mathrm{w}(\theta)$=$\Theta(1, \theta)$ is the temperature of the wall, and on the overall Nusselt number $\mathrm{Nu_m} = \frac{1}{2\pi}\int_0^{2\pi} \mathrm{N_u}(\theta)d\theta$. The increase of Pe means that the transfer by convection becomes important with respect to the thermal diffusion, consequently the overall Nusselt number also increases due to the decrease of Θ_w resulting from the lowering of the thickness of thermal boundary layer $\delta_{th} = \mathrm{O(Pe^{-1/2})}$. Figure 6 also compares the results found by the analytical and numerical solutions, showing clearly a very good agreement between the two methods both for $\mathrm{N_u}$ and $\mathrm{Nu_m}$. Finally, from the computational results we propose a useful practical correlation illustrated in red on Fig. 6, with coefficient of determination R-squared equal to 0.999974 in the interval $0.4 \le \mathrm{Pe} \le 200$: $\mathrm{Nu_m} = 1.73 \times \mathrm{Pe}^{0.505}$.

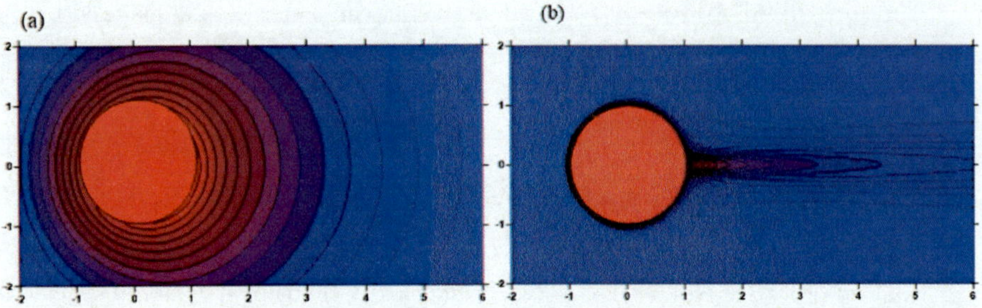

Fig. 5. The isotherms for (a) Pe = 0.1, and (b) Pe = 100.

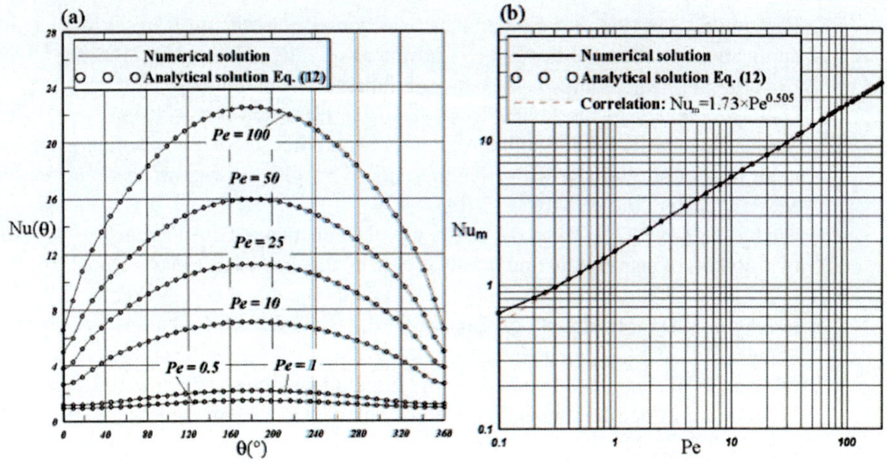

Fig. 6. Effect of Pe on (a) the local Nusselt and (b) the overall Nusselt numbers with comparison between analytical and numerical methods with correlation formula for Nu_m.

5 Conclusion

The solution of the fundamental problem of steady convection-diffusion in potential flow around a cylinder has obtained both analytically and numerically. The corresponding asymptotic solution for large distance from the cylinder in the thermal wake region is also determined. This one highlights a singular behavior of the solution in this region. This behavior is solved by introducing a new function which allowed us to obtain an accurate numerical solution presenting a maximum relative error less than 0.1% for the thermal characteristics compared to the exact analytical solution. This analytical solution is useful, since it has the interest to check not only the numerical solution of the energy equation for the simple case of potential flow but also to validate the numerical simulations of others more complicated problems considering flow over a cylinder with (NBC), such as viscous flow, slip flow, Magnetohydrodynamic flow, etc…

Acknowledgements. The authors gratefully acknowledge the helpful comments and suggestions of the reviewers, which have improved the article.

References

1. Kornev, K., Mukhamadullina, G.: Mathematical theory of freezing for flow in porous media. Proc. R. Soc. Lond. A **447**, 281–297 (1994)
2. Cummings, L.M., Hohlov, Y.E., Howison, S.D., Kornev, K.: Two-dimensional solidification and melting in potential flows. J. Fluid Mech. **378**, 1–18 (1999)
3. Choi, J., Margetis, D., Squires, T.M., Bazant, M.Z.: Steady advection–diffusion around finite absorbers in two-dimensional potential flows. J. Fluid Mech. **536**, 155–184 (2005)
4. Bazant, M.Z., Moffat, H.K.: Exact solutions of the Navier-Stokes equations having steady vortex structures. J. Fluid Mech. **541**, 55–64 (2005)

5. Boulais, E., Gervais, T.: Two-dimensional convection–diffusion in multipolar flows with applications in microfluidics and groundwater flow. Phys. Fluids **32**(12), 122001:1–12 (2020)
6. Erchiqui, F., Annasabi, Z.: 3D hybrid finite element enthalpy for anisotropic thermal conduction analysis. Int. J. Heat Mass. Transfer **136**, 1250–1264 (2019)
7. Serkar, D., Shah, K., Haji-Sheikh, A., Jain, A.: Analytical modeling of temperature distribution in an anisotropic cylinder with circumferentially varying convective heat transfer. Int. J. Heat Mass. Transfer **79**, 1027–1033 (2014)
8. Chang, Y., Kang, C.S., Chen, D.J.: The use of fundamental Green's functions for the solution of problems of heat conduction in anisotropic media. Int. J. Heat Mass. Transf. **16**, 1905–1918 (1973)
9. Polyanin, AD.: Handbook of Linear Partial Differential Equations for Engineers and Scientists. CRC, Chapman Hall (2002)
10. Ingham, D.B., Tang, T.A.: numerical investigation into the steady flow past a rotating circular cylinder at low and intermediate Reynolds numbers. J. Computat. Phys. **87**, 91–107 (1990)

Simulation of Combined Thermal Mixed Convection and Radiation in a Discretely Heated Lid-Driven Cavity using a Lattice Boltzmann Method

A. Daiz[1(✉)], A. Bahlaoui[1], I. Arroub[1], S. Belhouideg[1], A. Raji[2], and M. Hasnaoui[3]

[1] Polydisciplinary Faculty, Research Laboratory in Physics and Sciences for Engineers (LRPSI), Béni-Mellal, Morocco
daizabdelhak@gmail.com

[2] Faculty of Sciences and Technics, Energy and Materials Engineering Laboratory (LGEM), Béni-Mellal, Morocco

[3] Faculty of Sciences Semlalia, Laboratory of Fluid Mechanics and Energetics (LMFE), Marrakech, Morocco

Abstract. In this work, we examined numerically the heat transfer by mixed convection coupled with surface thermal radiation in a square cavity with a moving wall, using the Lattice Boltzmann method. The cavity is discretely heated from its left and lower walls with a hot temperature and cooled from its right wall, while the upper wall is adiabatic and moves from left to right at a constant velocity. The cooling medium is assumed to be air, which modeled as a radiatively transparent medium. The critical parameters in this study are the emissivity of the surfaces, ε, which varies between 0 and 1, and the number of Richardson, Ri, which varies from 0.1 to 10. The latter varies by varying the Reynolds number, Re, with a Rayleigh number, $Ra = 10^5$. The results obtained indicate that the dynamic structure is monocellular, which illustrated by the existence of a large cell generated by the combined effects of shear and buoyancy. In addition, it should be noted that the increase of the emissivity of the surfaces enhances the radiative heat exchange, reduces convective effect and improves the total heat exchange. Additionally, it is discovered that an increase in Ri causes a significant decrease in convective and global heat exchange, while just slight increase in radiative heat transfer.

Keywords: Numerical Study · Mixed Convection · Surface Radiation · Moving Wall · Discretely Heated · Lattice Boltzmann

1 Introduction

The problem of heat transfer by mixed convection coupled with radiation in a square cavity with moving walls has attracted an important attention from Scientists. This issue arises often in various technical applications, such as solar collectors, cooling of electrical equipments, thermal hydraulics in nuclear reactors and many others. However, due to its complexity, the effect of radiation has been neglected in most existing studies despite its

S. Aniss et al. (Eds.): CMM 2022, LNME, pp. 191–200, 2024.
https://doi.org/10.1007/978-3-031-46973-2_18

major impact on the total heat transfer [1, 2]. Since modest temperature differences are associated with remarkable radiative impacts, ignoring the latter becomes impractical.

During the past several decades a lot of work has been done to analyze the combined surface thermal radiation with natural convection effects in rectangular cavities maintained at different boundary conditions due to its wide variety of applications [3, 4]. According to the results, raising the wall emissivity and Rayleigh number enhances the total heat transfer rate for both isothermal and heat-producing heaters. In addition, in the presence of the radiation impact, the buoyancy force induced flow is favored in the case of an isothermal block, whereas it is delayed in the case of a heat-generating block. It is also discovered that surface radiation has a significant impact, in the cold surface emissivity fluctuation range (0–0.4), and that heat transfer (both convective and radiative) is highly dependent on the surfaces emissivity of the cavity. Recently, Rehhali et al. [5] explored numerically coupling natural convection and radiation heat transfer within a non-uniformly temperature heated tilted square enclosure by lattice Boltzmann. This study has shown that the surface radiation has a significant quantitative impact on overall heat transfer. Thus, the contribution of convection/radiation decreases/increases as Rayleigh number increases.

Mixed convection is often seen in appliances with very high output where forced or natural convection is not sufficient to dissipate all the necessary heat. At this point, a combination of natural and forced convection will often give the desired results [6, 7]. In the literature, the effects of radiation and mixed convection on dynamic behavior and thermal structure in moving walls cavities are relatively little investigated. Antar et al. [8] investigated the impact of radiation on mixed convection in a rectangular enclosure heating by the bottom wall and moving by a variable speed wall. The authors found that the Nusselt number values calculated at variable speed are between those obtained with (constant) speed profiles equal to V_{min} and V_{max}. Belmiloud and Chemloul [9] studied how radiation affected laminar mixed convection within a lid-driven square cavity with differential heating. Their findings demonstrate that radiation has an impact on both heat transfer components. Additionally, when the Richardson number rises, the Nusselt number, Nu, reduces. The impact of radiation on mixed convection in a lid-driven square cavity with variable heating was explored numerically by Dahani et al. [10]. They discovered that even at low Richardson numbers, radiation has a significant impact on overall heat transfer, and that this impact increases when the latter parameter is increased. Addini and Nassab [11] highlighted and underlined the radiation's role in lid-driven cavities containing gray materials that emit, absorb, and scatter radiation. They obtained that the radiative Nusselt number, evaluated on the hot wall, decreases progressively with increasing optical thickness of the medium.

The literature study indicates that there aren't enough research examining mixed convection and surface radiation in discretely heated lid-driven square enclosures. After that, further work will be needed to advance understanding in this area. Therefore, the principal objective of this work is to examine how surface radiation affects dynamic and thermal convection in a discretely heated square cavity with moving wall. The influence of various combinations of the governing factors, namely the Richardson number and the wall emissivity, on the thermal and dynamic flow fields as well as the thermal performance inside the cavity is investigated by performing a parametric analysis.

2 Description of the Problem and Mathematical Formulation

Figure 1 presents the schematic diagram of the flow configuration considered in this study. It consists of a lid-driven square cavity filled with air, whose Prandtl number is Pr = 0.71. The cavity is discretely heated on its left and bottom walls by at high temperature T'_H, and uniformly cooled on its right wall at low temperature T'_C. The upper wall is thermally insulated and moving from left to the right at a uniform speed, $u_{0'}$. The fluid characteristics are calculated at a low (cold) temperature. The internal surfaces that are in contact with the fluid are assumed to be as gray body, diffuse radiation emitters and reflectors with similar emissivities. The air is considered Newtonian, incompressible and obeys the Boussinesq approximation, while its flow is considered two-dimensional and laminar.

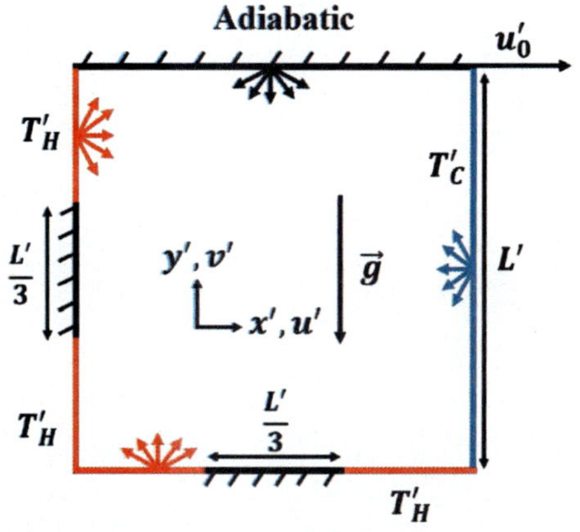

Fig. 1. Detailed description of the configuration under study

The LBM is a computational method that has been used to solve a variety of fluid flow and heat transfer issues at the microscopic level with great success. The LBM is based on the theory of kinetic energy of particles. In the lattice Boltzmann method (LBM), the solution is advanced in time and space through two steps collision and streaming. In the collision step, particle distributions are updated based on local equilibrium distributions and collision models. In the streaming step, particle distributions are propagated to neighboring lattice nodes based on predefined velocity directions. These steps are iteratively repeated to evolve the solution in time and space. The main equations of the Lattice Boltzmann model governs the studied problem is written in the Bhatnagar-Gross-Krook approximation. Thus, in the case of external force F, the lattice Boltzmann equation may be employed for both the flow and temperature fields, as shown below [12]:

$$f_k(r + c_k\Delta t, t + \Delta t) = f_k(r, t) - \frac{1}{\tau_f}\left(f_k(r, t) - f_k^{eq}(r, t)\right) + F_k \tag{1}$$

$$g_k(r + c_k\Delta t, t + \Delta t) = g_k(r, t) - \frac{1}{\tau_g}\left(g_k(r, t) - g_k^{eq}(r, t)\right) \tag{2}$$

where f_k^{eq} and g_k^{eq} are the equilibrium distribution functions for the momentum and energy equations, respectively and c_k is the microscopic velocity. τ_f and τ_g are the relaxation time; they are defined as an inverse of collision frequency ω_f and ω_g, and are related to the kinematic viscosity and thermal diffusivity respectively by $\left(\tau_f = \frac{1}{\omega_f} = 3\nu + 0.5\right)$ and $\left(\tau_g = \frac{1}{\omega_g} = 3\alpha + 0.5\right)$. In this work, a D2Q9 lattice model is applied to solve Eqs. (1) and (2). The D2Q9 is two-dimensional with nine directions.

Moreover, the equilibrium distribution functions f_k^{eq} and g_k^{eq} are expressed by:

$$f_k^{eq}(r,t) = w_k\rho\left[1 + \frac{c_k.u}{c_s^2} + \frac{1}{2}\frac{(c_k.u)^2}{c_s^4} - \frac{1}{2}\frac{u^2}{c_s^2}\right] \tag{3}$$

$$g_k^{eq}(r,t) = w_k T\left[1 + \frac{c_k.u}{c_s^2}\right] \tag{4}$$

where $c_s = 1/\sqrt{3}$ is the lattice speed of sound; u, ρ and T denote the macroscopic velocity, density and temperature, respectively.

In the D2Q9 lattice, the nine velocities, c_k, and weights of the velocity space discretization, w_k, are given as follows:

$$c_k = c\begin{cases} 0 & k = 0 \\ \left(\cos\left[(k-1)\frac{\pi}{2}\right]; \sin\left[(k-1)\frac{\pi}{2}\right]\right) & 1 \le k \le 4 \\ \sqrt{2}\left(\cos\left[(k-5)\frac{\pi}{2} + \frac{\pi}{4}\right]; \sin\left[(k-5)\frac{\pi}{2} + \frac{\pi}{4}\right]\right) & 5 \le k \le 8 \end{cases} \tag{5}$$

$$w_k = \begin{cases} 4/9 & k = 0 \\ 1/9 & 1 \le k \le 4 \\ 1/36 & 5 \le k \le 8 \end{cases} \tag{6}$$

The gravitational force appears in Eq. (1), F_k, with the application of the Boussinesq approximation, is calculated as follows:

$$F_k = 3w_k\rho g\beta c_{ky}(T - T_C) \tag{7}$$

where T_C (cold dimensionless temperature) is retained as dimensionless reference temperature; β is the coefficient of thermal expansion; c_{ky} is the component of the microscopic speed c_k in the y-axis direction.

In the case of the lattice Boltzmann method, the boundary conditions are not based on velocity or pressure for example, but rather on the distribution function. We will develop the different boundary conditions that we will use in our study.

The distribution functions of the particles density on the fixed surfaces of the cavity are expressed by using the classical bounce-back boundary conditions as follows:

$$\begin{cases} f_1 = f_3,\ f_5 = f_7 \text{ and } f_8 = f_6 & \text{on the left wall} \\ f_2 = f_4,\ f_5 = f_7 \text{ and } f_6 = f_8 & \text{on the bottom wall} \\ f_3 = f_1,\ f_6 = f_8 \text{ and } f_7 = f_5 & \text{on the right wall} \end{cases} \tag{8}$$

The distribution functions of the particles density on the top moving wall of the enclosure are evaluated as follows:

$$\begin{cases} \rho = f_0 + f_1 + f_3 + 2(f_2 + f_5 + f_6) \\ f_4 = f_2 \\ f_7 = f_5 + \frac{1}{2}(f_1 - f_3) - \frac{1}{2}\rho u_{0'} \\ f_8 = f_6 - \frac{1}{2}(f_1 - f_3) + \frac{1}{2}\rho u_{0'} \end{cases} \tag{9}$$

Now, the thermal boundary conditions are introduced:

a- For the adiabatic parts located on left and bottom walls:

$$g_{k,0} = \frac{4g_{k,1} - g_{k,2} - 2\Delta y \omega_k N_r Q_r}{3}, k = 1, \ldots, 8 \tag{10}$$

b- For the adiabatic top wall:

$$g_{k,M} = \frac{4g_{k,M-1} - g_{k,M-2} - 2\Delta y \omega_k N_r Q_r}{3}, k = 0, \ldots, 8 \tag{11}$$

c- For the heated parts on left wall:

$$\begin{cases} g_1 = T_H.(\omega_1 + \omega_3) - g_3 \\ g_5 = T_H.(\omega_5 + \omega_7) - g_7 \\ g_8 = T_H.(\omega_8 + \omega_6) - g_6 \end{cases} \tag{12}$$

d- For the heated parts on Bottom wall:

$$\begin{cases} g_2 = T_H.(\omega_2 + \omega_4) - g_4 \\ g_5 = T_H.(\omega_5 + \omega_7) - g_7 \\ g_6 = T_H.(\omega_6 + \omega_8) - g_8 \end{cases} \tag{13}$$

e- For the cold right wall:

$$\begin{cases} g_3 = T_C.(\omega_3 + \omega_1) - g_1 \\ g_7 = T_C.(\omega_7 + \omega_5) - g_5 \\ g_6 = T_C.(\omega_6 + \omega_8) - g_8 \end{cases} \tag{14}$$

The sum of the distribution functions at each position of the lattice indicates one of the macroscopic variables for each governing equation in the LBM.

$$\rho = \sum_k f_k \,;\ \rho u = \sum_k c_k f_k \,;\ T = \sum_k g_k \tag{15}$$

The radiosity method may be used to account for surface radiation and calculate the radiative flux on the cavity walls. The cavity's inside walls are subdivided into elementary surfaces centered on the network nodes. Each element surfaces is small enough to be regarded as isothermal. With the use of Hottel's [13] crossed string method, the view factors between the isothermal elementary surfaces were calculated. The non-dimensional radiosity and radiative heat flux for the ith element of the cavity, which is filled with a transparent fluid such as air, are given by the following equations:

$$J_i = \varepsilon_i \left(\frac{T_i}{T_r} + 1 \right)^4 + (1 - \varepsilon_i) \sum_{j=1}^{N} F_{ij} J_j \tag{16}$$

$$Q_r = J_i - I_i = \varepsilon_i \left[\left(\frac{T_i}{T_r} + 1 \right)^4 - \sum_{j=1}^{N} F_{ij} J_j \right] \tag{17}$$

where $J_i = J_i' / \sigma T_C'^4$ and $T_r = T_C' / (T_H' - T_C')$ are the non-dimensional radiosity and non-dimensional reference temperature, respectively.

The heat transfer exchanged by mixed convection and radiation through the cooled wall of the enclosure are assessed using the convective and radiative Nusselt numbers defined respectively as follows:

$$Nu_{Cv} = -\int_0^1 \left. \frac{\partial T}{\partial X} \right|_{X=1} dY \tag{18}$$

$$Nu_{Rd} = \int_0^1 N_r Q_r|_{X=1} dY \tag{19}$$

where $N_r = \sigma T_C'^4 L' \big/ \lambda (T_H' - T_C')$ and $Q_r = Q_r' / \sigma T_C'^4$ are the radiation-convection interaction ratio and the non-dimensional net radiation heat flux, respectively.

The total average Nusselt number is obtained by summing the convective and radiative components:

$$Nu = Nu_{Cv} + Nu_{Rd} \tag{20}$$

3 Numerical Validation and Grid Independence Test

The numerical code has been validated by contrasting our results with those obtained by Wang et al. [14] in the situation of coupled impacts of surface radiation and free convection in a differentially heated square cavity filled with air. In this investigation, a uniform 161 × 161 grid was employed. The convergence criterion is verified when the relative error between the new value and the old value of each variable is almost 10^{-8}. The comparative findings are presented in Table 1, in form of convective, radiative and total average Nusselt numbers calculated on the hot wall for $Ra = 10^4$ and $\varepsilon = 0.8$. The table demonstrates the maximum relative error between the mean Nusselt number values obtained by our code and those of the reference [14] is less than 0.8% in the worst case.

Table 1. Results of comparison of the numerical code for Ra = 10^4, ε = 0.8, T'_0 = 293.5 K and $\Delta T'$ = 10 K.

	Nu_{Cv}	Nu_{Rd}	Nu
Present work	2.246	2.382	4.628
Wang et al. [14]	2.249	2.401	4.650
Relative difference %	0.1	0.8	0.5

4 Results and Discussion

The purpose of this research is to investigate the influence of surface radiation on fluid flow and heat transfer in a lid-driven square cavity with discrete conditions imposed on two of its walls. The controlling parameters which are emissivity, ε, and Richardson number, Ri, were varied in the ranges $0 \le \varepsilon \le 1$ and $0.1 \le Ri \le 10$, respectively. The characteristics of the air are assessed at the cold temperature T'_C = 298.15 K. The radiation parameter N_r is calculated from the parameter Ra fixed at 10^5 and the air properties. The outcomes are illustrated in the form of streamlines, isotherms, and the average Nusselt numbers.

Figure 2 shows typically streamlines and isotherms, which characterize the flow structure and temperature patterns within the cavity, obtained numerically for various Richardson number and for ε = 0 (Fig. 2a) and ε = 1 (Fig. 2b). It is important to note that the Richardson number compares the magnitude of forced convection caused by the lid-driven to natural convection caused by buoyancy. In the case without radiation effect (ε = 0), Fig. 2a demonstrates that for Ri = 0.1 (Ri < 1), the forced flow generated by the movement of the entrained wall overcomes buoyancy effects for this small value of Ri. The streamlines in this figure show that the flow is dominated by a large main convective cell with its center is placed near to the middle of the cavity and that rotates clockwise. The formation of this is due to the shear effects caused by the moving wall and the Archimedean effects caused by the differences in temperature between the hot and cold walls. Also, it is seen a small counter clockwise secondary cell (negative cell), with a very faint intensity is also visible; it is imprisoned in a confined region towards the cavity's bottom right corner. The appearance of small cell can be explained by the combined effects of shear and buoyancy forces. By increasing Richardson to Ri = 1, the impact of natural convection caused by buoyancy forces is compared to that of forced convection caused by shear forces created by top wall movement. The streamlines show that the flow structure remains virtually unchanged. It is noted that the small cell observed for Ri = 0.1 disappears for the case of Ri = 1. Consequently, the flow structure becomes monocellular rotating clockwise. Further increase in the Richardson number up to Ri = 10 (Ri > 1 dominant natural convection regime), the flow within the cavity is mainly generated by the buoyancy forces caused by the temperature differences between the hot walls and the cold one. The increase in the Ri indicates that the dynamic structure undergoes limited qualitative changes in the shape of the main cell. It can also be noted that the intensity of the main cell could be improved according to Richardson number. The corresponding isotherms (in the right) are condensed at the level of the active walls implying an important convective heat transfer between the fluid and these walls. A fine thermal boundary layer is also observed near the active walls. Condensed isotherms provide evidence that the cold wall serves as a heat sink for all heat supplied by the heated wall. An increase in Richardson number from 0.1 to 10 shows that the boundary layer gets thicker, this thickening effect leads to a smaller thermal reaction between the active walls and the closed fluid. In addition, the thermal field shows that the boundary layer in the vicinity of the cold wall widens as emissivity increases for all convection regimes. This is because the emissivity reduces the convective heat transfer in favors to the radiation exchange.

When the radiation effect is taken into account ($\varepsilon = 1$), Fig. 2b shows that the radiation does not affect the dynamic structure of the flow for the two values of the Richardson number Ri = 0.1 and Ri = 1. On the other hand, for Ri = 10 the flow structure shows that the main cell undergoes changes. It should be mentioned that the main cell's intensity is subject to an increase or a decrease in the existence of surface radiation, relying on Ri. The corresponding isotherms presented in Fig. 2b (in the right) for $\varepsilon = 1$ and different Ri demonstrates that the presence of radiation makes isotherms more susceptible to changes with the Richardson number.

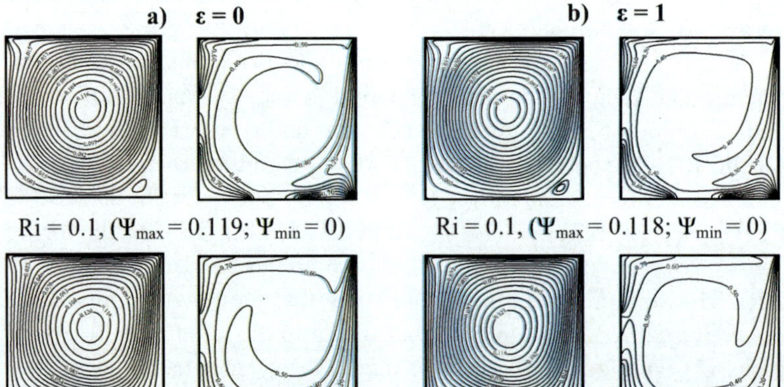

Fig. 2. Streamlines and isotherms obtained, for various values of Ri and two values of ε: a) $\varepsilon = 0$ and b) $\varepsilon = 1$.

Variations of the convective, radiative and total average Nusselt numbers, as function of ε, evaluated on the cold wall are shown in Figs. 3a–3c for various values of Ri. It can be seen from these figures that the convective component remains practically unaffected by increasing the emissivity, ε. But for high Ri, the effect of radiation leads to a slight decrease in convection. On the other hand, the radiative component's evolution as a function of ε is characterized by a linear monotonic increase. As a result of these variations, *Nu* follows the same tendency as Nu_{Rd} and is characterized by a drastic evolution with ε. In addition, by examining these figures, it should be observed that the radiative component becomes almost of the same order of magnitude as the convective one for ε = 1 in the case of Ri = 1. Moreover, the radiation effect dominates the mixed convection for ε ≥ 0.85 in the case of Ri = 10. Finally, by visualizing simultaneously these figures, we indicate that the radiative component is almost unchanged for any increase of Ri, but the convective and total heat transfers decrease notably with Ri.

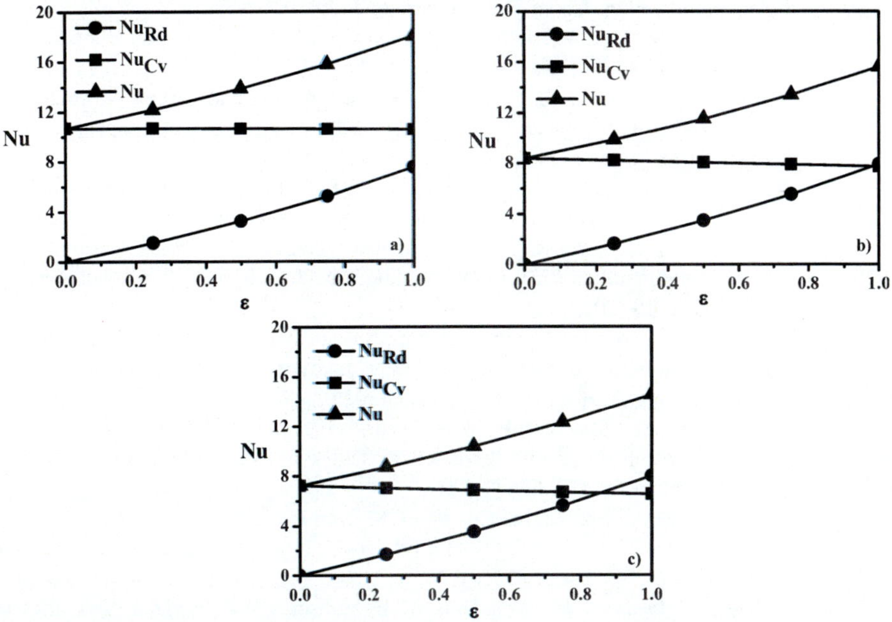

Fig. 3. Variations, with ε, of the mean Nusselt numbers on the cold wall for different values of Ri: a) Ri = 0.1, b) Ri = 1 and c) Ri = 10.

5 Conclusions

In this paper, we present a numerical study performed on the combined mixed convection and surface thermal radiation in a discretely heated lid-driven square cavity. The results obtained indicate that the flow structure is monocellular, characterized by the existence of a large cell generated by the combined effects of shear and buoyancy. This structure is not affected by the walls emissivity while the thermal field is slightly modified by the presence of radiation. In addition, it should be noted that the walls emissivity increases the radiative heat exchange, reduces relatively the convective component and improves the total heat exchange. It is also noted that the increase in Ri causes a decrease in Re, which is explained by a slow moving of the mobile wall and subsequently the reduction of the convective effect favored by the inertia of the wall. on the other hand, this predominance of natural convection relatively enhances radiation.

References

1. Arroub, I., Bahlaoui, A., Raji, A., Hasnaoui, M., Naïmi, M.: Cooling enhancement by nanofluid mixed convection inside a horizontal vented cavity submitted to sinusoidal heating. Eng. Comput. (Swansea, Wales) **35**, 1747–1773 (2018)
2. Arroub, I., Bahlaoui, A., Belhouideg, S., Raji, A., Hasnaoui, M.: Combined effects of inclination angle and imposed flow on mixed convective cooling inside a vented cavity crossed by nanofluids. Phys. Chem. Res. **11**, 631–642 (2023)
3. Sadaoui, D., Sahi, A., Djerrada, A., Mansouri, K.: Coupled radiation and natural convection within an inclined sinusoidal solar collector heated from below. Mech. Ind. **17**, 302 (2016)
4. Saravanan, S., Sivaraj, C.: Combined thermal radiation and natural convection in a cavity containing a discrete heater: effects of nature of heating and heater aspect ratio. Int. J. Heat Fluid Flow **66**, 1339–1351 (2017)
5. Rehhali, K., et al.: Lattice Boltzmann approach for natural convection and radiation in a tilted square cavity. J. Thermophys. Heat Transf.Thermophys. Heat Transf. **33**, 322–333 (2019)
6. Arroub, I., Bahlaoui, A., Raji, A., Hasnaoui, M., Naïmi, M.: Varying heating effect on mixed convection of nanofluids in a vented horizontal cavity with injection or suction. Heat Transf. Eng. **7632**, 1–18 (2019)
7. Arroub, I., Bahlaoui, A., Ezzaraa, K., Raji, A., Hasnaoui, M., Naïmi, M.: Effect of phase shift on mixed convection in a rectangular vented cavity filled with a nanofluid and submitted to periodic heating. J. Phys. Conf. Ser. **1226** (2019)
8. Antar, M.A., Ben-Mansour, R., Al-Dini, S.A.: The effect of thermal radiation on the heat transfer characteristics of lid-driven cavity with a moving surface. Int. J. Numer. Methods Heat Fluid Flow **24**, 679–696 (2014)
9. Belmiloud, M.A., Eddine, N., Chemloul, S.: Numerical study of mixed convection coupled to radiation in a square cavity with a lid-driven. Int. J. Mech. Aerospace, Ind. Mechatron. Manuf. Eng. **9**, 1690–1696 (2016)
10. Dahani, Y., Hasnaoui, M., Amahmid, A., El Mansouri, A., Hasnaoui, S.: Lattice Boltzmann simulation of combined effects of radiation and mixed convection in a lid-driven cavity with cooling and heating by sinusoidal temperature profiles on one side. Heat Transf. Eng. **41**, 433–448 (2020)
11. Addini, M.M., Gandjalikhan Nassab, A.: Combined mixed convection and radiation simulation of inclined lid driven cavity. Energy Equip. Syst. **6**, 261–277 (2018)
12. Mohamad, A.A.: Lattice Boltzmann Method Fundamentals and Engineering Applications with Computer Codes, 2nd edn. Springer, London (2011). https://doi.org/10.1007/978-0-85729-455-5
13. Hottel, H.C., Saroffim, A.F.: Radiative Heat Transfer. McGraw-Hill, New York (1967)
14. Wang, H., Xin, S., Le Quéré, P.: Étude Numérique Du Couplage De La Convection Naturelle Avec Le Rayonnement De Surfaces En Cavité Carrée Remplie D'Air. Comptes Rendus - Mec. **334**, 48–57 (2006)

Study of Thermal Conduction Within Two Orthotropic Mediums in Geometry Cylindrical Subjected to a Flux Varying Longitudinally on the Lateral Surface

M. Nikchi, R. Idmoussa, K. Zniber, J. Lahjomri, and A. Oubarra(✉)

Laboratory of Mechanics, Faculty of Science Aïn Chock, University Hassan II, 20100 Casablanca, Morocco
mohammed.nikchi-etu@etu.univh2c.ma, aoubarra@mail.com

Abstract. This article studies an exact solve for 2-D thermal conduction through cylinders with orthotropic properties, subjected to a prescribed temperature on both end sections and longitudinally varying heat flux over the whole lateral surface. The method of separation of variables is employed to integrate the orthotropic equation along with the boundary conditions, temperature continuity, and flux at the interface between the two mediums. The effectiveness of the obtained results is examined through the ratio of the principal thermal conductivity and the radial ratio of the thermal conductivity between the two mediums. The analysis demonstrates that the radial and axial profiles of temperature and flux are influenced by the specific values of the given problem data.

Keywords: Orthotropic Media · Radial and Axial Temperature · Radial and Axial Flux · Longitudinally Varying Heat Flux

1 Introduction

Composite orthotropic materials are commonly used in engineering and many investigators are focusing on their thermal property. It is very complicated to obtain the analytical solutions of the mixed boundary conditions, thus numerous computational methods are proposed to solve the problems listed, like FEM [1], FVM [2], FDM [3] and so on.

The transfer of heat in a cylindrical multi-layer is an important problem engineering [4] and [5]. The utilization of a material combination with variable thermomechanics property frequently aids in delivering performance features desired for engineered systems. The fuel rods in a nuclear reactor as an example, are frequently constructed as a multilayered cylinder with the nuclear reactor fuel housed in the ring region inside two tubes [6]. Additional uses where heat transfer in a layered cylinder is relevant include the storage of compressed hydrogen [7], super-conductive cables [8], transducers with piezoelectricity [9], and engineering structures [10]. [11] report a solution analytically in a cylindrical laminate. The influence of fiber arrangements of the orthotropic cylinder and thermal limit conditions on the unsteady conductive heat transfer of this orthotropic

S. Aniss et al. (Eds.): CMM 2022, LNME, pp. 201–212, 2024.
https://doi.org/10.1007/978-3-031-46973-2_19

cylinder is analyzed. [12] presented a study on determining the temperature distribution within a heat-generating cylinder under homogeneous boundary conditions. The authors proposed a method to estimate the thermal conductivity of the cylinder based on this distribution. They developed a code using the finite volume approach and employed the conjugate gradient method for optimization. Aimed to accurately model and analyze the temperature behavior within the cylinder. By obtaining the temperature distribution, they could then infer the thermal conductivity of the material. The finite volume method, a numerical technique commonly used for solving heat transfer problems, was implemented in their code to discretize the governing equations and obtain a numerical solution. [13] presented numerical analysis assisted by Python to model a 2D material domain and predict the temperature distribution. The investigation of orthotropic thermo-physical properties on heat conduction, focusing on Ti-4Al-4V alloy, provides insights for optimizing machining parameters and selecting appropriate cooling strategies. [14] presented a numerical solution of heat transfer within an orthotropic medium in cylindrical geometry, the heat equation for the orthotropic cylinder with corresponding boundary conditions is solved using the Implicit Alternating Direction (ADI) method. The results show that the isotherms depend on the values of the data problem. Moreover, the steady state time is rather affected by the ratio of the thermal conductivities of the network. The lower value of this time corresponds to higher values of the axial thermal conductivities of the media. [15] presented a thermoelastic heat conduction model for orthotropic materials, considering limited velocities of heat wave propagation. The model accounts for variations in the coefficient of thermal conductivity with temperature. The Laplace transform method is employed to obtain general solutions for the field variables equations.

The present paper studies the two-dimensional heat transfer within a cylinder formed by two orthotropic mediums. The cylinder is characterized by different temperatures at its left and right ends, while the side surface is subjected to a longitudinally varying heat flux. The primary contribution of this study lies in the development of an exact analytical solution for the problem at hand. The analysis involves solving the coupled conduction equations for the two orthotropic mediums, taking into account the appropriate boundary conditions, as well as ensuring continuity of temperatures and heat flux at the interface between the two mediums. The method employed to obtain the solution is the technique of separation of variables. By employing this analytical approach, it becomes possible to explore the thermal behaviors of both orthotropic mediums in-depth. Moreover, the study investigates the impact of various parameters, such as the principal thermal conductivity ratio and the ratio of the radial thermal conductivity of the two mediums, on the overall heat transfer characteristics.

2 Problem Formulation

Let us consider two cylindrical orthotropic mediums without heat source, of length L, and inner and outer radii respectively a and b, whose bases $z = 0$ and $z = L$ are kept respectively at different temperatures (T_l, T_r) while a radial heat flux varying longitudinally is applied on the lateral surface $r = b$ (Fig. 1).

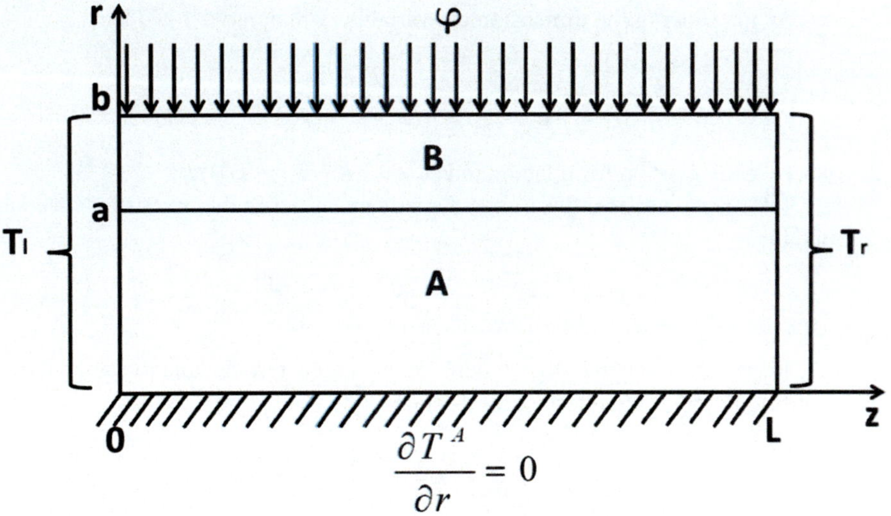

Fig. 1. Bidirectional representation of the two orthotropic mediums.

The heat transfer equations for orthotropic mediums in steady state are written:

$$div(\overline{\overline{K^i}}\overrightarrow{grad}T^i) = 0 \tag{1}$$

$$\begin{cases} \varphi_r = K_r^i \frac{\partial T^i}{\partial r} \\ \varphi_z = -K_z^i \frac{\partial T^i}{\partial z} \end{cases} \tag{2}$$

the index i represents the medium in which the study is conducted.

The determining equations of the thermal distribution are linked to the limit and interface conditions:

$$z = 0 \quad T^i(r, z = 0) = T_l \tag{3}$$

$$z = L \quad T^i(r, z = L) = T_r \tag{4}$$

$$r = 0 \quad \left.\frac{\partial T^A}{\partial r}\right|_{r=0} = 0 \tag{5}$$

$$r = a \quad T^A(r, z)\Big|_{r=a} = T^B(r, z)\Big|_{r=a} \tag{6}$$

$$r = a \quad \left.\frac{\partial T^A}{\partial r}\right|_{r=a} = \frac{K_r^B}{K_r^A}\left.\frac{\partial T^B}{\partial r}\right|_{r=a} \tag{7}$$

$$r = b \quad \left.\frac{\partial T^B}{\partial r}\right|_{r=b} = \frac{\varphi(z)}{K_r^B} \tag{8}$$

By introducing the dimensionless variables which are defined by:

$$Z = \frac{z}{b}, R = \frac{r}{b}, \theta = \frac{T^i - T_r}{T_l - T_r}, \Phi^i_{R,Z} = \frac{\varphi^i_{r,z}}{\varphi_{\text{Ref}}} \tag{9}$$

where, $G = L/b$ the form factor and $\varphi_{ref} = (K_z^B(T_r - T_l))/L$

Taking into account these characteristic quantities in the heat diffusion equations we obtain:

$$\frac{\partial^2 \theta^i}{\partial R^2} + \frac{1}{R}\frac{\partial \theta^i}{\partial R} + \upsilon^2 \frac{\partial^2 \theta^i}{\partial Z^2} = 0 \tag{10}$$

Taking into account the dimensionless characteristic quantities in the boundary conditions and at the interfaces we obtain:

$$Z = 0 \quad \theta^i(R, Z = 0) = 1 \tag{11}$$

$$Z = G \quad \theta^i(R, Z = G) = 0 \tag{12}$$

$$R = 0 \quad \left.\frac{\partial \theta^A}{\partial R}\right|_{R=0} = 0 \tag{13}$$

$$R = \xi \quad \theta^A(R = \xi, Z) = \theta^B(R = \xi, Z) \tag{14}$$

$$R = \xi \quad \left.\frac{\partial \theta^A}{\partial R}\right|_{R=\xi} = \frac{1}{\delta}\left.\frac{\partial \theta^B}{\partial R}\right|_{R=\xi} \tag{15}$$

$$R = 1 \quad \left.\frac{\partial \theta^B}{\partial R}\right|_{R=1} = \frac{\upsilon^B}{G}\Phi(Z) \tag{16}$$

With: $\upsilon^2 = K_z/K_r$ principal thermal conductivity ratio, $\delta = K_r^A/K_r^B$ Radial thermal conductivity ratio and $\xi = a/b$.

Solving the dimensionless system of equations provides access to the temperature and flux profile in cylindrical coordinates. They will be determined analytically by the variables separation method.

3 Problem Solving

The dimensionless system of equations, expressed by the heat conduction equations, the boundary and interface conditions, reflecting the thermal behavior of the two cylinders, is solved by the method of separations of variables in steady state.

We apply the variable separation method for each medium, by calculation we according to Newman's condition at the external lateral surface, the imposed flux varies as a function of Z in a sinusoidal way $\Phi(Z) = \sin(\pi Z/G)$, Consequently the general expression of $\theta^A(R, Z)$ and $\theta^B(R, Z)$ in this case is written in the form:

$$\Theta^A(R, Z) = 1 - Z/G + \sum_{n=1} \frac{\upsilon}{n\pi}.\frac{1}{I_1(\lambda\upsilon)}\left[1 + C.\left(\frac{K_1(\lambda\upsilon)}{I_1(\lambda\upsilon)} + \frac{K_0(\lambda\upsilon\xi)}{I_0(\lambda\upsilon\xi)}\right)\right]. \sin(\lambda Z).I_0(\lambda\upsilon R) \tag{17}$$

$$\Theta^B(R,Z) = 1 - Z/G + \sum_{n=1} \frac{\upsilon}{n\pi} \cdot \frac{1}{I_1(\lambda\upsilon)} \left[I_0(\lambda\upsilon R) + C. \left(\frac{K_1(\lambda\upsilon)}{I_1(\lambda\upsilon)} . I_0(\lambda\upsilon R) + K_0(\lambda\upsilon R) \right) \right] . \sin(\lambda Z) \quad (18)$$

I_0, I_1, K_0 and K_1: Are Bessel functions.

With: $C = \frac{\left(\frac{1}{\delta} - 1\right)}{\frac{K_1(\lambda\upsilon)}{I_1(\lambda\upsilon)} + \frac{K_0(\lambda\upsilon\xi)}{I_0(\lambda\upsilon\xi)} - \frac{1}{\delta}\left(\frac{K_1(\lambda\upsilon)}{I_1(\lambda\upsilon)} - \frac{K_1(\lambda\upsilon\xi)}{I_1(\lambda\upsilon\xi)}\right)}$.

Equations (19)–(22) give the general expressions for the axial and radial heat fluxes for each medium.

$$\Phi_R^A(R,Z) = \sum_{n=1} \frac{\delta}{I_1(\lambda\upsilon)} \left[1 + C. \left(\frac{K_1(\lambda\upsilon)}{I_1(\lambda\upsilon)} + \frac{K_0(\lambda\upsilon\xi)}{I_0(\lambda\upsilon\xi)} \right) \right] . \sin(\lambda Z) . I_1(\lambda\upsilon R) \quad (19)$$

$$\Phi_R^B(R,Z) = \sum_{n=1} \frac{1}{I_1(\lambda\upsilon)} \left[I_1(\lambda\upsilon R) + C. \left(\frac{K_1(\lambda\upsilon)}{I_1(\lambda\upsilon)} . I_1(\lambda\upsilon R) - K_1(\lambda\upsilon R) \right) \right] . \sin(\lambda Z) \quad (20)$$

$$\Phi_Z^A(R,Z) = \frac{1}{G} + \sum_{n=1} \frac{\upsilon^2}{G} \cdot \frac{1}{I_1(\lambda\upsilon)} \left[1 + C. \left(\frac{K_1(\lambda\upsilon)}{I_1(\lambda\upsilon)} + \frac{K_0(\lambda\upsilon\xi)}{I_0(\lambda\upsilon\xi)} \right) \right] . \cos(\lambda Z) . I_0(\lambda\upsilon R) \quad (21)$$

$$\Phi_Z^B(R,Z) = \frac{1}{G} + \sum_{n=1} \frac{\upsilon^2}{G} \cdot \frac{1}{I_1(\lambda\upsilon)} \left[I_0(\lambda\upsilon R) + C. \left(\frac{K_1(\lambda\upsilon)}{I_1(\lambda\upsilon)} . I_0(\lambda\upsilon R) + K_0(\lambda\upsilon R) \right) \right] . \cos(\lambda Z) \quad (22)$$

The temperatures and flux, axial and radial in each medium are established by a calculation iterative, comparing the consecutive difference of all internal temperatures among two orders of series and using a convergence criterion, based on:

$$\left| \Theta^{\mathrm{A}}(R,\ Z,\ n+1) - \Theta^{\mathrm{A}}(R,\ Z,\ n) \right| + \left| \Theta^{\mathrm{B}}(R,\ Z,\ n+1) - \Theta^{\mathrm{B}}(R,\ Z,\ n) \right| < \varepsilon \quad (23)$$

4 Results and Discussion

The radial and axial temperature and flux profiles will be represented. The results are relative to the case of orthotropic media in cylindrical geometry, of constant factor of form G (G = 2), for values of the ratio of thermal conductivities = υ0.5, 1, 2 and values of the ratio of radial thermal conductivities δ = 0.5, 1, 2 and of sinusoidal imposed flux. The position of the interface between the mediums is taken to be equal to 0.8. The Figs. 2, 3 and 4 show the temperature radial contour in the mediums corresponding to a fixed value of the radial thermal conductivity ratio coupled with different main thermal conductivity ratio values. Hence, for a certain determined value of δ, the temperature radial shape rises from the middle to the outer side surface of the cylinder. The increase of the ratio of the principal thermal conductivity decreases the thermal level in the first medium and raises it in the second medium. Moreover, it is noticed that for a constant value of the ratio of the principal thermal conductivity, the increasing of results in the same findings with a global reduction of the level of heat. Globally, it seems that the 2nd medium is more sensible to the impact of υ. Moreover, we note a specific radius relating to the reversal of the second medium temperature regarding υ. Curves Figs. 5, 6 and 7 present the radial flux of different curvatures associated with the mediums. The radial flux increases from the center to the outer end of the outer cylinder. The radial flux profile increases in the mediums corresponding to the ratio of the radial thermal conductivity δ. It is noted that when the main thermal conductivity ratio υ decreases, the radial flow profile becomes higher.

For all Figs. 8, 9 and 10 extreme axial temperatures correspond to boundary conditions 1 and 0 respectively. Note that the shape of axial temperature depends on the values of υ and δ. The more the thermal conductivity ratio υ is increased, the axial temperature profile becomes weaker for the different values of the radial thermal conductivity ratio for medium A.

On the other hand, for medium B the axial temperature profile becomes greater when the ratio thermal conductivities υ increase due to the flux imposed on the lateral surface of medium B.

Figures 11, 12 and 13 show that the axial flux of medium A increases as the ratio of thermal conductivities increase for values of z which are less than 1, and conversely for values of Z greater than 1. Note that for values of Z less than 1, and for low thermal conductivity ratio, the axial flux of medium B becomes more dominant. For the opposite case where Z greater than 1, the axial flux of medium B becomes weak.

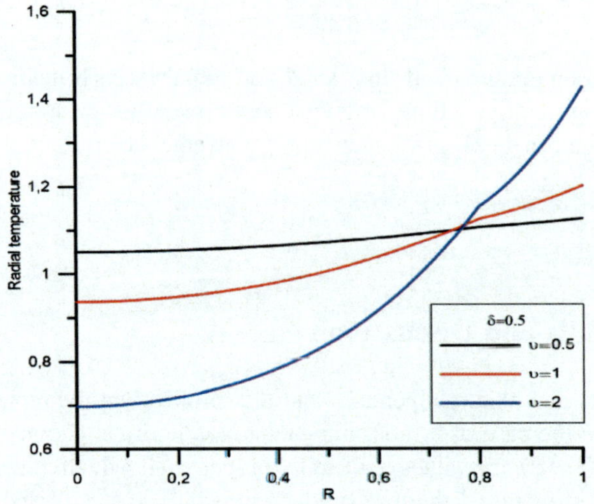

Fig. 2. Central radial temperature for $\delta = 0.5$.

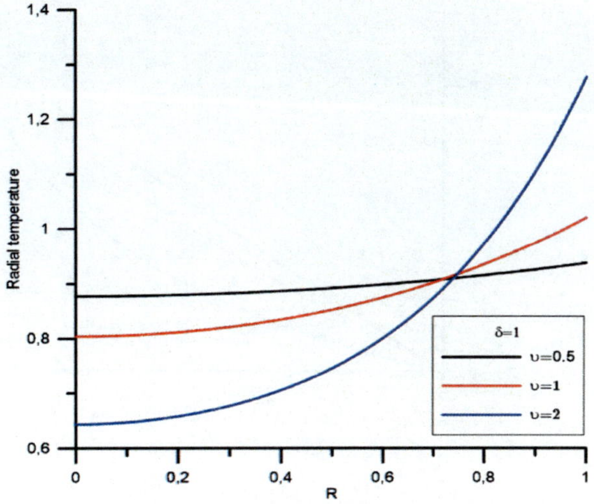

Fig. 3. Central radial temperature for $\delta = 1$.

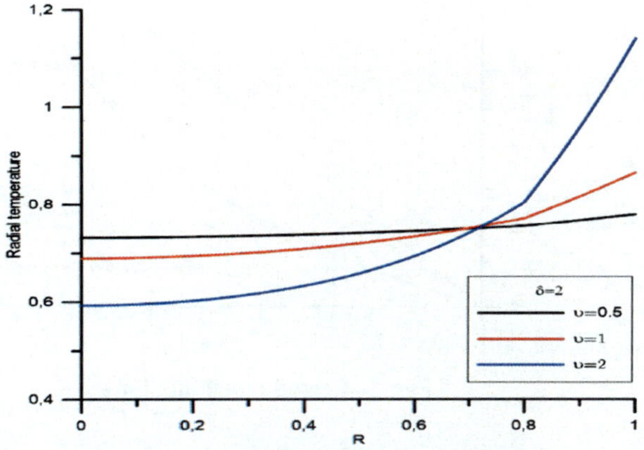

Fig. 4. Central radial temperature for $\delta = 2$.

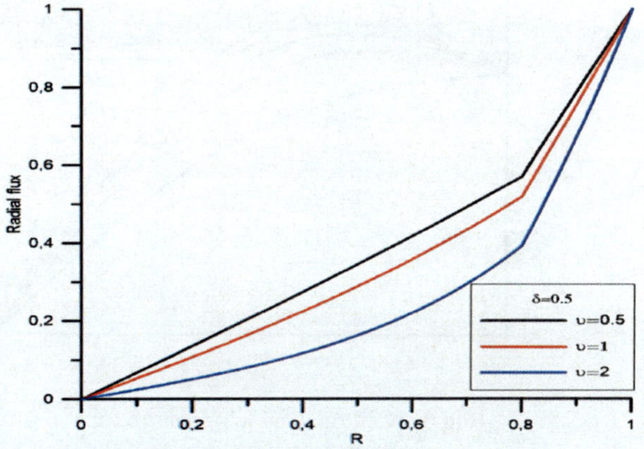

Fig. 5. Central radial flux for $\delta = 0.5$.

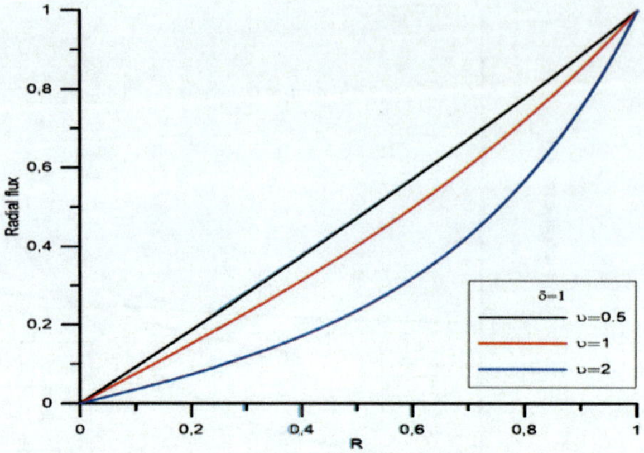

Fig. 6. Central radial flux for δ = 1.

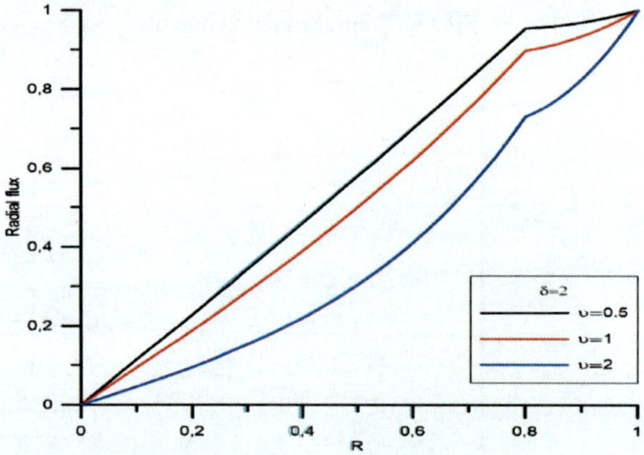

Fig. 7. Central radial flux for δ = 2.

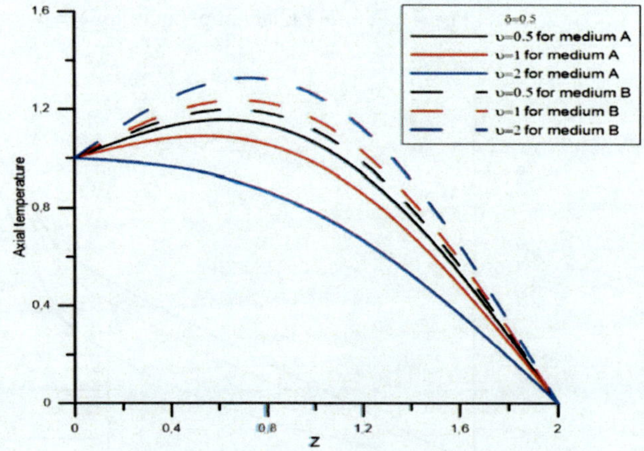

Fig. 8. Central axial temperature for δ = 0.5.

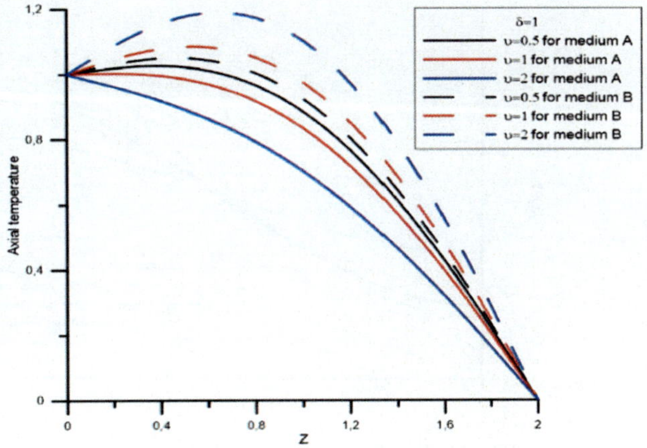

Fig. 9. Central axial temperature for δ = 1.

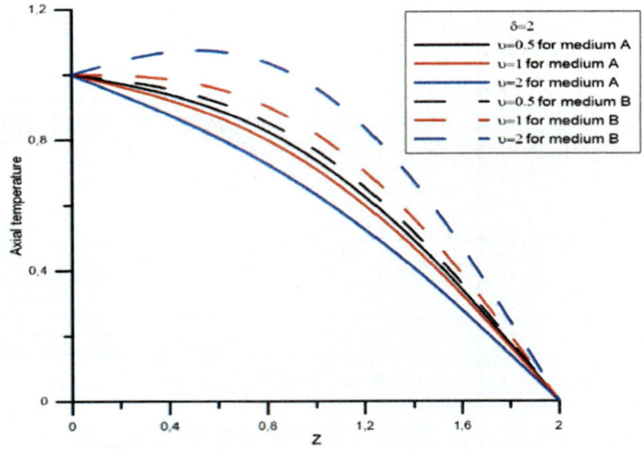

Fig. 10. Central axial temperature for δ = 2.

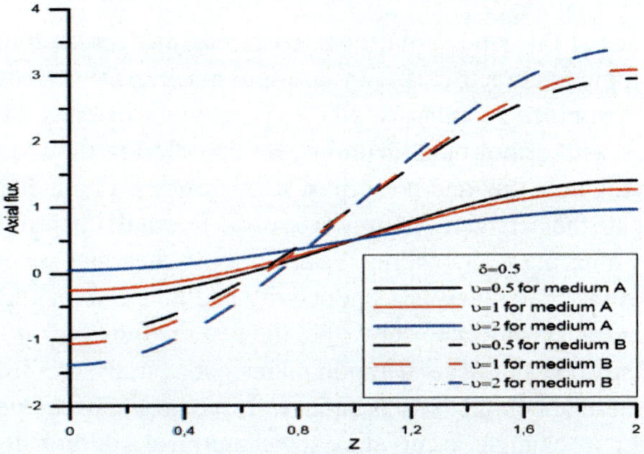

Fig. 11. Central axial flux for δ = 0.5.

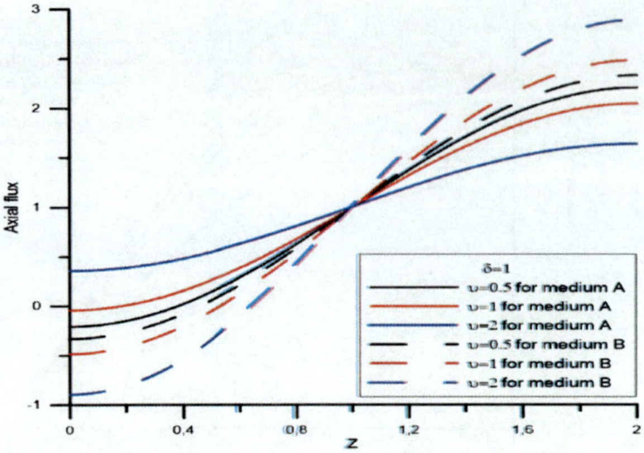

Fig. 12. Central axial flux for $\delta = 1$.

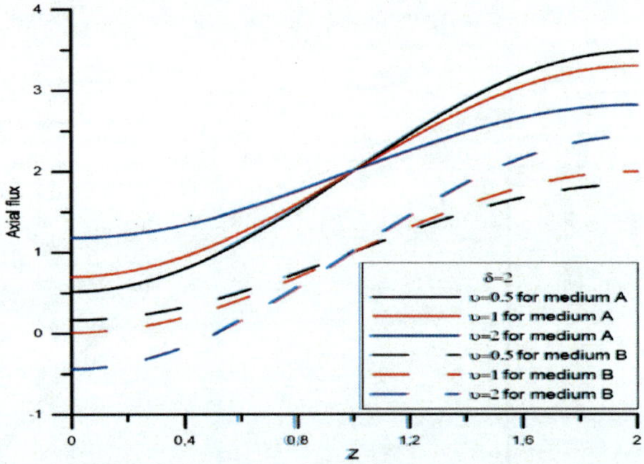

Fig. 13. Central axial flux for $\delta = 2$.

5 Conclusion

The findings of this study contribute to our ongoing research on heat transfer phenomena in complex geometries and orthotropic materials. Previous studies have primarily focused on simpler geometries, whereas this work extends the analysis to cylindrical geometries with orthotropic mediums. By considering the specific case of a longitudinally varying heat flux and prescribed temperature on both end sections, we were able to investigate the behavior of temperature and thermal flux profiles in this configuration. This study aims at understanding the orthotropic character and in particular the effect of the ratio of the radial thermal conductivity and the principal thermal conductivity ratio on the heat transfer mechanism within the two mediums.

Our study revealed several important insights. Firstly, we observed that when the ratio of radial thermal conductivity is equal to 1, the profile of temperature and flux remains regular without changes in curvature at the interface. Additionally, we found that the level of radial fluxes decreases when the ratio of principal thermal conductivity is very high.

This observation highlights the significance of the principal thermal conductivity ratio in influencing the radial heat transfer. Furthermore, we demonstrated that an increase in the ratio of radial thermal conductivity leads to a higher radial flux profile. This finding suggests the importance of considering the orthotropic behavior of the medium when analyzing heat transfer in such systems. Furthermore, we observed that an increase in the values of the ratio of principal thermal conductivity leads to an increase in axial flux. This result emphasizes the role of the principal thermal conductivity ratio in governing the heat transfer mechanism within the mediums. Lastly, our study demonstrated that by raising the ratio of radial thermal conductivity, the effect of the flux applied to the lateral side surface is reduced. These results expand the existing knowledge in the field and provide valuable insights for researchers and engineers working on heat transfer analysis and design in similar configurations.

References

1. Pashah, S., Arif, A.F.M., Zubair, S.M.: Study of orthotropic pin fin performance through axisymmetric thermal non-dimensional finite element. Appl. Therm. Eng. **31**, 376–384 (2011). https://doi.org/10.1016/j.applthermaleng.2010.09.028
2. Demuth, C., Mendes, M.A.A., Ray, S., Trimis, D.: Performance of thermal lattice Boltzmann and finite volume methods for the solution of heat conduction equation in 2D and 3D composite media with inclined and curved interfaces. Int. J. Heat Mass Transf. **77**, 979–994 (2014). https://doi.org/10.1016/j.ijheatmasstransfer.2014.05.051
3. Mosayebidorcheh, S., Farzinpoor, M., Ganji, D.D.: Transient thermal analysis of longitudinal fins with internal heat generation considering temperature dependent properties and different fin profiles. Energy Convers. Manag. **86**, 365–370 (2014). https://doi.org/10.1016/j.enconman.2014.05.03
4. Hahn, D.W., Ozisik, M.N.: Heat Conduction, 3rd edn. Wiley, Hoboken (2012)
5. Kakaç, S., Yener, Y.: Heat Conduction, 2nd edn. NASA STI/Recon. Technical report A 86, 21310 (1985)
6. Fenech, H.: Heat Transfer and Fluid Flow in Nuclear Systems. Elsevier, Amsterdam (2013)
7. Manahan, S.: Environmental Chemistry (Novena Edi). United States of America (2010)
8. Olsen, S., Traeholt, C., Kuhle, A., Tonnesen, O., Daumling, M., Oestergaard, J.: Loss and inductance investigations in a 4-layer superconducting prototype cable conductor. IEEE Trans. Appl. Supercond. **9**, 833–836 (1999). https://doi.org/10.1109/77.783426
9. Wang, J., Qin, L., Li, W., Song, W.: Parametric analysis and optimization of radially layered cylindrical piezoceramic epoxy composite transducers. Micromachines **9**, 585 (2018). https://doi.org/10.3390/mi9110585
10. Cavalcante, M., Marques, S., Pindera, M.J.: Transient thermomechanical analysis of a layered cylinder by the parametric finite-volume theory. J. Therm. Stresses **32**, 112–134 (2008). https://doi.org/10.1080/01495730802540783
11. Norouzi, M., Rezaei, N.S.M., Kayhani, M.H., Shariati, M., Karimi, D.M., Naghavi, M.S.: Exact solution of unsteady conductive heat transfer in cylindrical composite laminates. J. Heat Transf. **134** (2012). https://doi.org/10.1115/1.4006009
12. Patil, S., Chintamani, S., Kumar, R., Dennis, B.H.: Determination of orthotropic thermal conductivity in heat generating cylinder. In: ASME International Mechanical Engineering Congress and Exposition, American Society of Mechanical Engineers Digital Collection (2016). https://doi.org/10.1115/IMECE2016-67918

13. Ranjan, A., Kumar, A., Roy, A.K., Kumar, K.: Python assisted numerical analysis of heat conduction for an orthotropic material. Adv. Mater. Process. Technol. **8**, 2014–2028 (2022). https://doi.org/10.1080/2374068X.2022.2031561
14. Nikchi, M., Hamza, H., Lahjomri, J., Zniber, K., Oubarra, A.: Numerical solution of heat transfer by conduction inside orthotropic solid materials in cylindrical geometry. Comput. Therm. Sci. **14**, 81–100 (2022). https://doi.org/10.1615/ComputThermalScien.2022043168
15. Abouelregal, A.E., Ahmad, H., Yavuz, M., Nofal, T.A., Alsulami, M.D.: An orthotropic thermo-viscoelastic infinite medium with a cylindrical cavity of temperature dependent properties via MGT thermoelasticity. Open Phys. **20**, 1127–1141 (2022). https://doi.org/10.1515/phys-2022-0143

Improved Heat Transfer by Using a Channel with Wavy Surface

S. Elalaoui(✉), A. Oubarra, J. Lahjomri, and K. Zniber

Laboratory of Mechanics, Faculty of Sciences Ain Chock, Hassan II University of Casablanca, 20000 Casablanca, Morocco
soaad.elalaoui-etu@etu.univh2c.ma

Abstract. The work consists of a numerical study of a forced convection of an incompressible, laminar flow of a Newtonian fluid within a channel with symmetrical wavy walls of semi-infinite length.

The temperature profile and the distribution of the local Nusselt number are determined numerically using the Alternating Direction Implicit method. We present the effect of certain parameters on the radial and axial temperature profiles and on the Nusselt number.

Keywords: ADI · Flow · Laminar · Forced convection · Heat transfer · Corrugated channel

1 Introduction

Due to the importance of improving heat transfer in different domains such as, for example, chemical reactors or nuclear systems, there are several works on flows in pipes or channels with variable cross-section as reported in the literature. Among these systems, there are in particular ones with corrugated walls used for their importance in the industrial field. Many studies were reported in the literature dealing with such systems.

A numerical study on the effect of geometrical parameters was conducted by **Mohammed et al.** [1]. The upper and lower walls of the corrugated channel are subjected to a constant heat flux. The governing equations are solved by the finite volume method. They concluded that the use of a corrugated channel is a suitable method to increase the thermal efficiency and achieve a higher compactness of the heat exchanger. A detailed numerical study of heat transfer of a fluid flowing in a corrugated channel with Reynolds number (Re = 2000–5500), and corrugated plate angles V ($\theta = 20°$, 40°, 60°), with constant heat flow ($q'' = 580, 830, 1090\ w/m^2$), was presented by **Yang et al.** [2]. The numerical results have been validated with data reported by **Naphon** [3]. They concluded that increasing the angles of corrugated plates can improve the heat transfer. A numerical study of transient flow with Reynolds number varying from 200 to 3000 in three dimensions, to examine the geometrical effect on the thermo-hydraulic characteristics of periodic cross-corrugated channels, was presented by **Liu and Niu** [4]. They

S. Aniss et al. (Eds.): CMM 2022, LNME, pp. 213–224, 2024.
https://doi.org/10.1007/978-3-031-46973-2_20

concluded that the vertex angle has a strong impact on the heat transfer and pressure drop in the corrugated channel.

In this present work, we studied numerically the thermal behavior for a Newtonian fluid in a channel with corrugated walls. In the first part, we describe the problem and its mathematical formulation, i.e., the energy equation as well as the imposed boundary conditions and the necessary assumptions that make it easier for us to solve the problem. In the second part, we presented the numerical solution of the basic equation using the ADI method, with a calculation algorithm and the validation of the calculation code developed using the FORTRAN language. The third part is reserved for the presentation of the influence of the geometry of the channel on the structure of the flow and on the heat transfer rate by presenting the parameters influence on the phenomenon of the laminar forced convection inside a corrugated channel and the different results of the numerical study with necessary interpretations. Finally, a general conclusion that summarizes the present work and the perspectives is drawn.

2 Geometric Configuration and Formulation of the Problem

The system studied is illustrated by a channel with symmetrical wavy walls of semi-infinite length, of axis of symmetry Δ, of radius $R(x)$ designating the radial distance from a point of the surface to the axis of symmetry, (Fig. 1), in which circulates a Newtonian fluid whose velocity profile is that of Poiseuille for x < 0. The flow is supposed crossing a very long distance before entering at x = 0, so that it is completely developed in the upstream regime, while on downstream one only a perturbation appeaser due to corrugations as reported in the work of reference [7].

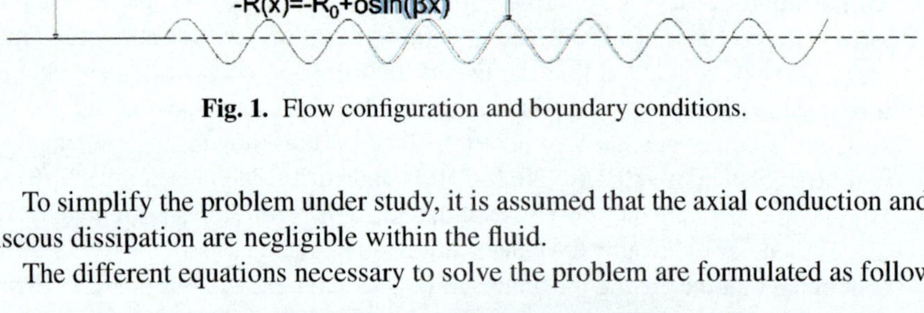

Fig. 1. Flow configuration and boundary conditions.

To simplify the problem under study, it is assumed that the axial conduction and the viscous dissipation are negligible within the fluid.

The different equations necessary to solve the problem are formulated as follows:

$$\left[\frac{\partial T}{\partial t} + u\frac{\partial T}{\partial x} + v\frac{\partial T}{\partial y}\right] = \alpha\left(\frac{\partial^2 T}{\partial y^2}\right) \quad (1)$$

where:

$\alpha = \frac{k}{\rho c_p}$ and the boundary conditions shown in Fig. 1.

According to the article by **Castellões et al.** [5], for a two-dimensional Newtonian incompressible laminar flow in a corrugated channel with an amplitude less than or equal to 0.1 of the corrugated walls ($\delta \leq 0.1$), we have for x > 0: (the same process used by the work of reference [6])

$$u(x, y) = \frac{3R_0 u_m}{2[\delta \sin(\beta x) - R_0]^3}\left[y^2 - (R_0 - \delta \sin(\beta x))^2\right] \quad (2)$$

$$v(x, y) = \frac{3}{2}\left[\frac{R_0 u_m \beta \delta \cos(\beta x) y\left[y^2 - (R_0 - \delta \sin(\beta x))^2\right]}{[-R_0 + \delta \sin(\beta x)]^4}\right] \quad (3)$$

It should be noted that (2) and (3) satisfy well:

$$div\, \vec{u} = \frac{\partial u}{\partial x} + \frac{\partial v}{\partial y} = 0,\ \left.\frac{\partial u}{\partial y}\right|_{y=0} = 0,\ v(x, 0) = 0, u(x, R(x)) = 0, v(x, R(x)) = 0 \quad (4)$$

And for $\delta = 0$ we have:

$$u(x, y) = \frac{3}{2} u_m \left(1 - \frac{y^2}{R_0^2}\right),\ v(x, y) = 0 \quad (5)$$

which is the classical form of the speed of Hagen-Poiseuille flow and indicating the properties of continuity for x = 0.

A quantity of practical importance, the local Nusselt number is defined by:

$$Nu(x, t) = \frac{-D_h k \left.\frac{\partial T}{\partial n}\right|_{x,R(x),t}}{k\big(T_m(x, t) - T_p(x, t)\big)} \quad (6)$$

Let us now define new variables to transform the effect of the corrugated wall from the boundary conditions into equations. The transformed coordinates are:

$$\begin{cases} \eta = \dfrac{y}{R(x)} \\ \xi = \dfrac{x}{PeR_0} \end{cases} \quad (7)$$

Using these new variables, we notice that η varies between 0 and 1 and the domain becomes rectangular. This transformation is introduced in order to solve numerically the energy equation by the ADI method on a rectangular domain.

After performing the domain transformations, a dimensionless analysis is now performed. Thus, we define the following dimensionless quantities:

$$\theta = \frac{T - T_e}{T_p - T_e},\ U(\xi, \eta) = \frac{u(x, y)}{u_m},\ V(\xi, \eta) = \frac{v(x, y)}{u_m},\ \tau = \frac{\alpha t}{R_0^2} \quad (8)$$

With u_m and R_0 represent the mean speed and the average half-distance respectively.

Taking into account of the dimensionless quantities defined above, and the dimensionless pulsation and the occlusion parameters $\omega = R_0 P_e \beta$ and $\varepsilon = \frac{\delta}{R_0}$, respectively, the previous Eqs. (1) and (6) become:

$$\begin{aligned} &\frac{\alpha}{R_0^2}\frac{\partial\theta}{\partial\tau} + \frac{3}{2}\frac{R_0 u_m}{(R_0 - \delta\sin(\omega\xi))}\left(1-\eta^2\right)\left[\frac{\eta\delta\beta\cos(\omega\xi)}{R_0 - \delta\sin(\omega\xi)}\frac{\partial\theta}{\partial\eta} + \frac{1}{R_0 Pe}\frac{\partial\theta}{\partial\xi}\right] \\ &-\frac{3}{2}\left[\frac{R_0 u_m \beta\delta\cos(\omega\xi)\eta\left(1-\eta^2\right)}{(R_0 - \delta\sin(\omega\xi))}\right]\left(\frac{1}{R_0 - \delta\sin(\omega\xi)}\right)\frac{\partial\theta}{\partial\eta} = \alpha\frac{1}{(R_0 - \delta\sin(\omega\xi))^2}\frac{\partial^2\theta}{\partial\eta^2} \end{aligned} \tag{9}$$

$$Nu(\xi,\tau) = \frac{4\sqrt{1 + \frac{\varepsilon^2\omega^2}{pe^2}\cos^2(\omega\xi)}\left.\frac{\partial\theta}{\partial\eta}\right|_{\xi,1,\tau}}{[1-\theta_m(\xi,\tau)]} \tag{10}$$

In order to confine the domain of the numerical resolution, we use the following transformation from infinite to finite domain limit values.

$$\xi = \frac{\xi'}{1-\xi'} \Rightarrow \xi' = \frac{\xi}{\xi+1} \tag{11}$$

By admitting the transformation expressed by Eq. (1), the previous Eqs. (9) and (10) become:

$$\frac{\partial\theta}{\partial\tau} + \frac{3}{2}\left[\frac{\left(1-\eta^2\right)\left(1-\xi'\right)^2}{1-\varepsilon\sin\left(\omega\frac{\xi'}{1-\xi'}\right)}\right]\frac{\partial\theta}{\partial\xi'} = \frac{1}{\left(1-\varepsilon\sin\left(\omega\frac{\xi'}{1-\xi'}\right)\right)^2}\frac{\partial^2\theta}{\partial\eta^2} \tag{12}$$

$$Nu\left(\xi',\tau\right) = \frac{4\sqrt{1+\frac{\varepsilon^2\omega^2}{pe^2}\cos^2\left(\omega\frac{\xi'}{1-\xi'}\right)}\left.\frac{\partial\theta}{\partial\eta}\right|_{\xi',1,\tau}}{[1-\theta_m(\xi',\tau)]} \tag{13}$$

With the following initial condition and boundary conditions:

- Initial condition:

$$\theta\left(\xi',\eta,\tau=0\right) = \theta_i \tag{14}$$

- At the entrance of the canal:

$$\theta\left(\xi'=0,\eta,\tau\right) = 0 \tag{15}$$

- Symmetry condition:

$$\frac{\partial\theta}{\partial\eta}\left(\xi',\eta=0,\tau\right) = 0 \tag{16}$$

- Condition on corrugated wall:

$$\theta\left(\xi',\eta=R_1(x),\tau\right) = 1 \tag{17}$$

The numerical resolution of (12) will now be performed using the initial and boundary conditions (14)–(17).

3 Numerical Method

The resolution method is that of ADI (Alternating Direction Implicit), which consists of iterating the implicit equation system in two time steps.

1st time step $\tau \to \tau + \frac{1}{2}$

$$A(i)\theta_{i-1,j}^{k+\frac{1}{2}} + B(i)\theta_{i,j}^{k+\frac{1}{2}} + C(i)\theta_{i+1,j}^{k+\frac{1}{2}} = D(i) \tag{18}$$

With:

$$A(i) = -\frac{3\Delta\tau}{8\Delta\xi'}\left[\frac{\left(1-\eta^2(i)\right)\left(1-\xi'(i)\right)^2}{1-\varepsilon\sin\left(\omega\frac{\xi'(i)}{1-\xi'(i)}\right)}\right] \tag{19}$$

$$B(i) = 1 \tag{20}$$

$$C(i) = \frac{3\Delta\tau}{8\Delta\xi'}\left[\frac{\left(1-\eta^2(i)\right)\left(1-\xi'(i)\right)^2}{1-\varepsilon\sin\left(\omega\frac{\xi'(i)}{1-\xi'(i)}\right)}\right] \tag{21}$$

$$D(i) = \theta_{i,j}^{k} + \frac{\Delta\tau}{2\Delta\eta^2}\left[\frac{1}{\left(1-\varepsilon\sin\left(\omega\frac{\xi'(i)}{1-\xi'(i)}\right)\right)^2}\right]\left[\theta_{i,j-1}^{k} - 2\theta_{i,j}^{k} + \theta_{i,j+1}^{k}\right] \tag{22}$$

2nd time step $\tau + \frac{1}{2} \to \tau + 1$

$$A(j)\theta_{i,j-1}^{k+1} + B(j)\theta_{i,j}^{k+1} + C(j)\theta_{i,j+1}^{k+1} = D(j) \tag{23}$$

With:

$$A(j) = -\frac{\Delta\tau}{2\Delta\eta^2}\left[\frac{1}{\left(1-\varepsilon\sin\left(\omega\frac{\xi'(i)}{1-\xi'(i)}\right)\right)^2}\right] \tag{24}$$

$$B(j) = 1 + \frac{\Delta\tau}{\Delta\eta^2}\left[\frac{1}{\left(1-\varepsilon\sin\left(\omega\frac{\xi'(i)}{1-\xi'(i)}\right)\right)^2}\right] \tag{25}$$

$$C(j) = -\frac{\Delta\tau}{2\Delta\eta^2}\left[\frac{1}{\left(1-\varepsilon\sin\left(\omega\frac{\xi'(i)}{1-\xi'(i)}\right)\right)^2}\right] \tag{26}$$

$$D(j) = \theta_{i,j}^{k+\frac{1}{2}} - \frac{3\Delta\tau}{8\Delta\xi'}\left[\frac{\left(1-\eta^2(j)\right)\left(1-\xi'(i)\right)^2}{1-\varepsilon\sin\left(\omega\frac{\xi'(i)}{1-\xi'(i)}\right)}\right]\left[\theta_{i+1,j}^{k+\frac{1}{2}} - \theta_{i-1,j}^{k+\frac{1}{2}}\right] \tag{27}$$

4 Results and Discussion

Figure 2 presents the validation between of the numerical solution and another solution obtained with Maple software. The solutions are presented for given values of the parameters. A concordance between Maple code results with the boundary conditions is well respected. Moreover, we notice more clearly that if we increase ξ' the temperature levels also increase, and the domain for which the temperature remains constant is reduced, which shows that the transfer is more and more efficient on a given section as we progress along the channel. Far from the entrance, the temperature varies slightly on the whole section, which shows that the developed regime is close to one, the dimensionless wall temperature.

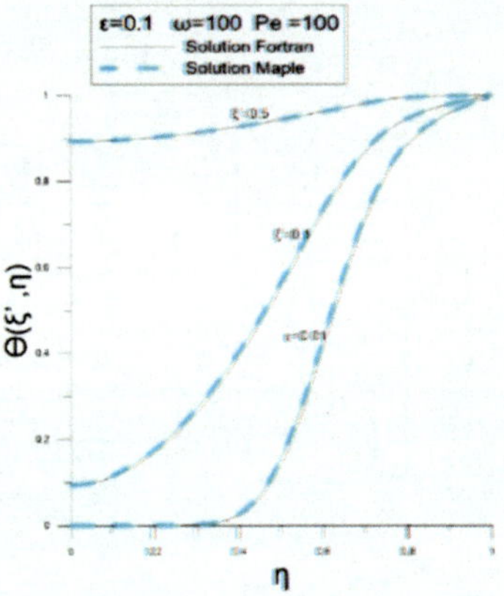

Fig. 2. Steady state comparison between the numerical solution (Fortran) and the numerical solution (Maple) for different values of the dimensionless axial position

Figure 3 shows the variation of the radial temperature at a section $\xi' = 0.01$ for different meshes applied to the corrugated channel. It is clear that the temperature profiles are almost identical for the different meshes studied, especially the meshes (301, 301) and (401, 401). For our calculations, we retain the mesh (301 $\times$ 301) with a time step of ($\Delta\tau = 10^{-4}$). It should be noted that this mesh is influenced by the variables and the parameters of the present study. The profiles in Fig. 2 are practically the same for different grid sizes. Finally, we can add that the other criterion to support our choice of the mesh is the comparison of our different results with those obtained by using the Maple code.

Figures 4, 5 and 6 show the effect of the amplitude ε of the corrugated wall for different ξ' on the radial temperature. Indeed, we note that by increasing the amplitude ε and/or the values of η and ξ' ($\xi' = 0.05$, $\xi' = 0.1$, $\xi' = 0.5$), the radial temperature continues to increase globally and strongly with η for small values of ξ'. This shows that the heat transfer progresses with these parameters, which explains among other things the importance of the choice of a corrugated wall and its amplitude. The effect of ε begins to fade for large values of ξ' ($\xi' = 0.5$ or $\xi = 1$) which corresponds to a large

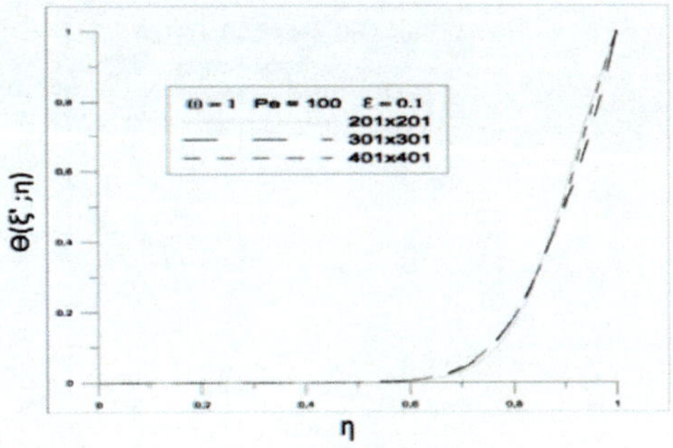

Fig. 3. Test of the mesh used for $\omega = 100$

value of the axial distance (x) which is normal because the temperature begins to reach a uniform developed regime.

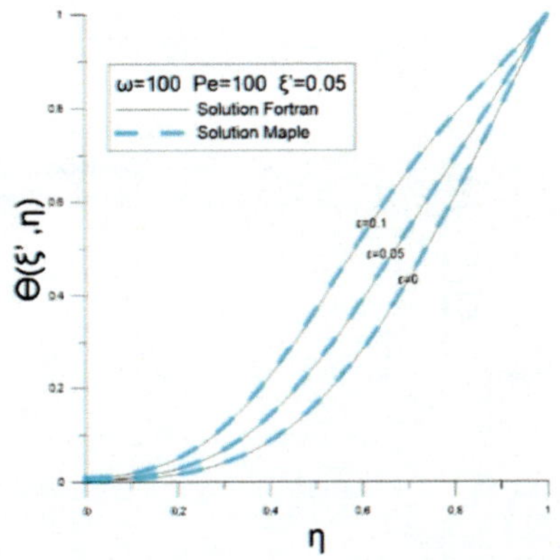

Fig. 4. Evolution of the radial temperature: Influence of amplitude. $\xi' = 0.05$, for $\omega = 100$

In Fig. 7, we clearly see a substantial increase in temperature for small values of ξ' and much lower beyond certain values of the axial distance ξ' ($\xi' = 0.1$, $\xi' = 0.5$). Moreover, we can state that the influence starts tardily for the smaller values of ξ'.

Figures 8, 9 and 10 show the curves of the axial temperature for different amplitudes $\varepsilon = 0$; $\varepsilon = 0.05$ and $\varepsilon = 0.1$ and for $\eta = 0.1$; $\eta = 0.5$ and $\eta = 0.8$; respectively. The comparison between these 3 figures shows that the smaller η is, the more the curves separate, showing that $\theta(\xi', \eta)$ increases with ε and that when one approaches the corrugated wall ($\eta = 0.8$), the curves corresponding to $\varepsilon = 0.05$ and $\varepsilon = 0.1$ tighten and oscillate around a variable axial mean value, which is undoubtedly close to that represented for $\varepsilon = 0$ (flat wall). We can also make the connection with the cases of flat wall ($\varepsilon = 0$; $R = R_0$) and corrugated wall ($\varepsilon \neq 0$; $R = R_0 - \delta \sin(\beta x)$).

Overall, we can say that the exchange improves with ε and that the value of ξ' to reach $\theta(\xi', \eta) = 1$ seems to be close to 0.75 and independent of η.

On Figs. 11 and 12, we observe the evolution of the axial temperature at different η for $\varepsilon = 0.1$ and $\omega = 100$ or $\omega = 500$, respectively. Globally and for the two plots, and as

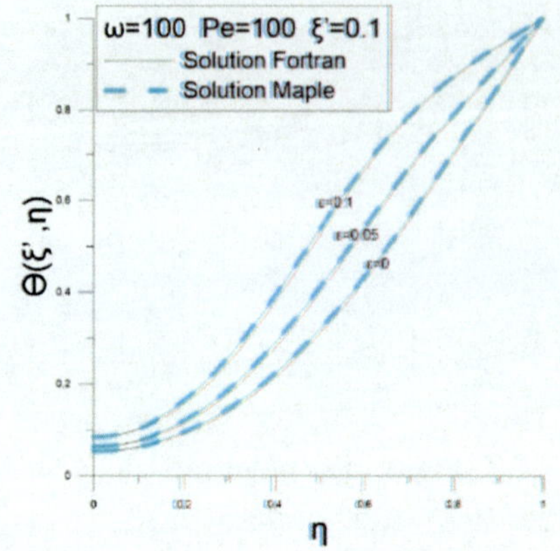

Fig. 5. Evolution of the radial temperature: Influence of amplitude. $\xi' = 0.1$, for $\omega = 100$

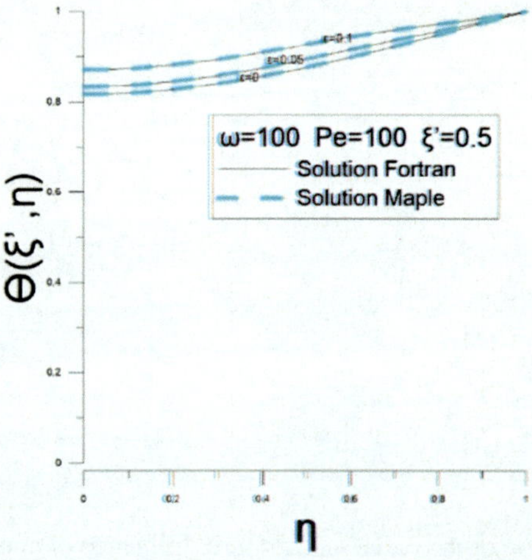

Fig. 6. Evolution of the radial temperature: Influence of amplitude $\xi' = 0.5$ for $\omega = 100$

it was foreseeable, the temperature increases with ξ' for the different η while oscillating in a damped way around an axial average value increasing with η before reaching the constant value 1 of the developed thermal regime. The amplitude of the oscillations of the temperature increases when approaching the wall, which was also foreseeable, and the frequency of the oscillations follows that of the speed which itself follows that of the walls ($\frac{2\pi}{\omega} \approx 0.0628 = \frac{1}{f}$). The passage from $\boldsymbol{\omega} = 100$ to $\boldsymbol{\omega} = 500$ seems apparently to affect only the frequency per spatial period which becomes important as shown in Figs. 11 and 12.

Figure 13 represents the evolution of the local Nusselt number for $\omega = 1$ and $\varepsilon = 0$; $\varepsilon = 0.05$ and $\varepsilon = 0.1$. Two main behaviors are observed. The first one is the thermal inlet for which a strong decrease of Nu is observed followed by a second for which Nu is a periodic function of period $\frac{2\pi}{\omega} = \frac{2\pi}{100} = 0.0628$, which is identical to that of $\theta(\xi', \eta)$, corresponding to the established periodic regime of the Nusselt number which starts at

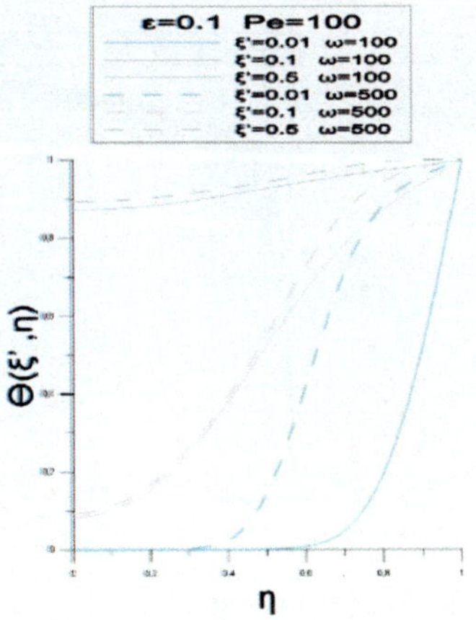

Fig. 7. Evolution of the radial temperature: Influence of the dimensionless pulsation for $\omega = 100$ and $\omega = 500$

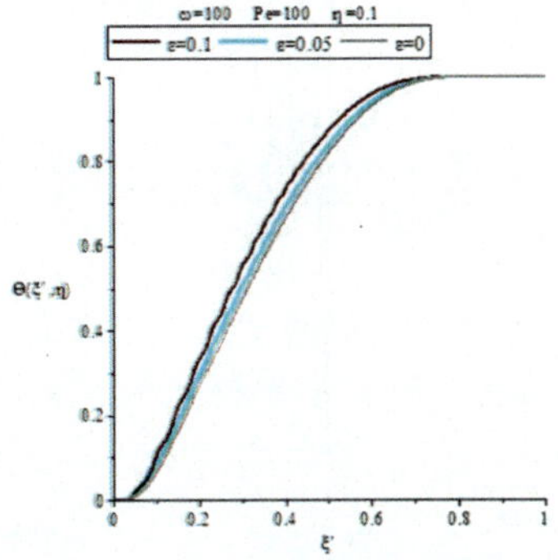

Fig. 8. Evolution of the axial temperature: Influence of the amplitude.$\eta = 0.1$ and $\omega = 100$.

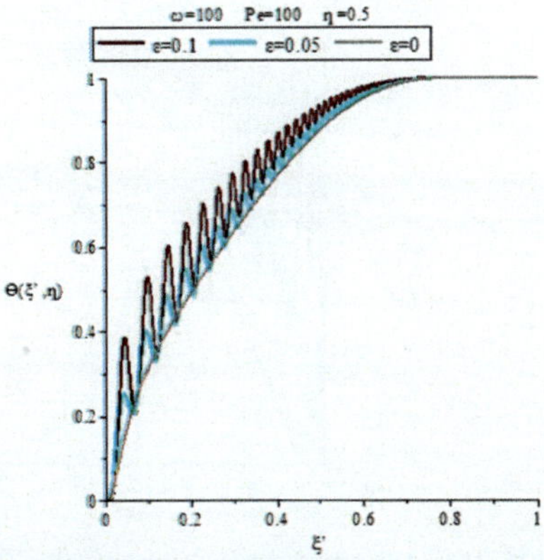

Fig. 9. Evolution of the axial temperature: Influence of the amplitude.$\eta = 0.5$ and $\omega = 100$.

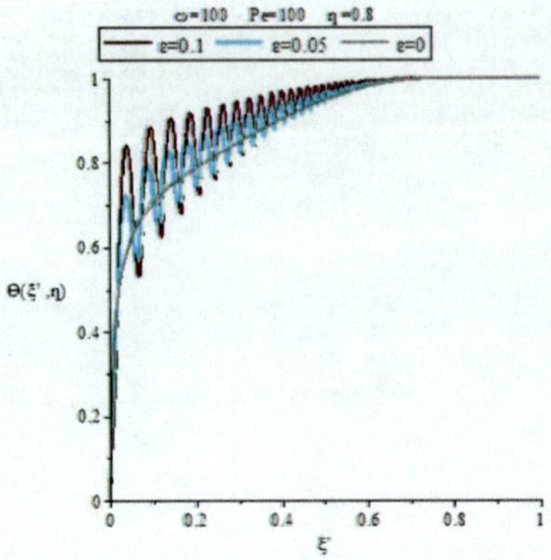

Fig. 10. Evolution of the axial temperature: Influence of the amplitude.$\eta = 0.8$ and $\omega = 100$.

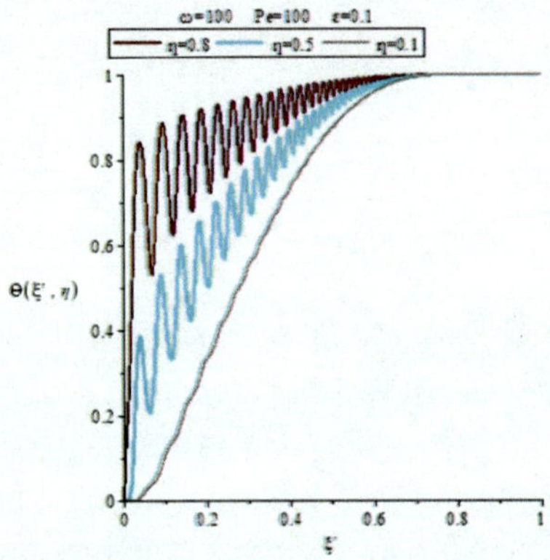

Fig. 11. Evolution of the axial temperature for $\omega = 100$

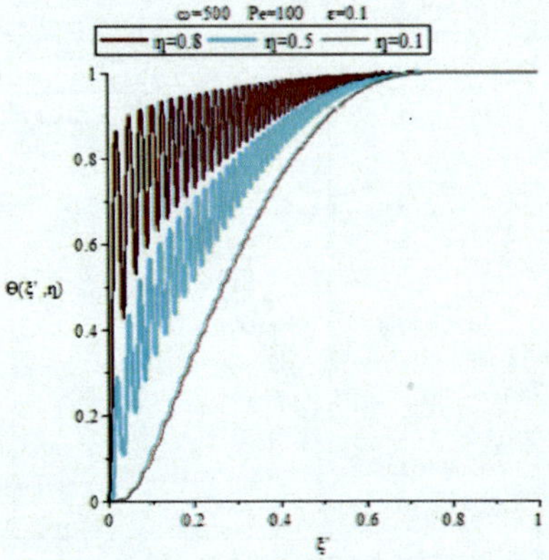

Fig. 12. Evolution of the axial temperature for $\omega = 500$

about $\xi' \approx 0.1$ for each ε. We also notice that the oscillation amplitude of Nu increases with ε and the same behavior was observed in [7], and that the mean value seems to be close to the asymptotic one for $\varepsilon = 0$. The Nusselt value is approximately 7.5, which is in agreement with the value reported in [5, 7, 8], and [9]. Moreover, we see clearly in Fig. 13 that the Nu increases at the entrance with ε, Globally, one can say that this is the mean value in an interval $[0.\xi']$ increases with this parameter.

Finally, let us note that the increase of the amplitude of oscillation of Nu would be undoubtedly at the origin of the improvement of the thermal exchange observed previously on the representations of $\theta(\xi', \eta)$ for different ε.

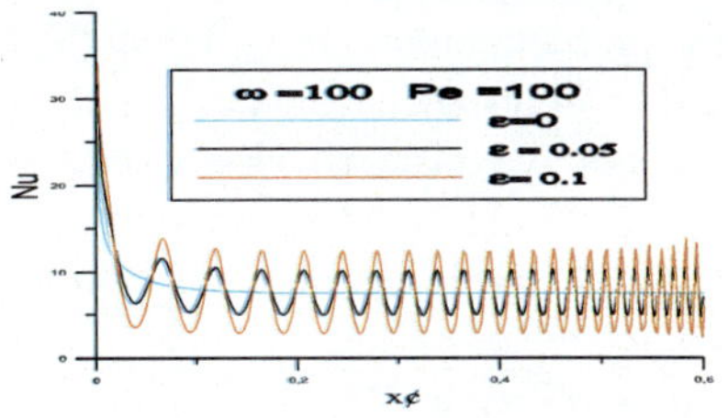

Fig. 13. Evolution of the local Nusselt number for $\omega = 100$: Influence of the amplitude

From Fig. 14, one can observe that the average value of the asymptotic Nusselt number $Nu_{as}(\approx 7.5)$ seems to be not strongly dependent of ω, and that the thermal inlet length increases (from $\approx$0.1 to >0.4) when ω decreases which shows that even this parameter allows us to improve the thermal exchange, and this is in agreement with the previous observations made for $\theta(\xi', \eta)$. Moreover, we see clearly in Fig. 14 that the Nu increases at the entrance with ω, Globally, one can say that this is the mean value in an interval $[0.\xi']$ increases with this parameter.

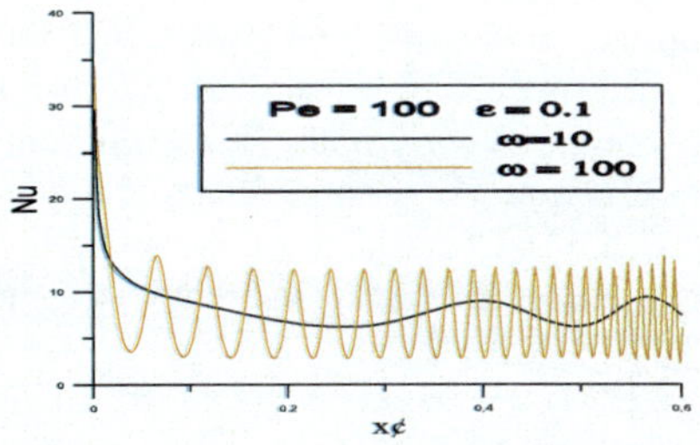

Fig. 14. Evolution of the local Nusselt number for $\omega = 100$ and for $\omega = 10$: Influence of the pulsation

5 Conclusions

The main objective of the presented work is to verify if the channel geometry can have an amplifying effect on heat transfer rate.

The results obtained show that the corrugated channel promotes heat transfer by convection, but according to the scientific literature in this field, it inevitably causes pressure drops. Despite the minimal pressure losses in the case of the smooth tube, the latter is not necessarily appreciable for heat transfer applications.

The presence of corrugations will under certain conditions give a better heat transfer performance and the system can be optimized.

The heat transfer performance increases with increasing wall amplitude and pulsation compared to a straight wall.

Finally, several points can be developed in the future based on this work. Among the most important, in perspective, we recommend in this area, the following points:

- To study the same problem taking into account the axial conduction within the fluid.
- To study the problem in turbulent regime.
- To extend the problem by considering the viscous dissipation in the fluid.
- Include other non-Newtonian fluid models.
- To apply other thermal conditions such as a heat flux instead of a temperature to the walls.
- Study the unsteady problem in 3D.

References

1. Mohammed, H., Abbas, K., Sheriff, J.: Influence of geometrical parameters and forced convective heat transfer in transversely corrugated circular tubes. Heat Mass Transf. **44**, 116–126 (2013)
2. Yang, Y., Chen, P.: Numerical simulation of fluid flow and heat transfer characteristics in channel with V corrugated plates. Heat Mass Transf. **46**, 437–445 (2010)
3. Naphon, P.: Heat transfer characteristics and pressure drop in channel with V corrugated upper and lower plates. Energy Convers. Manage. **48**, 1516–1524 (2007)
4. Liu, X., Niu, J.: Effects of geometrical parameters on the thermo hydraulic characteristics of periodic cross-corrugated channels. Heat Mass Transf. **84**, 542–549 (2015)
5. Castellões, F.V., Quaresma, J.N.N., Cotta, R.M.: Convective heat transfer enhancement in low Reynolds number flows with wavy walls. Heat Mass Transf. **53**, 2022–2034 (2010)
6. Ozisik, M., Trepp, C., Egolf, H.: Laminar forced convection in converging or diverging planar symmetric ducts. Heat Mass Transf. **25**, 1477–1480 (1982)
7. Wang, C., Chen, C.: Forced convection in a wavy-wall channel. Heat Mass Transf. **45**, 2587–2595 (2002)
8. Jambal, O., Shigechi, T., Davaa, G., Momoki, S.: Effects of viscous dissipation and fluid axial heat conduction on heat transfer for non-Newtonian fluids in ducts with uniform wall temperature. Heat Mass Transf. **32**, 1165–1173 (2005)
9. Lahjomri, J., Oubarra, A., Alemany, A.: Heat transfer by laminar Hartmann flow in thermal entrance region with a step change in wall temperatures: the Graetz problem extended. Heat Mass Transf. **45**, 1127–1148 (2002)

Modified Thermo-Physical Properties and Thermal Behavior of Plasterboards with Different Mixing Water Ratios

M. Touil(✉), O. Rahmoun, A. Lachheb, M. Dlimi, A. Charef, and R. Saadani

Laboratory of Advanced Materials Studies and Applications, FS-EST, Moulay Ismail University, Meknes, Morocco
mohamed.touil.as@gmail.com

Abstract. The building sector is one of the most energy-intensive fields in the world. Hence, enhancing the production techniques of traditional materials seems to offer a viable solution for tackling this challenge effectively. However, plaster is considered one of the most used construction substances thanks to its advantageous characteristics such as its aesthetic aspect, its fire resistance and its thermo-acoustic insulation qualities. The present paper includes in its entirety an experimental study aiming to evaluate the effect of the mixing water ratio on the thermo-physical properties, and the thermal behavior of the plaster matrix. Otherwise four blending proportions were investigated (0.5, 0.6, 0.7 and 0.8). The obtained results showed that increasing the water quantity used during the preparation process allows reducing simultaneously the plaster thermal conductivity from 0.415 to 0.359 W/m.K and its density from 1398.4 to 1096.4 kg/m^3. Furthermore, an appreciable improvement of the thermal behavior has been noted. Thus gains of approximately 24.47 and 15.68% were achieved for the time lag and the temperature attenuation factor, respectively. On the other hand, this process has adversely affected the mechanical performance of plaster. Therefore, by transitioning from a water-to-mix ratio of 0.5 to 0.8, the mechanical strength decreases from 5.8 to 2.1 MPa, respectively.

Keywords: Plaster · Mixing water ratio · Thermo-physical properties · Thermal behavior · Bending strength

1 Introduction

Nowadays, the building sector has the greatest potential to improve the energy efficiency since it directly interferes with the external environment. In Morocco, it is considered as one of the largest consumers of primary energy, with approximately 25% of the total energy [1]. Moreover, the heating and cooling demands essential to ensure occupants' thermal comfort significantly contribute to the electrical peak. In this regard, many steps have been taken by the country to establish a passive energy efficiency code, which deals with the upgrade of the building insulation. However, among the major elements implemented in this respect is the development of the thermal characteristics and the choice of materials with low energy consumption.

S. Aniss et al. (Eds.): CMM 2022, LNME, pp. 225–234, 2024.
https://doi.org/10.1007/978-3-031-46973-2_21

Plaster is considered one of the conventional substances widely used in construction in several decorative applications, to benefit from its aesthetic appearance. In addition, its advantageous traits compared to other materials, such as its thermo-physical properties and its resistance to fire makes it a highly valued material and justifies the interest granted to it [2]. Despite its effective qualities in terms of insulation and regulation of humidity. The plaster remains a substance that exhibits a worrying fragility and can undergo more enhancement. This explains the origin of the research works all over the world to strengthen it. Moreover, several innovative procedures have brought a wide knowledge of the methods, and the conditions of its reinforcement [3–5].

Two previous research articles conducted by our research laboratory also addressed the improvement of plaster's thermal aspect [6,7]. The authors reinforced the plaster matrix with alfa fibers and coffee grounds under varying water-to-mix ratios. The main findings demonstrated that increasing the percentage of additives enhanced the thermo-physical properties of the construction material. Particularly noteworthy was the impressive improvement observed with the increase in the water-to-mix ratio. With a 1% to 4% increase in alfa fiber content, the thermal conductivity decreased by 30.40% and 42.69%, and the thermal diffusivity decreased by 33.3% to 46.04%, for composites prepared with W/P ratios of 0.5 and 0.7, respectively. Moreover, composites reinforced with 8% coffee grounds also experienced significant reductions in thermal conductivity (21.17% to 34.78%) and thermal diffusivity (29.98% to 46.83%) for W/P ratios of 0.5 and 0.7, respectively.

In light of this context, the primary objective of this study is to investigate the influence of four water-to-mix ratios (0.5, 0.6, 0.7, and 0.8) on the thermo-physical properties and thermal behavior of plaster. To evaluate its performance in comparison to other techniques, the selected characteristics for analysis include thermal conductivity, density, decrement factor, and time lag. Additionally, a brief analysis of mechanical performance is also provided to better comprehend the impact of water quantity on the mechanical strength of plaster.

2 Materials and Methods

2.1 Plaster Matrix and Sample Preparation

The raw material used in this experiment is a plaster extracted from gypsum rocks near the city of Safi, Morocco. It is a porous compound of calcium sulfate dihydrate, industrially manufactured from gypsum, whose chemical formula is $CaSO_4.2H_2O$. When this latter is heated to 100 °C, it loses its water content and is transformed into calcium sulfate hemihydrate $2CaSO_4.H_2O$. The solubility of plaster in water produced solid pieces. Its chemical composition is presented in the Table 1 [8].

Table 1. The chemical composition of the plaster in %.

SiO_2	R_2O_3a	CaO	MgO	Na_2O-K_2O	SO_3	H_2O	CO_2 and volatiles
0.46	0.17	37.61	0.12	53.66	0.03	6.16	1.59

As already mentioned, four mixing water ratios (W/P) were investigated to prepare the samples. However, the production process was based on several steps. Firstly, the required amount of water was added in a container, followed by a gradual dispersion of the plaster powder on the water surface. Thereafter the paste was blended manually until obtaining a homogeneous mixture. Then, the latter was poured into a mold of $25 * 25 * 4$ cm^3. Afterward it underwent a forced drying in an oven at a temperature of 50 ± 1 °C, until removing all their water content. Finally, the manufactured products are stored under ambient conditions for 28 days. It is also worth noting that, in order to improve and validate the accuracy of the experimental results, each model was manufactured in three samples under the same ambient conditions, and each sample was experimentally tested three times to reduce sensitivity to variations and random errors associated with sample preparation.

2.2 Experimental Setup and Measurement Techniques

The measurement of the conductivity and the evaluation of the samples' thermal behavior were identified using a highly insulated model house of $40 * 40 * 40$ cm^3 (Fig. 1) [9].

Fig. 1. Experimental equipment.

This chamber has replaceable lateral walls with a square opening on each side of $21 * 21$ cm^2 to allow studying several types of materials. Indeed, the investigated plasters were fixed in the inside face in a way to seal the aperture. The internal temperature was increased by means of a 100 W electric bulb connected to a thermo-regulated thermostat. It produced a steady state interior temperature of about 50 °C. After several hours, the heat flow was stationary, and the temperature of the chamber (T_b), the inner and outer sample faces (T_{in}) and (T_{ex}), and that of the laboratory (T_l) were recorded using thermocouples. The thermal conductivity was calculated through Fourier's law (Eq. 1).

$$\lambda = \frac{P}{S} \frac{e}{(T_{in} - T_{ex})} \tag{1}$$

where S and e are respectively the surface and the sample thickness in m.

The $\left(\frac{\phi}{S}\right)$ value is determined by the average thermal flow of the inner and outer air surfaces of the box, through the formulas (Eq. 2) and (Eq. 3).

$$\left(\frac{\phi}{S}\right)_{in} = h(T_b - T_{in}) \tag{2}$$

$$\left(\frac{\phi}{S}\right)_{ex} = h(T_{ex} - T_l) \tag{3}$$

where h is the heat transfer coefficient in the case of a natural movement of air in contact with a wall. It has a value of 8.1 in W/m^2.K [10].

As for the bulk density, it was measured after a complete drying of the samples. It is calculated by Eq. 4, where the weight (M in g) was determined by a precision balance of ±0.1 g, and the volume (V in m^3) was determined from the samples' dimensions.

$$\rho = \frac{M}{V} \tag{4}$$

The thermal behavior was evaluated through the time lag (TL_ϕ) and the decrement factor (f). The first one represents the period needed for the propagation of a thermal excitation along the sample thickness. While the second one describes the reduction ratio of the outside temperature amplitude to the interior one. The method of determining these two properties was performed by exposing the inner faces of the materials to periodic powers of 80 W, for 90 min each. The highest and lowest temperatures (T_{ex} and T_{in}) of each side are thus recorded and used for the calculations (Eq. 5 and Eq. 6).

$$TL_\phi = T_{in\_max} - T_{ex\_max} \tag{5}$$

$$f = \frac{T_{ex\_max} - T_{ex\_min}}{T_{in\_max} - T_{in\_min}} \tag{6}$$

Before starting the measurement process, it is necessary to verify the validity of the experimental methodology. For this purpose, a test was conducted on two commercially available samples of wood and polyurethane, both of which have well-known thermo-physical properties. Table 2 presents the thermal conductivity of each material, determined experimentally in this study and validated through the two-box method [3] and literature. The results are approximately the same, demonstrating the reliability of the measurement cell.

Table 2. Validation of the experimental process by measuring the thermal conductivity (W/m.K) of wood and polyurethane.

Material	This research	Two-box method	Literature
Wood	0.128 ± 0.004	0.122 ± 0.002	0.12 [9]
Polyurethane	0.034 ± 0.003	0.031 ± 0.001	0.030 [11]

3 Results and Discussion

3.1 Analysis of the Thermo-Physical Properties

Figure 2 illustrates the variation of the plaster thermal conductivity as a function of the mixing water ratio. From its analysis, it appears clear that increasing the W/P rate induces a significant reduction of the concerned physical property. However, by changing the consistency of preparation from 0.5 to 0.8, the thermal conductivity decreases from 0.401 to 0.359 W/m.K, which represents a considerable gain of 10.47%. This finding demonstrates that a plaster matrix formulated with a high-mixing proportion of water has a good insulating capacity.

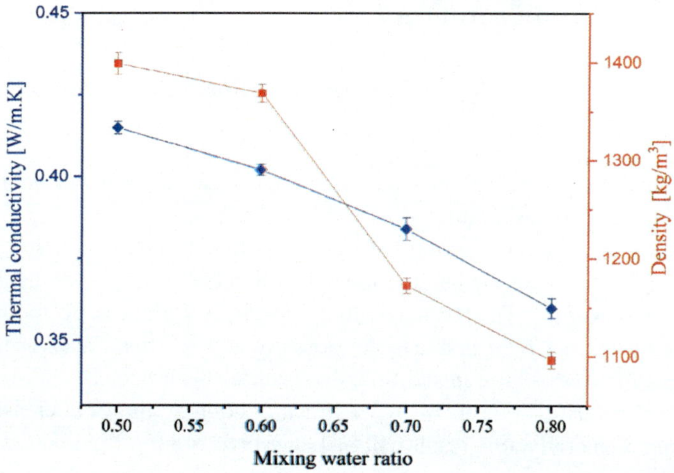

Fig. 2. Thermal conductivity and density as a function of the mixing water ratio.

This phenomenon is explained by the enhanced crystallization degree of the plaster particles, associated with the higher W/P ratio. As a consequence, several pores are formed, constituting the main cause of this observed effect, as already mentioned in the articles by Touil et al. [6,7] (Sect. 1). Figure 3 shows microscopic images of the two models prepared with W/P ratios of 0.5 and 0.7. However, it appears that the plaster hydration process with low water to mix ratios leads to the production of insufficiently crystallized specimens. By contrast, increasing the water content generates a structure composed of randomly oriented crystals in the form of needles or plates (approx. 1 to 3 μm). As a result, the crystalline structure interlock and create voids, thus increasing the porosity of the hardened plaster. As a fluid with low thermal properties (lambda = 0.026 W/m.K), the air that fills the created cavities improves the insulating capacity of the plaster [12].

a. With (W/P: 0.5) b. With (W/P: 0.7)

Fig. 3. SEM images of some neat plasters treated.

The extent of the mixing water ratio not only affects the thermal profile of the plaster matrix, but also has a direct impact on its lightness. However, Fig. 2 depicts the density variation as a function of the W/P proportion. From its evaluation, it can be seen that the maximization of the latter induces a significant reduction of the density from 1398.4 to 1096.4 Kg/m^3. This corresponds to a gain in lightness of about 21.6%. The main cause of this behavior is due to the porosity actions, that create empty holes inside the sample microstructure. In addition, this profile is validated by the study of Del Río et al. [13], who verified that the increase in water content allows decreasing the plaster bulk density. Generally, this result will be beneficial for large-scale buildings to reduce their weight.

3.2 Assessment of the Thermal Behavior

Before proceeding with any building construction, it is essential to carry out a thermal study on the materials' performance. Concretely, this kind of assessment allows obtaining a precise understanding of the material's behavior and will enable making the right design choices to meet the needs in terms of comfort. Indeed, to link with the objective of this article, this section aims to present the effect of the mixing water ratio on the plaster thermal profile and to characterize its impact on the thermal inertia properties. The Fig. 4 presents an example of the evolution of interior and exterior surface temperatures over time for each model. The recorded amplitudes represent the basic data for calculating the decrement factor and time shift, as mentioned in Sect. 2.2. It is worth noting that these tests were replicated three times for each sample to enhance the sensitivity of thermal inertia measurements.

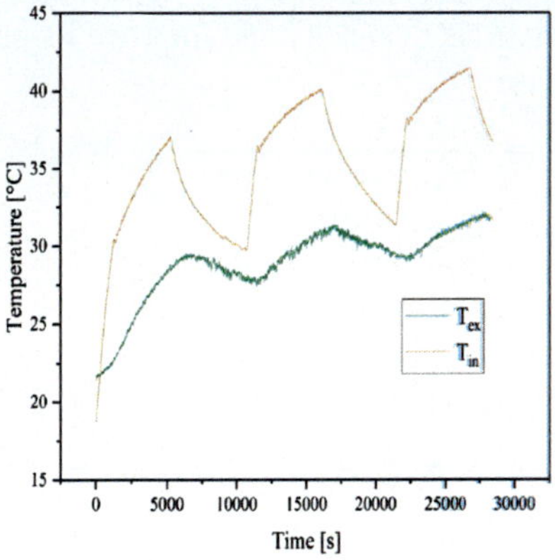

a. With (W/P: 0.5)

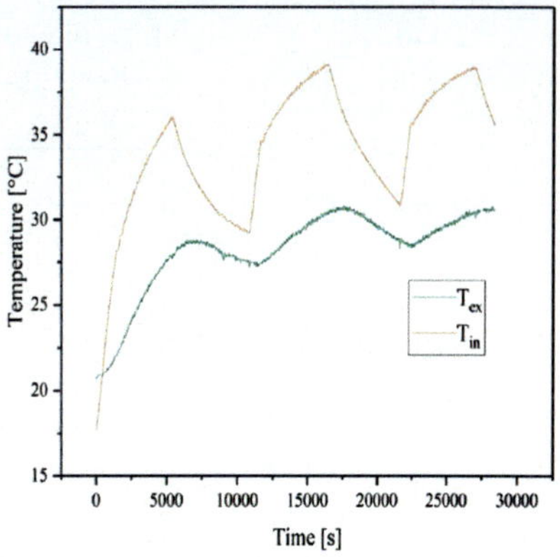

b. With (W/P: 0.6)

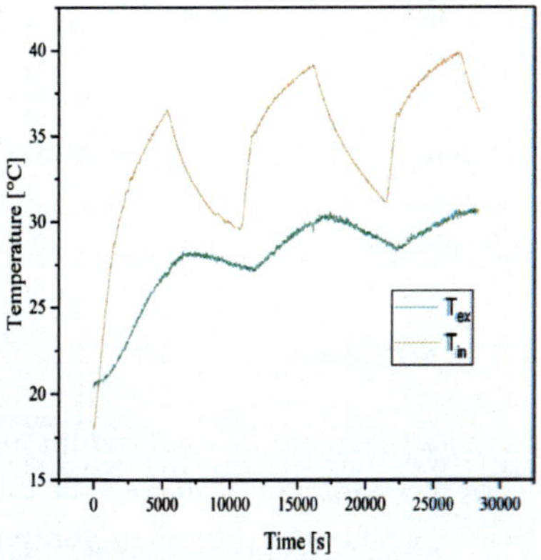

a. With (W/P: 0.7)

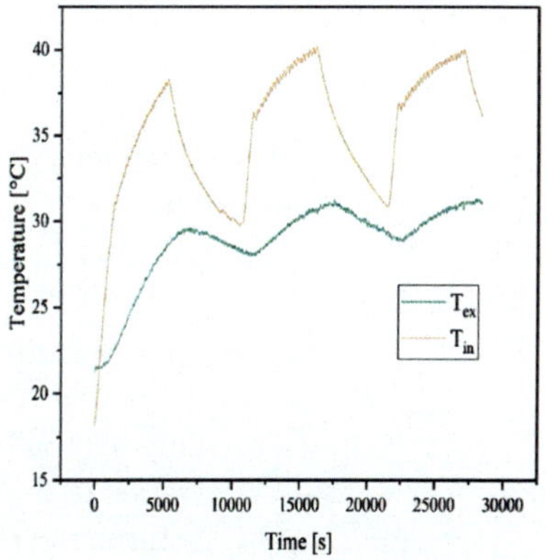

b. With (W/P: 0.8)

Fig. 4. Example of indoor temperature variation and its response on the outdoor side

Figure 5 illustrates the variation of the decrement factor and phase shift as a function of the water proportion, used in the preparation of the plaster. Through its analysis, it can be seen that there is a specific relationship between the water rate and the two calculated performance parameters. It can be noted that the damping factor was reduced from 0.402 to 0.339, which represents a significant attenuation of the internal temperature amplitude by approximately 15.68%. Another interesting observation is the delay of the external thermal response to the heat flow throughout the material's inner face. This reflects a considerable improvement of the time lag by 24.47% at the highest mixing proportion.

Similarly to the thermo-physical properties, this finding reveals the positive influence of increasing the W/P ratio on the plaster insulating performance and its heat storage ability. This effect can be explained by the enhancement of density and thermal conductivity, which provide sufficient intrinsic heat capacity and improve its inertia.

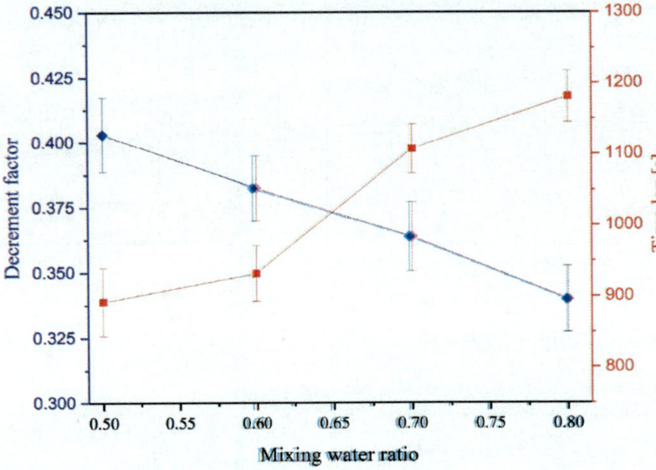

Fig. 5. Decrement factor and time shift as a function of the mixing water ratio.

4 Mechanical Performance: A Brief Analysis

In the previous study by Touil et al. [7], three materials manufactured with three different water-to-mix ratios (0.5, 0.6, and 0.7) were mechanically tested under three-point bending loads. To further expand the experimental plan and provide a comprehensive understanding of the mechanical behavior of all the models presented in this research, three additional samples were produced with a water-to-mix ratio of 0.8. The Table 3 illustrates the complete set of mechanical resistance results obtained for each material.

Table 3. Results of mechanical strength.

W/P	0.5	0.6	0.7	0.8
Strength [MPa]	5.8 ± 0.02	4.7 ± 0.035	3.89 ± 0.024	2.1 ± 0.027

The bending strength measures the material's ability to resist deformation or fracture under external forces applied during flexion. However, materials made with a W/P ratio of 0.5 demonstrate higher mechanical resistance, indicating a greater capacity to withstand loads without breaking. As mentioned previously, increasing the W/P ratio enhances the thermal behavior of plaster but adversely affects its mechanical performance. This is primarily due to the presence of pores within the plaster's structure, which weaken the material. However, based on the standards set by European Standard EN 13279-2 [14], the minimum required mechanical resistance for plaster should be at least 2 MPa. The strength of the specimens made with a W/P ratio of 0.8 is still sufficient.

5 Conclusion

This paper presents an experimental study on the development of the thermo-physical properties, and the thermal behavior of the plaster matrix. The idea was to improve its insulating capacity and its thermal inertia, in order to smooth the thermal flows and to attenuate the summer extreme temperatures. Four proportions of mixing water ratios (0.5, 0.6, 0.7 and 0.8) were investigated to measure the influence of this parameter on the physical characteristics concerned by this research. The following conclusions were drawn:

- Increasing the blending water rate reduced the plaster's ability to conduct heat and limits the thermal agitation of its particles. Consequently, an optimal thermal conductivity of 0.359 W/m.K can be obtained using a mixing ratio of 0.8; this illustrates a gain of 10.47% compared to the material prepared with a dry consistency (0.5). This technique represents an alternative and efficient solution to diminish the thermal exchange inside gypsum materials.
- Meanwhile, it has been demonstrated that the exploitation of very high water quantities has allowed to significantly decrease the plaster density. However, a considerable lightness benefit of 21.6% was obtained by increasing the mixing ratio from 0.5 to 0.8. Indeed, this result reports an effective method to reduce the buildings' weight and to facilitate the transport and the construction activities.
- Another interesting remark is the significant impact of the blending water proportion on the attenuation and the temperature phase shift. In fact, by switching from a dry consistency (0.5) to a fluid one (0.8), the plaster decrement factor was diminished by approximately 15.68%, while a net rise of 24.47% of the time lag was observed. The estimation of these two characteristics allowed describing the favorable relationship between the high mixing ratio and the plaster's thermal inertia and demonstrating its effective behavior under extreme thermal loads.
- Increasing the W/P ratio enhances thermal behavior but compromises mechanical performance due to weakened plaster structure from pores. However, specimens with a W/P ratio of 0.8 still meet minimum mechanical resistance standards.

References

1. National Agency for the Development of Renewable Energy and Energy Efficiency (ADEREE): Thermal regulation of the construction in Morocco, Morocco. http://www.aderee.ma
2. Şahin, S., Karaman, S.: The properties of expanded polystyrene - pumice-Gypsum blocks as a building material. J. Tekirdag Agric. Faculty **9**, 51–56 (2019). https://hdl.handle.net/20.500.11776/1665
3. Touil, M., Rahmoun, O., Iken, O., El Harti, K., Saadani, R., Rahmoune, M.: Exploring the thermal behaviour and thermo-mechanical properties of Ferula Communis reinforced plaster and mortar composites: an integrated experimental and numerical approach. Energy Conves. Manage. **289**, 117119 (2023). https://doi.org/10.1016/j.enconman.2023.117119
4. Gutiérrez-González, S., Gadea, J., Rodríguez, A., Junco, C., Calderón, V.: Lightweight plaster materials with enhanced thermal properties made with polyurethane foam wastes. Constr. Build. Mater. **28**, 653–658 (2011). https://doi.org/10.1016/j.conbuildmat.2011.10.055
5. Wang, Z.: Experimental investigation on the effect of temperature on thermal conductivity of synthetic rock samples. IOP Conf. Ser. Earth Environ. Sci. **570**, 032066 (2020). https://doi.org/10.1088/1755-1315/570/3/032066
6. Touil, M., Lachheb, A., Saadani, R., Kabiri, M.R., Rahmoune, M.: A new experimental strategy assessing the optimal thermo-mechanical properties of plaster composites containing Alfa fibers. Energy Build. **262**, 111984 (2022). https://doi.org/10.1016/j.enbuild.2022.111984
7. Touil, M., Lachheb, A., Saadani, R., Sanbi, M., Talidi, A., Rahmoune, M.: Experimental investigation on the combined effect of the water mixing ratio and the addition of spent coffee grounds on plaster's thermo-mechanical properties. Therm. Sci. Eng. Prog. **36**, 101488 (2022). https://doi.org/10.1016/j.tsep.2022.101488
8. Touil, M., Lachheb, A., Saadani, R., Sanbi, M., Rahmoune, M.: Influence of the mixing water rate on the plaster thermophysical properties: experimental and dynamic study. E3S Web Conf. **336**, 00012 (2022). https://doi.org/10.1051/e3sconf/202233600012
9. Laboratory Experiments, Physics: P2360300 PHYWE Series of Publications. HYWE SYSTEME GMBH & Co. KG, Göttingen, Germany
10. Navacerrada, M.A., Fernández, P., Díaz, C., Pedrero, A.: Thermal and acoustic properties of aluminium foams manufactured by the infiltration process. Appl. Acoust. **47**, 456–501 (2013). https://doi.org/10.1016/j.apacoust.2012.10.006
11. Papadopoulos, A.M.: State of the art in thermal insulation materials and aims for future developments. Energy Build. **37**, 77–86 (2005). https://doi.org/10.1016/j.enbuild.2004.05.006
12. Tan, Y., Longtin, J.P., Sampath, S.: Modeling thermal conductivity of thermal spray coatings: comparing predictions to experiments. J. Therm. Spray Technol. **15**, 545–552 (2006). https://doi.org/10.1361/105996306X147216
13. Del Río, M., Fernández Olivares, F.: Lightened plaster: alternative solutions to cellular solids addition. Constr. Mater. **275**, 65–77 (2004). https://doi.org/10.3989/mc.2004.v54.i275.248
14. European Standard EN 13279-2. Gypsum binders and plasters for building - Part 2: Test methods (2014)

Numerical Analysis of Mixed Convection in a Ventilated Square Cavity Containing a Circular Heat-Generating Cylinder

R. Hidki, L. El Moutaouakil, M. Boukendil(✉), Z. Charqui, Z. Zrikem, and A. Abdelbaki

LMFE, Department of Physics, Faculty of Sciences Semlalia, Cadi Ayyad University, B.P. 2390, Marrakesh, Morocco

m.boukendil@uca.ac.ma

Abstract. This work aims to investigate the complex phenomenon of mixed convection in a vented cavity containing a circular heat-generating cylinder, which holds significant relevance for electronic device cooling applications. To achieve this objective, a comprehensive numerical study is conducted using the finite volume technique. The mathematical model is discretized, and the resulting equations are solved numerically employing a dedicated numerical code based on the SIMPLE algorithm. In this study, the effects of various parameters on the flow and thermal field are examined. These parameters include the Richardson number ($0.1 \leq Ri \leq 10$), Reynolds number ($50 \leq Re \leq 500$), thermal conductivity ratios ($0.1 \leq K \leq 100$), and cylinder position. The Prandtl number is fixed at 0.71, and the cylinder diameter is set at 0.4. It is found that the Reynolds number plays a significant role in influencing the flow patterns. Increasing the Reynolds number leads to enhanced cooling effectiveness of the circular cylinder. This suggests that higher flow rates contribute to better heat transfer and improved cooling performance. Additionally, the investigation demonstrates that an increase in thermal conductivity ratios yields substantial improvements in the overall performance of the cooling system.

Keywords: Mixed convection · Ventilated cavity · Finite volume method · Numerical simulation · Heat-generating cylinder

1 Introduction

Natural or mixed convection in cavities equipped with heating cylinders has garnered significant attention from numerous scientific researchers due to its wide range of practical applications. These applications include air conditioning and heating of buildings, cooling of electronic devices, and heat exchangers. Extensive studies on this topic can be found in the existing literature [1–5].

Chamkha et al. [6] conducted a numerical investigation of mixed convection in a vented cavity with an isothermal square cylinder. The authors explored the influence of the outlet position for various Richardson (*Ri*) and Reynolds (*Re*) numbers. Their

S. Aniss et al. (Eds.): CMM 2022, LNME, pp. 235–245, 2024.
https://doi.org/10.1007/978-3-031-46973-2_22

findings revealed that positioning the outlet port at the top resulted in the minimum temperature within the cavity. In another study, Chatterjee and Mishra [7] examined the scenario of two isothermal circular cylinders in a ventilated cavity. Their analysis indicated that the Richardson number had a negligible effect on the mean Nusselt number. Gupta et al. [8] performed a numerical investigation on mixed convection within a differentially heated cavity containing a centrally located circular conductive cylinder. They demonstrated that, for low Ri values, increasing the cylinder's conductivity led to a decrease in the Nusselt number and an increase in the mean temperature. This same configuration was also reconsidered by [9, 10] but with a square conductive cylinder. Mamun et al. [11] investigated a hollow circular cylinder placed inside a vented square enclosure. Their findings revealed that, for $Ri = 0$, thermal conductivity had no impact on the flow. Ahammad et al. [12] studied a vented cavity heated from below, considering a square heat-generating cylinder. Their results showed that the Nusselt number reached its maximum for small cylinder sizes.

The literature review highlights the significance of studying mixed convection in ventilated cavities with heating cylinders, given its relevance to various practical applications. However, it is evident that most existing research primarily focuses on isothermal or simple conductive cylinders. Comparatively, fewer studies have explored the case of heat-generating cylinders. Therefore, this research aims to investigate the air cooling of a circular cylinder that uniformly generates heat when placed inside a vented cavity. In this study, the cavity is cooled through the left side and is insulated from the other walls. The research aims to analyze the influence of Richardson, Reynolds numbers, thermal conductivity ratios, as well as the cylinder's position on the dynamic and thermal characteristics of the system. Ultimately, the findings from this study can be applied to the design of cooling systems that generate a small amount of heat.

2 Mathematical Model

Figure 1 depicts the geometry of the physical problem under consideration. It highlights a circular cylinder placed inside a square air-vented cavity ($Pr = 0.71$). The size of openings (inlet and outlet) is set to equal $h_i = h_o = 0.1L$. The cylinder generates a uniform volume power $Q(W/m^3)$ and has a diameter of $D = d/L = 0.4$. All the cavity's walls are perfectly insulated except for the left one, which is maintained at a cold temperature. In addition, the inflow and outflow ports are located on the left and right walls, respectively.

The Boussinesq approximation is employed in our study to account for the buoyancy effect by considering the density variation with temperature. This approximation is valid for our problem since the temperature difference (ΔT) is relatively small compared to the absolute temperature (T), typically below 0.1. By making this assumption, we can simplify the governing equations while capturing the significant effects of buoyancy-driven flows and thermal convection. The other physical properties of air are assumed to be constant throughout the system.

By applying these approximations to the continuity, momentum, and energy equations, the following dimensionless system is obtained:

$$\frac{\partial U}{\partial X} + \frac{\partial V}{\partial Y} = 0 \tag{1}$$

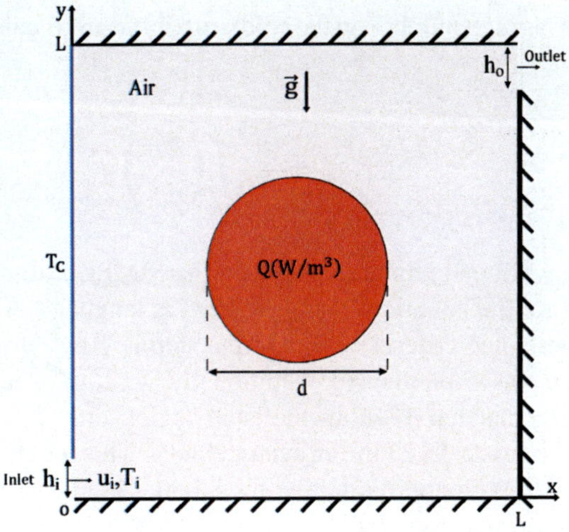

Fig. 1. Schematic configuration.

$$U\frac{\partial U}{\partial X} + V\frac{\partial U}{\partial Y} = -\frac{\partial P}{\partial X} + \frac{1}{Re}\nabla^2 \mathrm{U} \tag{2}$$

$$U\frac{\partial V}{\partial X} + V\frac{\partial V}{\partial Y} = -\frac{\partial P}{\partial Y} + \frac{1}{Re}\nabla^2 \mathrm{V} + Ri\theta \tag{3}$$

$$U\frac{\partial \theta}{\partial X} + V\frac{\partial \theta}{\partial Y} = \frac{1}{RePr}\nabla^2 \theta \tag{4}$$

$$\frac{K}{RePr}\nabla^2 \theta_s + \frac{K}{RePr} = 0 \tag{5}$$

In Eqs. (1)–(5), the dimensionless variables and parameters are specified as follows:

$$(U, V) = \frac{(u, v)}{u_i}, \theta = \frac{(T - T_C)}{\Delta T}, \theta_s = \frac{(T_s - T_C)}{\Delta T}, \Delta T = \frac{QL^2}{k_s}, (X, Y) = \frac{(x, y)}{L} \tag{6}$$

$$Re = \frac{u_i L}{\upsilon}, Pr = \frac{\upsilon_f}{\alpha_f}, Ra = \frac{g\beta L^3 \Delta T}{\upsilon_f \alpha_f}, Ri = \frac{Ra}{Re^2 Pr}, K = \frac{k_s}{k_f} \tag{7}$$

Equations (1)–(5) are solved using the following dimensionless boundary conditions:

- At the inlet: $U = 1, V = 0, \theta = 0$
- On all solid walls: $U = V = 0$
- On all adiabatic walls: $\frac{\partial \theta}{\partial n} = 0$
- On the cold wall: $\theta = 0$
- At the outlet: $P = 0$
- On the circular cylinder surfaces: $\theta_f = \theta_s$, $\frac{\partial \theta_f}{\partial n} = K\frac{\partial \theta_s}{\partial n}$

The mean Nusselt number on the cold wall (left wall) is calculated using the following equation:

$$Nu_m = \frac{1}{0.9}\int_{0.1}^{1}\left(\frac{\partial\theta_f}{\partial X}\right)_{X=0} dY \tag{8}$$

The finite volume technique based on the SIMPLE algorithm has been utilized to solve the differential equations using the Fortran language. All variables were subjected to 10^{-4} convergence criteria, and the time step is 10^{-5}. To enhance accuracy, a finer grid resolution was implemented. Additionally, we employed an approach where a control volume is considered within the solid region only if its center P lies within the proper curved boundaries of the circular cylinder. This methodology effectively reduces errors associated with approximating the circular boundaries, ensuring a more precise representation of the geometry.

A mesh test was conducted to evaluate the maximum temperature within the cavity, with the cylinder placed at the center and varying Reynolds and Richardson numbers values. According to Table 1, it is clear that the 150 × 150 mesh yields good results in terms of accuracy and computational time.

Table 1. Grid test on θ_{max} for $Re = 50, 500$ and $Ri = 0.1, 10$ with $K = 1$.

Grid	$\theta_{max} \times 10^2$			
	50 × 50	100 × 100	150 × 150	200 × 200
$Re = 50$, $Ri = 0.1$	2.5137	3.0140	3.2019	3.2143
$Re = 500$, $Ri = 0.1$	2.2525	2.5007	2.4707	2.4859
$Re = 50$, $Ri = 10$	2.5585	3.1828	3.6380	3.6540
$Re = 50\ 0$,$Ri = 10$	2.0853	2.4036	2.5060	2.4906

The obtained results have been validated by comparing them with those of Gupta et al. [8]. Figure 2 compares streamlines and isotherms for $Ri = 5$, $Re = 100$, $K = 5$, $D = 0.4$, and $Pr = 0.71$. The results indicate a good agreement between our data and those presented by Gupta et al. [8]. Furthermore, additional validation can be found in the works of [5, 13–15].

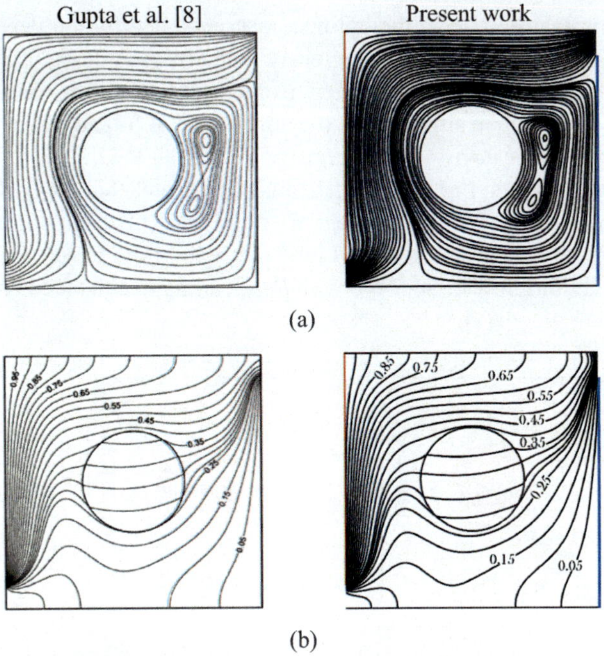

Fig. 2. (a) Streamlines and (b) isotherms.

3 Results and Discussion

The results are provided in this part in the form of isotherms, streamlines, and cylinder maximum temperature. For $Pr = 0.71$ and $D = 0.4$, the effects of Re ($50 \leq Re \leq 500$), Ri ($0.1 \leq Ri \leq 10$), K ($0.1 \leq K \leq 100$), and cylinder's position are investigated.

3.1 Streamlines and Isotherms

Figure 3 presents the streamlines for various combinations of (Ri, Re), with the circular cylinder positioned at the center of the cavity and $K = 1$. When $Re = 50$, regardless of the Ri value, the streamlines indicate that the external air uniformly fills the entire cavity, exhibiting symmetry with respect to the line connecting the inlet and outlet ports. This suggests that the cold external air entering the cavity does not interact significantly with the warmer inner air. However, a notable change in the streamlines is observed as Re increases to 500. In this case, the main flow is situated in the right half of the cavity, with a large vortex occupying the remaining region and enclosing the heat-generating cylinder. This observation demonstrates that the inlet air starts to interact with the internal air. Additionally, it can be observed that the effect of Ri on the streamlines is negligible for all considered values of Re.

Under the same conditions, Fig. 4 illustrates the isotherms and mean heat transfer for different combinations of (Ri, Re). The figure demonstrates that, for $Re = 50$, the isotherms are distributed throughout the domain, with more pronounced thermal gradients observed below the heat-generating cylinder. Increasing Ri ($Re = 50$) leads to a slight temperature rise in the cylinder, accompanied by the development of a thermal plume above the cylinder, as previously noted by Ahammad et al. [12]. For $Re = 500$, the air in the cavity becomes nearly isothermal, except for the region surrounding the cylinder. This can be attributed to the dominance of forced convection over natural convection within the cavity. In addition, for $Re = 500$, the influence of Ri is not significant, as explained earlier. Furthermore, the figure indicates that the Nusselt number rises with increasing values of both the Ri and Re numbers. This behavior can be attributed to the combined effects of an increased heat rate within the cavity (associated with Ri) and intensified fluid flow (associated with Re), leading to enhanced convective heat transfer.

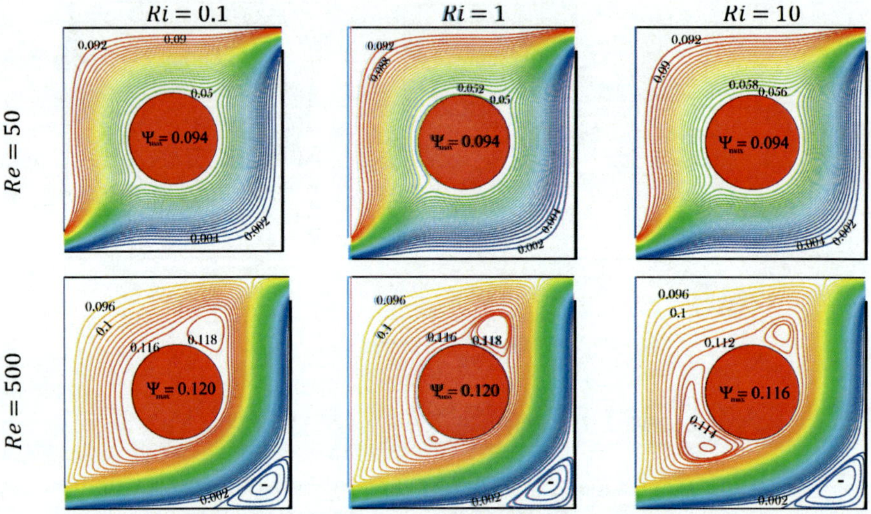

Fig. 3. Streamlines for different values of (Ri, Re).

For $Re = 200$, $Ri = 10$, and $K = 1$, Fig. 5 presents the variations of streamlines (a) and isotherms (b) for different positions of the cylinder. The streamlines reveal that when the cylinder is positioned at ($X = 0.5$, $Y = 0.25$) or ($X = 0.75$, $Y = 0.5$), a vortex is formed above the inlet ports due to the frictional force between the inlet air and the cylinder. When the cylinder is placed at the center ($X = Y = 0.5$), two vortices are generated above the inlet, and the primary airflow occupies most of the cavity. Lastly, when the cylinder is positioned near the cold wall ($X = 0.25$, $Y = 0.5$), the flow intensifies around the cylinder with two small vortices formed above and below it. This observation suggests that the heat-generating body is effectively cooled.

Regarding the isotherms (Fig. 5b), they exhibit a higher concentration within the heat-generating cylinder. The effect of cylinder position indicates that placing the cylinder closer to the cold wall or at the middle of the cavity results in better cooling of the cylinder. The Nusselt number displays a notable dependence on the position of the cylinder within the cavity. Specifically, the Nusselt number reaches its minimum value when the cylinder is positioned near the left wall. Conversely, the maximum Nusselt number is achieved when the cylinder is close to the cold wall. This proximity enhances

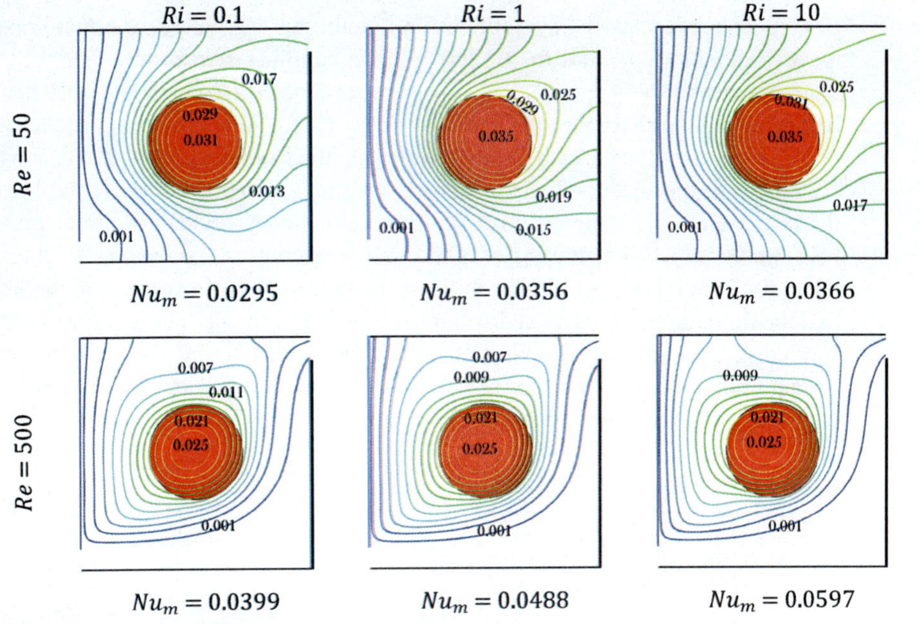

Fig. 4. Isotherms for different values of (Ri, Re).

flow dynamics, leading to stronger convective heat transfer and increased heat extraction from the cylinder.

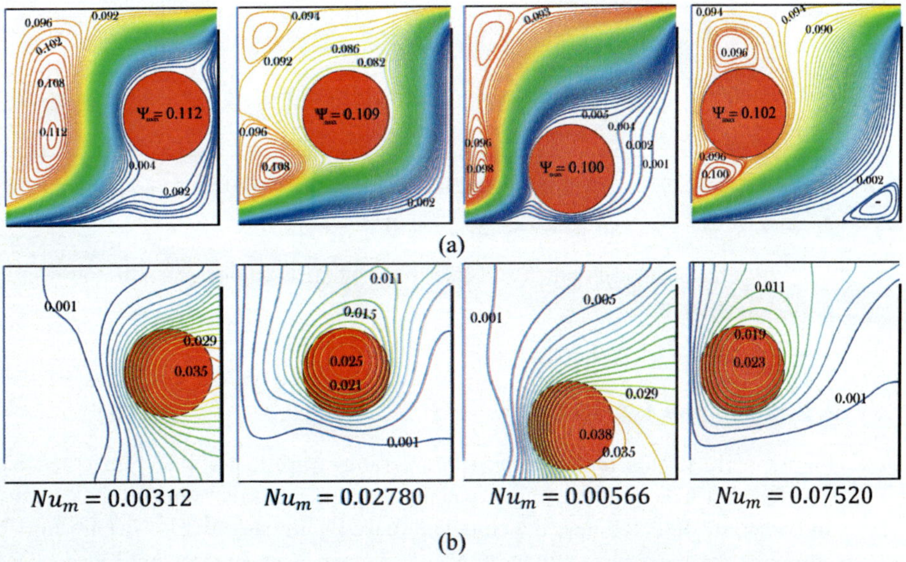

Fig. 5. Impact of the cylinder's position on (a) streamlines and (b) isotherms.

Figure 6 presents the streamlines and isotherms for various thermal conductivity ratios (with $Ri = 10$ and $Re = 200$). The streamlines remain unaffected by thermal conductivity as the same amount of heat is evacuated ($Ri = 10$) within the system, regardless of the thermal conductivity values. This observation is also reflected in the constant value of the Nusselt number. However, the isotherms demonstrate considerable sensitivity to thermal conductivity. The isotherms accumulate in a circular form within the heat-generating cylinder for lower thermal conductivity values, indicating that the heat source impedes heat transfer in the cavity. In contrast, increasing the thermal conductivity from 0.1 to 5 gradually shifts the isotherms away from the cylinder, extending into the fluid region. This behavior signifies an enhancement in the cooling system's performance. Consequently, for $K \geq 10$, the cylinder becomes nearly isothermal, reaching a maximum temperature value. This finding highlights the crucial role of higher thermal conductivity in improving the overall cooling efficiency of the system.

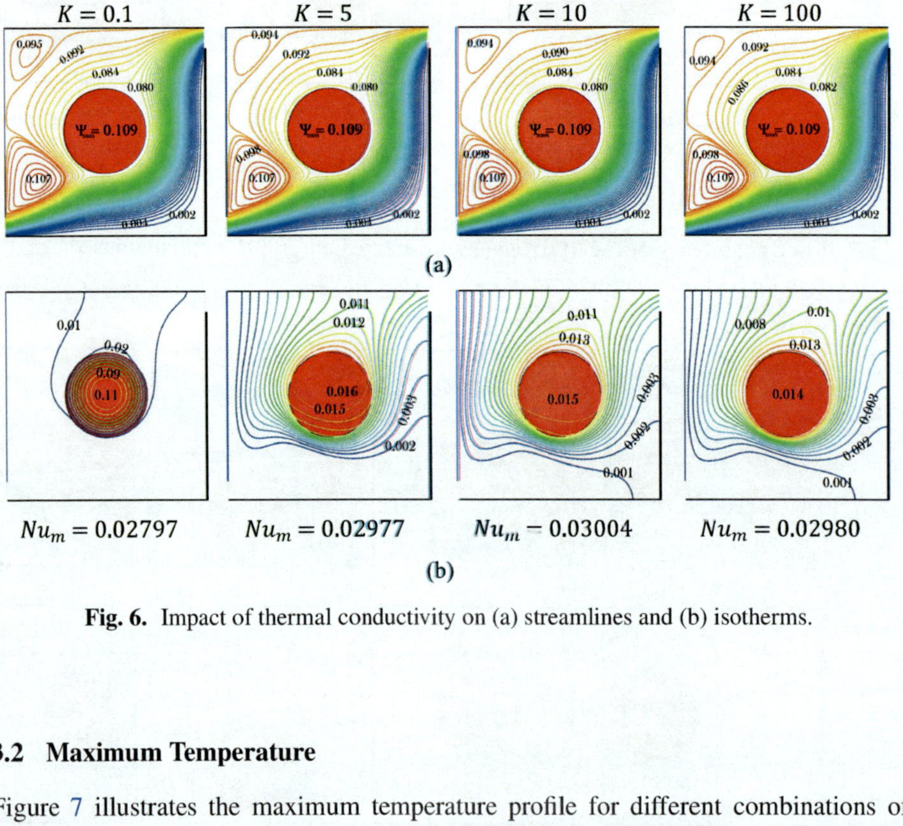

Fig. 6. Impact of thermal conductivity on (a) streamlines and (b) isotherms.

3.2 Maximum Temperature

Figure 7 illustrates the maximum temperature profile for different combinations of (Ri, Re) and $K = 1$ in two cases: when the cylinder is positioned at the center (case A) and when it is located near the cold wall (case B). In case A (Fig. 7a), for low Reynolds numbers, the maximum temperature θ_{max} increases as Ri rises until it reaches its peak at $Ri = 5$. Subsequently, in the $5 \leq Ri \leq 10$ range, θ_{max} decreases, which aligns with the observation made by [12]. The increase in Re weakens this trend. When Ri is held constant, θ_{max} decreases rapidly as Re increases. When Re varies from 50 to 200, θ_{max}

decreases by 33% of its initial value. It is noteworthy that the lowest temperature is achieved for high values of (*Ri*, *Re*). In case B (Fig. 7b), temperature variations are observed only for low values of the Richardson number. The results also indicate that the optimal position for effective cooling is when the cylinder is closest to the cold wall. This preference is attributed to the significant contribution of the cold wall in removing heat from the system.

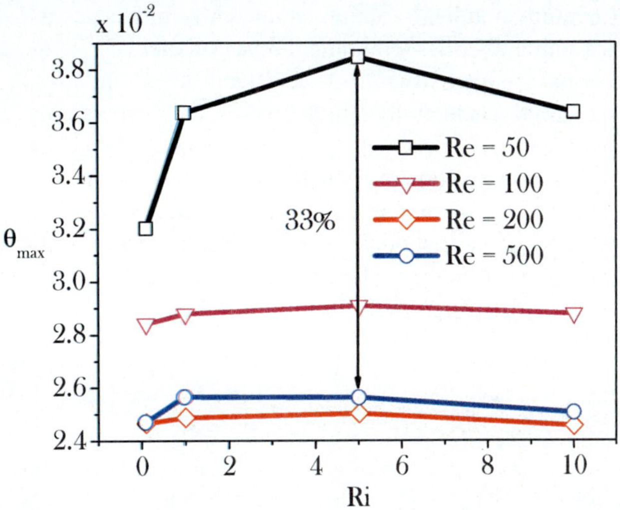

(a)

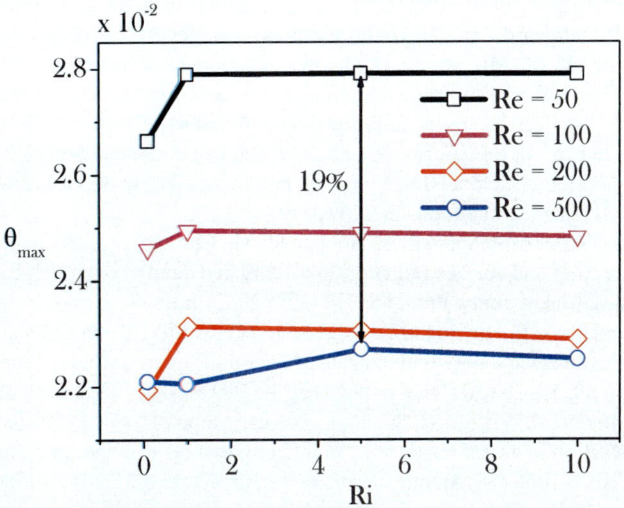

(b)

Fig. 7. Maximum temperature for different combinations of (*Ri*, *Re*): (a). Case A, (b). Case B.

4 Conclusion

This study presents a numerical investigation using the Finite Volume Method to simulate the cooling of a circular heat-generating cylinder through mixed convection. The cylinder is positioned within a square cavity ventilated with air. The results are presented considering various Reynolds numbers, Richardson numbers, thermal conductivity ratios, and cylinder positions. The key findings can be summarized as follows:

- The Reynolds number variation significantly influences the streamlines, while the Richardson number has a minimal impact on them.
- The thermal conductivity ratio affects the distribution of isotherms within the cavity. As the thermal conductivity ratio increases, the temperature of the cylinder decreases notably.
- The highest combination of Richardson and Reynolds numbers results in a minimum value for the maximum temperature within the cavity.
- The position of the cylinder has a significant impact on the thermal and dynamic characteristics of the flow. Placing the cylinder closest to the cold wall facilitates efficient heat extraction.

References

1. Boutina, L., Bessaïh, R.: Numerical simulation of mixed convection air-cooling of electronic components mounted in an inclined channel. Appl. Therm. Eng. **31**, 2052–2062 (2011). https://doi.org/10.1016/j.applthermaleng.2011.03.021
2. Gupta, N., Nayak, A.K., Malik, S.: Conjugate heat and species transport in an air filled ventilated enclosure with a thermo-contaminated block. Int. J. Heat Mass Transf. **117**, 388–411 (2018). https://doi.org/10.1016/j.ijheatmasstransfer.2017.10.028
3. Hssain, M.A., Mir, R., El Hammami, Y.: Numerical simulation of the cooling of heated electronic blocks in horizontal channel by mixed convection of nanofluids. J. Nanomater. **2020**, 1–11 (2020). https://doi.org/10.1155/2020/4187074
4. Hamouche, A., Bessaïh, R.: Mixed convection air cooling of protruding heat sources mounted in a horizontal channel. Int. Commun. Heat Mass Transf. **36**, 841–849 (2009). https://doi.org/10.1016/j.icheatmasstransfer.2009.04.009
5. Hidki, R., El Moutaouakil, L., Boukendil, M., Charqui, Z., Zrikem, Z., Abdelbaki, A.: Mixed convection and surface radiation in a ventilated cavity containing two heat-generating solid bodies. Mater. Today Proc. **66**, 318–324 (2022). https://doi.org/10.1016/j.matpr.2022.05.404
6. Chamkha, A.J., Hussain, S.H., Abd-Amer, Q.R.: Mixed convection heat transfer of air inside a square vented cavity with a heated horizontal square cylinder. Numer. Heat Transf. Part A Appl. **59**, 58–79 (2011). https://doi.org/10.1080/10407782.2011.541216
7. Chatterjee, D., Mishra, R.: Numerical investigation of transient magnetohydrodynamic mixed convection in a ventilated cavity containing two heated circular cylinders. Heat Transf. Eng. **39**, 1052–1066 (2018). https://doi.org/10.1080/01457632.2017.1358487
8. Gupta, S.K., Chatterjee, D., Mondal, B.: Investigation of mixed convection in a ventilated cavity in the presence of a heat conducting circular cylinder. Numer. Heat Transf. Part A Appl. **67**, 52–74 (2015). https://doi.org/10.1080/10407782.2014.916113
9. Rahman, M.M., Alim, M.A., Saha, S., Chowdhury, M.K.: Effect of the presence of a heat conducting horizontal square block on mixed convection inside a vented square cavity. Nonlinear Anal. Model. Control. **14**, 531–548 (2009). https://doi.org/10.15388/na.2009.14.4.14472

10. Rahman, M.M., Alim, M., Saha, S., Chowdhury, M.: A numerical study of mixed convection in a square cavity with a heat conducting square cylinder at different locations. J. Mech. Eng. **39**, 78–85 (2008). https://doi.org/10.3329/jme.v39i2.1850
11. Mamun, M.A.H., Rahman, M.M., Billah, M.M., Saidur, R.: A numerical study on the effect of a heated hollow cylinder on mixed convection in a ventilated cavity. Int. Commun. Heat Mass Transf. **37**, 1326–1334 (2010). https://doi.org/10.1016/j.icheatmasstransfer.2010.07.019
12. Ahammad, M.U., Rahman, M.M., Rahman, M.L.: Mixed convection flow and heat transfer behavior inside a vented enclosure in the presence of heat generating obstacle. Int. J. Innov. Appl. Stud. **3**, 967–978 (2013)
13. Hidki, R., El Moutaouakil, L., Charqui, Z., Boukendil, M., Zrikem, Z.: Natural convection in a square cavity containing two heat-generating cylinders with different geometries. Mater. Today Proc. **45**, 7415–7423 (2021). https://doi.org/10.1016/J.MATPR.2021.01.626
14. Hidki, R., Moutaouakil, L.E., Boukendil, M., Charqui, Z., Abdelbaki, A.: Natural convection and surface radiation in an inclined square cavity with two heat-generating blocks. In: Vasant, P., Zelinka, I., Weber, G.-W. (eds.) ICO 2021. LNNS, vol. 371, pp. 948–957. Springer, Cham (2022). https://doi.org/10.1007/978-3-030-93247-3_90
15. Hidki, R., El Moutaouakil, L., Boukendil, M., Charqui, Z., Abdelbaki, A.: Effect of thermal radiation on natural convection in an air-filled cavity with an inclined heat-generating elliptical body. In: Ben Ahmed, M., Boudhir, A.A., Karaş, İR., Jain, V., Mellouli, S. (eds.) SCA 2021. LNNS, vol. 393, pp. 293–303. Springer, Cham (2022). https://doi.org/10.1007/978-3-030-94191-8_23

Effect of Fineness and Ash Replacement Rate on Mortar Physical Properties: An Experimental Investigation

A. Lachheb[1] (✉), M. Touil[1], Y. Bandadi[1], R. Saadani[1], A. Sbaibi[2], and M. Rahmoune[1]

[1] Laboratoire d'Etude des Matériaux Avancées et Applications, FSM-ESTM, Moulay Ismail University of Meknes, Meknes, Morocco
a.lachheb@umi.ac.ma

[2] Faculté des Sciences et Techniques Mohammedia, Université of Hassan II, Casablanca, Morocco

Abstract. The construction industry is a large consumer of cement which is used in concrete and mortar manufacture. Cement substitution would reduce its economic and environmental cost. This work is part of the local materials development such as cement and residual ash from thermal power plants. Our objective is to study the effect of ash fineness and replacement rate on the thermomechanical characteristics of mortar. For this reason, the ash was ground into two different fineness values of 4500 cm^2/g and 6500 cm^2/g and used as replacement rate ranging from 10 up to 50% by weight of cement. In a first step, the thermal properties of ash mortars have been experimentally determined in terms of thermal conductivity and diffusivity using the box method, subsequently, thermomechanical characterizations were performed using the H10KL bench to estimate the ash mortars mechanical parameters. Test results show that the use of ash in the mortar resulted in a significant improvement in thermomechanical properties and a substantial decrease in water absorption. This subsequently contributes to the reduction of building energy consumption, CO_2 emissions in the atmosphere and annual quantities of ash produced by thermal power stations.

Keywords: Ash · cement · mortar · box method · H10KL bench · thermomechanical properties · energy consumption

1 Introduction

Nowadays, waste recycling and its use in the construction sector is becoming a major trend to improve the buildings energy efficiency. This trend aims to reduce the high energy consumption of the building sector worldwide.

In this perspective, Morocco has implemented an energy efficiency code in buildings, which includes improving materials thermal characteristics. Given the fact that the building sector is the second largest energy consumer, accounting for 33% of the country's total energy consumption [1].

S. Aniss et al. (Eds.): CMM 2022, LNME, pp. 246–254, 2024.
https://doi.org/10.1007/978-3-031-46973-2_23

The role of construction materials in the building envelope is paramount. According to objective or individual criteria, their choice has a direct impact on the thermal envelope quality [2]. As a result, several studies focus on the research and development of new building materials [3, 4].

The physical characteristics of building materials are key factors in the building's energy performance. And the characterization of new materials, by different measurement methods, has been the subject of many studies [5, 6].

On the other hand, the cement manufacturing process is one of the most energy consuming and environmentally polluting processes. In order to remedy these problems, many researches have been carried out on the subject and show that ash can be incorporated as aggregates in concrete and mortar, and can replace cement thanks to its physical and chemical properties. Some of them are interested in the characterization of concrete ashes [7, 8]. Others have focused on the use of ash as a substitute material for cement in the production of cementitious materials [9, 10].

In Morocco, more than 467.000 t of fly ash are produced annually by thermal power plants. The ashes of JORF are the only ones in Morocco likely to be used in cement and concrete, because they are evacuated by dry way unlike the other plants which evacuate them by wet way.

In this paper, we study the effect of ashes from the JORF Lasfar thermal power plant on the thermophysical characteristics of cement mortar, through experimental characterization using the box method and the H10KL bench.

Overall, this work highlights an innovative waste management method that can improve the thermophysical properties of building materials and save energy in the construction sector.

For this purpose, the ashes were ground to two different fineness values of 4500 cm^2/g and 6500 cm^2/g to substitute the cement with percentages that vary between 0% and 50% by weight of cement. Initially, the water absorption and retention capacity of these samples is determined through water absorption tests. Subsequently, thermomechanical characterizations were performed to estimate the ash mortars thermophysical and mechanical parameters. Finally, a comparative study of all these results will allow the selection of the most effective and adequate dosage and fineness to be incorporated in the mortar in order to integrate it in the building envelope.

2 Methods and Materials

2.1 Materials Preparation

The samples are prepared at ambient temperature, initially the sand is deposited, the cement and ash are poured, then the three components are mixed in a dry state to homogenize the mixture. Then the heap is dug into the crater and the necessary amount of water is poured in regularly until the mixture reaches the right consistency, then it is mixed for 2 min to obtain a cement mortar ready for use. This cement mortar is poured into molds of dimensions $[27 \times 27 \times 4]\ cm^3$ for thermal characterization and $[16 \times 4 \times 4]\ cm^3$ for mechanical characterization and then dried in an oven maintained at 70 °C.

In order to study the influence of the dosage and fineness of the ash on the physical properties of the mortar, a series of mixes with two values of fineness and a different level of cement replacement by ash were designed. The water-cement mass ratio was thus 0.5 for all mortar mixes.

2.2 Water Absorption

Water absorption capacity of processed materials is determined by measuring the mass increase of a sample dried in an oven at 70 °C (M_D) after being immersed in water for 24 h and surface dried (M_A).

The ratio of mass increase to dry sample mass, expressed as a percentage, is defined as Water absorption (W_{Abs}), which is:

$$W_{Abs} = \left(\frac{M_D - M_A}{M_D}\right) \cdot 100\% \tag{1}$$

2.3 Thermal Properties Measurement

The thermal characteristics of building materials are decisive factors in the energy performance of buildings.

The thermal characterization of our samples is performed by the box method (Fig. 1).

Fig. 1. EI700 box method.

The thermal conductivity (λ) of ash mortars with different percentages was measured using the first box of the method after obtaining the steady state. It is calculated by the following equation:

$$\lambda = \frac{e}{S \cdot (T_H - T_C)}\left[\frac{V^2}{R} - c \cdot (T_B - T_A)\right] \tag{2}$$

Where e is the sample thickness (m), U is the heating resistor voltage (V), R is the value of the heating resistor (Ohm), C is the coefficient of the Box 1 loss (W/ °C) and S is the sample area (m^2). T_A and T_B are the ambient and the inside box temperatures (°C); T_H and T_C are the hot and cold faces temperatures respectively (°C).

While, the diffusivity (α) is measured by the flash method in the second box of the experimental device. It is estimated by the average of the three values calculated by the following expressions [1]:

$$\alpha_1 = \frac{e^2}{t_{5/6}^2}\left[1.15t_{5/6} - 1.25t_{2/3}\right]$$
$$\alpha_2 = \frac{e^2}{t_{5/6}^2}\left[0.76t_{5/6} - 0.926t_{1/2}\right] \quad (3)$$
$$\alpha_3 = \frac{e^2}{t_{5/6}^2}\left[0.618t_{5/6} - 0.862t_{1/3}\right]$$

Where e is the sample thickness (m) and t the times for which we have 1/2, 1/3, 2/3, and 5/6 of the maximum value of temperature (s).

2.4 Mechanical Properties Measurement

The mechanical characterization of the samples is performed by the H10KL bench characterized by a maximum load capacity of 10 KN (Fig. 2).

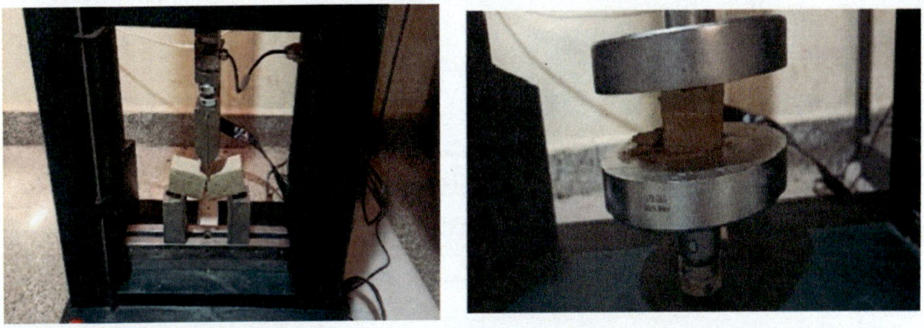

Fig. 2. H10KL bench for measuring compressive and bending strengths.

The 3-point bending tests were performed by applying an increasing load to the middle of the specimen, held on two supports with a loading rate of 50 N/s, until it broke into two separate pieces. The determination of the maximum bending strength (S_b) is deduced from the following equation:

$$S_b = \frac{3}{2}\frac{FL}{e^3} \quad (4)$$

Where e and L are the sample thickness and length (m).

The half-prisms of the specimen obtained after flexural failure will be broken in compression. The compressive strength (S_c) was calculated by dividing the force (F) by the specimen area (S).

$$S_c = \frac{F}{S} \quad (5)$$

3 Results and Discussions

For each sample, we carried out three series of measurements to test the repeatability of the experimental methods used. Mean values and standard deviations were calculated and reported in this work. The results obtained were used to construct polynomial models, enabling the physical properties of mortars to be correlated with the ash dosage rate.

3.1 Water Absorption

Water absorption is used to determine the amount of water absorbed under specific conditions. Figure 3 shows the mean values of water absorption for simple cement mortars and ash mortars with different fineness values and ash replacement rates.

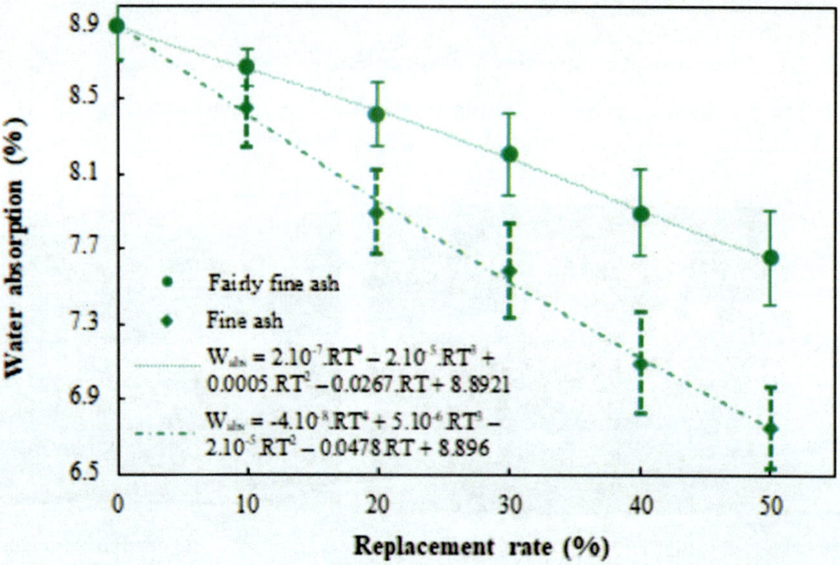

Fig. 3. Average water absorption evolution of mortars as a function of dosage and ash fineness.

As shown in Fig. 3, the water absorption of ash mortars is lower than that of simple cement mortar. This indicates that the incorporation of ash results in a reduction in water absorption.

In addition, the water absorption of ash mortars decreased with an increase in the ash replacement rate. This is reflected in the decrease in porous particles.

The results also show that the reduction is improved with finer ash due to the smoother surface of finer ash particles.

3.2 Thermal Properties

The dosage and fineness effect of ash on cement mortar thermal properties has been studied particularly on the thermal conductivity and diffusivity.

From these results, it can be seen that the higher the ash percentage, the lower the ash mortar conductivity and diffusivity. This shows that the more ash the composite contains, the less heat it can transmit, making it important in terms of thermal insulation (Figs. 4 and 5).

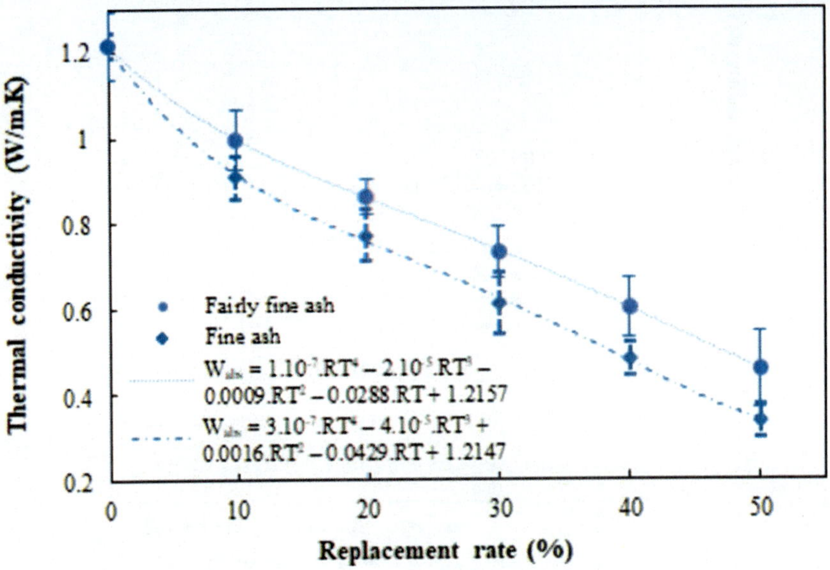

Fig. 4. Dosage and ash fineness effect on the average thermal conductivity of mortar.

These results also show that the ash mortars thermal properties decrease with the ash fineness. This is due to the fact that the increase in fineness is accompanied by an increase in the porosity, which results in a significant air content within the porous medium, which has lower thermal properties than the solid matrix.

3.3 Bending and Compressive Strength

The ash effect on the cement mortar mechanical properties, namely bending and compressive strength, were analyzed.

The results obtained show that the replacement of a part of the cement by the ash can improve in a sensitive way the mechanical resistances to the compression and to the flexion especially for substitution values of about 30%. Beyond this value, slight improvements in strength are observed.

It is also shown that the finer the sample, the higher the mechanical performance. Indeed, the thinner elements gather more active material and have a higher setting speed (Figs. 6 and 7).

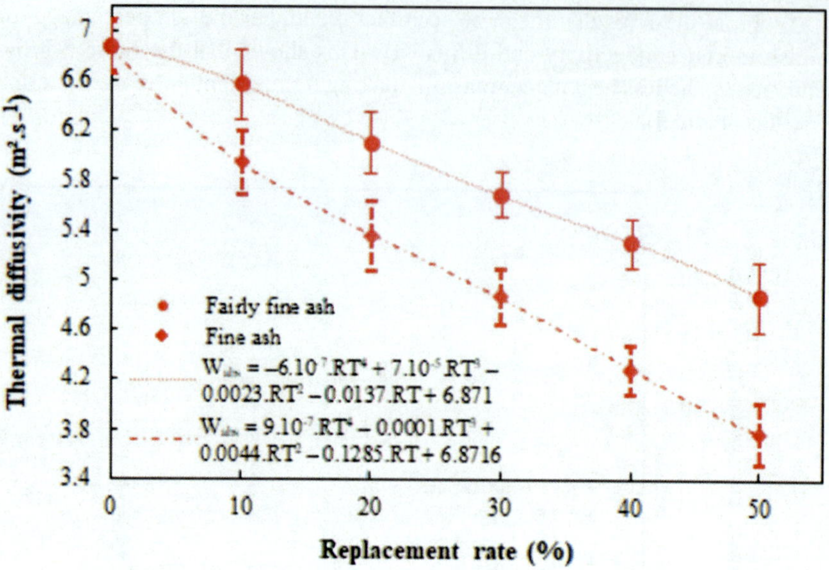

Fig. 5. Dosage and ash fineness effect on the average thermal diffusivity of mortar.

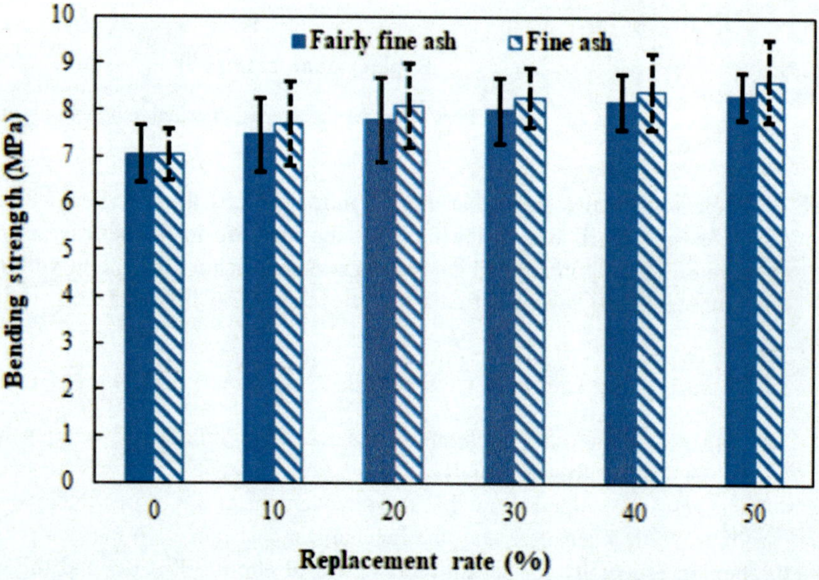

Fig. 6. Variation in average bending strength of ash mortars as a function of ash fineness and dosage rate.

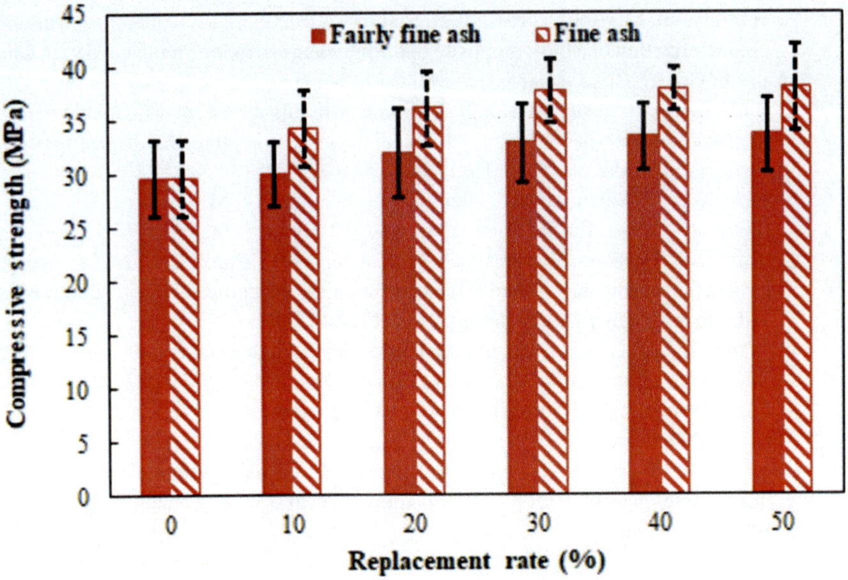

Fig. 7. Average compressive strength variation of ash mortars as a function of ash fineness and dosage rate.

4 Conclusion

This study presents the results of an experimental study aimed at assessing the influence fineness and ash replacement rate on mortar thermophysical properties. Several conclusions can be drawn from these results:

- Ash mortar is a promising composite compared to conventional mortar. Increasing the percentage of ashes in mortar formulation improves its insulating capacity by reducing its thermophysiques properties.
- The ash fineness has a significant influence on the cement mortar properties. Increasing the fine ash percentage in the mortar formulation improves its thermomechanical properties.
- The integration of ash mortar into the building envelope contributes to significant energy savings as it reduces CO_2 emissions into the environment and recycles ashes produced from thermal power plants.

References

1. Lachheb, A., et al.: Thermal insulation improvement in construction materials by adding spent coffee grounds: an experimental and simulation study. J. Clean. Prod. **209**, 1411–1419 (2019)
2. Lachheb, A., Saadani, R., Rahmoune, M.: Improving energy efficiency of Moroccan buildings through energy efficient envelope design: a case study. Int. J. Energy Environ. Econ. **26**(2), 115–141 (2018)

3. Boumhaout, M., Boukhattem, L., Hamdi, H., Benhamou, B., Nouh, F.A.: Thermomechanical characterization of a bio-composite building material: mortar reinforced with date palm fibers mesh. Constr. Build. Mater. **135**, 241–250 (2017)
4. Sudarsan, J.S., Vaishampayan, S., Parija, P.: Making a case for sustainable building materials to promote carbon neutrality in Indian scenario. Clean Technol. Environ. Policy **24**, 1609–1617 (2022). https://doi.org/10.1007/s10098-021-02251-4
5. Pezeshki, Z., Soleimani, A., Darabi, A., Mazinani, S.M.: Thermal transport in: building materials. Constr. Build. Mater. **181**, 238–252 (2018)
6. Touil, M., Lachheb, A., Saadani, R., Kabiri, M.R., Rahmoune, M.: A new experimental strategy assessing the optimal thermo-mechanical properties of plaster composites containing Alfa fibers. Energy and Buildings **262**, 111984 (2022)
7. Wang, W., Lu, C., Li, Y., Li, Q.: An investigation on thermal conductivity of fly ash concrete after elevated temperature exposure. Constr. Build. Mater. **148**, 148–154 (2017)
8. Xie, H., et al.: Characterization and use of biomass power plant ash in sulfoaluminate cementitious materials. Constr. Build. Mater. **325**, 126667 (2022)
9. Rivera, F., Martínez, P., Castro, J., López, M.: Massive volume fly-ash concrete: a more sustainable material with fly ash replacing cement and aggregates. Cement Concr. Compos. **63**, 104–112 (2015)
10. Kumar, P.C., Shanthala, T., Aparna, K., Babu, S.V.: Experimental investigation on the combined effect of fly ash and eggshell powder as partial replacement of cement. In: Kondraivendhan, B., Modhera, C.D., Matsagar, V. (eds.) Sustainable Building Materials and Construction: Select Proceedings of ICSBMC 2021, vol. 222, pp. 371–378. Springer, Singapore (2022). https://doi.org/10.1007/978-981-16-8496-8_46

Author Index

S. Aniss et al. (Eds.): CMM 2022, LNME, pp. 255–256, 2024.
https://doi.org/10.1007/978-3-031-46973-2